Nanophytopathology

Editors

Graciela Avila-Quezada
Facultad de Ciencias Agrotecnologicas
Universidad Autonoma de Chihuahua
Chihuahua, Mexico

Mahendra Rai
Department of Biotechnology
SGB Amravati University
Amravati, Maharashtra, India

CRC Press is an imprint of the
Taylor & Francis Group, an **informa** business
A SCIENCE PUBLISHERS BOOK

Cover illustrations reproduced by kind courtesy of Elanit Rubalcava Avila and Yossef Rubalcava Avila, Chihuahua City, Mexico.

First edition published 2023
by CRC Press
6000 Broken Sound Parkway NW, Suite 300, Boca Raton, FL 33487-2742

and by CRC Press
4 Park Square, Milton Park, Abingdon, Oxon, OX14 4RN

CRC Press is an imprint of Taylor & Francis Group, LLC

Library of Congress Cataloging-in-Publication Data (applied for)

ISBN: 978-1-032-38323-1 (hbk)
ISBN: 978-1-032-38328-6 (pbk)
ISBN: 978-1-003-34451-3 (ebk)

DOI: 10.1201/9781003344513

Typeset in Times New Roman
by Radiant Productions

Preface

The excessive use of chemical pesticides and fertilizers has polluted the ecosystems. The chemicals used are hazardous to microbes and other living beings as much of the chemicals are lost and drained into the water bodies; consequently, the aquatic flora and fauna are at high risk. Moreover, many fungicides/pesticides are unsafe for humans. Not only this, there are many other fungicides that are not effective in the management of various plant diseases caused by pathogens and pests because the pathogens and pests have developed resistance to the fungicides. Considering these facts, European Commission has decided to cut 50% of pesticides by 2030. In this case, there is a need for alternative technologies such as biological control or the application of smart nanotechnological tools such as nanobiosensor in the detection of plant diseases, delivery of fungicides, and therapy for the diseases caused by plant pathogens and pests. The use of nanomaterials will minimize the huge amount of application of pesticides, fungicides, and fertilizers, thereby reducing environmental pollution. This technology is eco-friendly, economically viable, and useful for sustainable crop production.

The book on "Nanophytopathology" discusses the role of nanomaterials including inorganic nanoparticles of copper, zinc, titanium, silver, carbon materials, and organic and biodegradable nanoparticles such as sulfur, chitosan, and pullulan, among others. The authors are experts and experienced in their respective fields.

The book is divided into three sections: Section I of the book deals with Nanophytopathology as a new and emerging science, and also discusses the perspectives of nanophytopathology in plant protection; Section II encompasses the role of nanomaterials in the management of different plant diseases and insect pests; while Section III incorporates the new and emerging technologies such as CRISPER cas9 and nitric oxide in the management of plant diseases.

This book provides up-to-date, research-based literature for postgraduate students and research scholars. It would be beneficial for agriculture scientists, nanotechnologists, microbiologists, green chemistry experts, and biotechnologists. It will also enrich academicians and researchers of agro-industries.

We thank the authors for providing updated and innovative contributions on the subject of their corresponding chapters. Their hard and sincere work will undoubtedly augment and update the readers' knowledge concerning nanophytopathology as a novel and evolving science. Further, we would like to express our sincere thanks to the publishers and authors of the chapters whose research work has been mentioned in the book.

Finally, Mahendra Rai thankfully acknowledges the financial support rendered by the Polish National Agency for Academic Exchange (NAWA) (Project No. PPN/ULM/2019/1/00117/U/00001) to visit the Department of Microbiology, Nicolaus Copernicus University, Toruń, Poland.

Graciela Avila-Quezada, Mexico
Mahendra Rai, India

Contents

Section I
General

Chapter 1

Nanophytopathology
A New and Emerging Science

Graciela Avila-Quezada,[1,*] *Mahendra Rai*,[2,3] *Nuvia Orduño-Cruz*,[1] *Patricia Rivas-Valencia*,[4] *Patrycja Golińska*,[3] *Laila Muñoz-Castellanos*,[1] *Denisse Mercado-Meza*[1] and *Hilda Saenz-Hidalgo*[1,5]

Introduction

To feed the almost 7.7 billion people who currently inhabit the planet, the 8 billion inhabitants estimated for 2030, more than 9.7 billion who will inhabit it in 2050, and nearly 11 billion around 2100 (UN 2019), it will be necessary to produce more food sustainably and apply new ways of delivering agricultural inputs to plants. Food is a necessity for all, which has implications that involve health, the economy, migration, and conflicts.

In addition to this challenge, 33% of the world's soil presents moderate to high levels of degradation, and more than 80% of the mineral supply is not used by agricultural crops (UNEP 2016), but fertilizers are lost in the environment by leaching, volatilization or denitrification (Avila-Quezada and Rai 2022).

Losses in the agricultural sector due to pests and diseases affect up to 40% of yield (FAO 2021). Phytopathogens are responsible for billions of dollars of agricultural crop losses. FAO estimates that each year, plant diseases cost the world economy around US$220 billion (FAO 2019). These data show that new strategies

[1] Universidad Autonoma de Chihuahua, Escorza 900, col. Centro, Chihuahua 31000, Mexico.
[2] Nanobiotechnology Lab., Department of Biotechnology, SGB Amravati University, Amravati-444 602, Maharashtra, India.
[3] Department of Microbiology, Nicolaus Copernicus University, Torun, Poland.
[4] Instituto Nacional de Investigaciones Forestales, Agrícolas y Pecuarias, Campo Experimental Valle de Mexico, Km 13.5 Carretera Mexico-Texcoco, Coatlinchan, Texcoco, 56230, Estado de Mexico, Mexico.
[5] Centro de Investigacion en Alimentacion y Desarrollo AC. av. 4a Sur 3820, Delicias, 33089, Chihuahua, Mexico.
* Corresponding author: gdavila@uach.mx

for managing plant diseases are necessary. The potential of nanotechnology, in particular the use of nanoparticles (NPs), may ensure production if it is put into integrated crop management.

As per historical data, nanostructured materials date back to prehistoric technology, such as rock paintings in the Chauvet-Pont-d'Arc cave in France, where nanographene was used. The ancient Egyptians used soot (carbon nanomaterials) from oil lamps to make opaque black pigments for writing on papyrus. In addition, with chemical methods, they elaborated palladium disulfide nanoparticles (PdS_2NPs) as hair dye ingredients (Barhoum et al. 2022). The ancient Mayans manufactured pigments such as Maya blue for ceramics and wall painting, which have lasted despite the humidity and are resistant to corrosion. This dye is composed of nanomaterials chemically combined with clay nanopores (Gettens 1962). Between the 4th and 17th centuries, the brightly colored stained-glass windows of European cathedrals were made from gold NPs. To make the ceramics shine, Islamic people used NPs of silver or copper (Barhoum et al. 2022). Damascus saber blades also contain carbon nanotubes and cementite nanowires. Nowadays, with electron microscopes, it has been possible to know matter of nano sizes 1 to 100 nm and then manipulate them.

To positively affect food production and thus supply the growing population, each individual technology, including nanotechnology, must converge, transferring knowledge to the others. Nano-enabled agrochemicals such as nanofertilizers and nanopesticides have recently been developed (Avila-Quezada et al. 2021, Avila-Quezada and Rai 2022). These nanoproducts could offer benefits such as greater efficiency, durability, and reduction of the amount currently being applied. For instance, metallic nanoparticles including metal or metal oxides have strong antimicrobial activity due to Reactive Oxygen Species (ROS) production, membrane interaction, and ATP depletion (Slavin et al. 2017); however, metal oxide NPs have shown antimicrobial activity against many microbes because of surface ROS production (Behera et al. 2019). Also, nanoformulations such as emulsions or microcapsules are functional by synchronizing the demand of the crop and the release of the required inputs (Kalia et al. 2020). In addition, nanosensors and biosensors can monitor pathogens throughout the agricultural production chain (Kaushal and Wani 2017). This information will be useful for making decisions about when and how much pesticide to apply. As a result, excessive use of agrochemical products will be avoided.

Thus, nanophytopathology as a new and emerging field for food production will play a key role in the application of NPs for the management of crop pathogens. The trend is to be more efficient in agriculture, with sustainable practices that require fewer inputs, generate less waste and reduce pollution (Rai and Ingle 2012).

Losses in crops due to plant pathogens

Recent studies have estimated the percentages of losses caused by emerging or re-emerging pests and diseases, in the five main crops that provide food for the world population: wheat (21.5%), rice (30.0%), corn (22.5%), potato (17.2%) and soy (21.4%) (Savary et al. 2019).

Currently, the production and feeding schemes of the general population have changed since the COVID-19 pandemic in 2020, the confinement and the consequent reduction of non-essential activities and in some countries the total cessation of activities, increased the demand for agricultural products (Rivas-Valencia et al. 2021); in addition, the growing population (Figure 1) will require food.

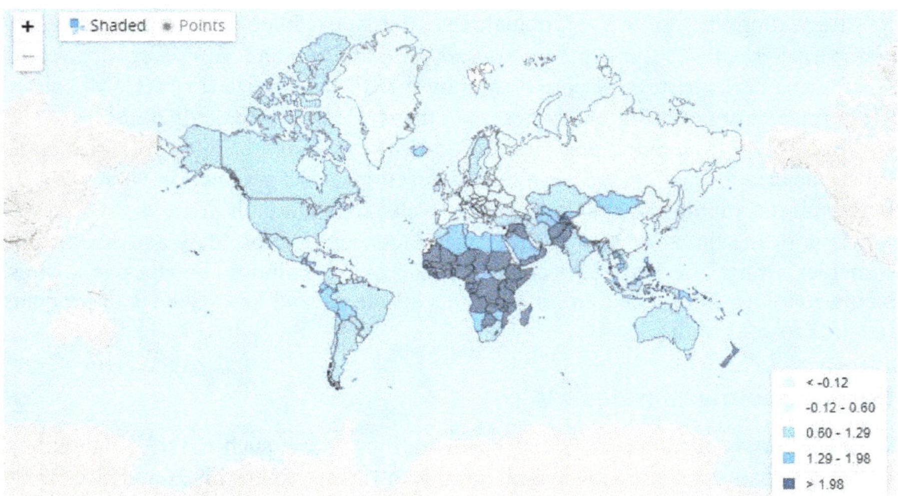

Figure 1: Annual population growth (%). World Bank. https://datos.bancomundial.org/indicator/SP.POP.GROW?end=2020&start=1961&type=shaded&view=map&year=2020.

Feeding such a large population will be a great challenge for farm production when the demand for food will increase between 60 to 120% (FAO 2019). In addition, other factors are relevant and will determine the success of food production. For instance, the availability of water and climate change directly affect the water reserves, reducing them. Besides, the international trade flow of agricultural products can distribute pathogens around the world.

Examples of diseases showing economic and social damage are potato late blight caused by *Phytophthora infestans*, which caused the Irish famine in 1840; coffee rust in Ceylon caused by *Hemileia vastatrix* in the 1860s; and the Great Famine in Bengal in 1943 caused by *Helminthosporium oryzae* (Schumann 1991, Avila-Quezada et al. 2016).

Chestnut disease caused by *Cryphonectria parasitica*, wiped out the American chestnut (*Castanea dentata*) in the 1950s, killing 80% of chestnut trees (Schumann 1991, Clark et al. 2019).

Puccinia graminis rust (Saunders et al. 2019) and *Puccinia striiformis* (Liu et al. 2017), in addition to *Puccinia graminis* f. sp. *tritici* Ug99 which affects wheat, cause losses up to 70%; this rust was a new virulent strain identified in wheat in Uganda in 1999, thus designated as Ug99 (Singh et al. 2011, Avila-Quezada et al. 2018).

Moreover, a new strain of *P. infestans* has rapidly displaced other strains of late blight (Cooke et al. 2012). Sudden wilting of olive trees caused by *Xylella fastidiosa* subsp. *pauca* has destroyed millions of olive trees in Puglia, Italy, and also threatens other European and Mediterranean countries. This is an example of how a pathogen can affect a crop and also the landscape of a region (Schneider et al. 2020). In California and Oregon, in the United States of America, as well as in other areas, *Phytophthora ramorum* causes sudden oak death and poses a threat to forest ecosystems (Rizzo et al. 2005).

Regarding the groups of nematodes, the most important is the sedentary endoparasites of the genera *Heterodera*, *Globodera*, and *Meloidogyne*. About viruses, the citrus tristeza disease caused by the Citrus Tristeza Virus (CTV) caused the death of more than 300 million trees in the world (Moreno et al. 2008).

Recently, Huanglongbing (HLB) caused by the bacterium *Candidatus* Liberibacter spp. has caused significant losses in citrus production (Bové 2012). Vector insects (aphids and psyllids, respectively) transmit both diseases.

Due to the threat of plant pathogens to food production, new and sustainable strategies must be implemented to ensure agricultural production. Thus, nanotechnology has much to offer due to its efficiency and low volumes of products required to control the disease.

Diversity in nanomaterials

As it is known, nanomaterials can be of various forms such as NPs, nanotubes, fullerenes, nanofibers, nanowhiskers, and nanosheets (Nurfatihah and Siddiquee 2019), with various useful properties for the management of phytopathogens; consequently, a low amount of nanoencapsulated active ingredient or NPs is applied, compared to the large amounts of conventional agrochemicals used in modern agriculture (Kumar et al. 2022).

NPs are characterized by having good encapsulation properties and being effective in releasing active ingredients (ai) and released in response to particular triggers (Huang et al. 2018). Dendrimers are active ingredient delivery systems, which are a class of polymer-coated particles. These can be used as sensors, and catalysts, in addition to administering active ingredients, because they are non-toxic and biodegradable (Kowalczuk et al. 2014).

Regarding control, organic nanomaterials are used as carriers for pesticides. These nanoencapsulates are micelles, liposomes or nanospheres based on lipids, proteins, and polysaccharides. Lipid-based nanomaterials are among the most widely applied because they can be produced from natural ingredients and have the ability to encapsulate compounds with different solubilities (Esfanjani et al. 2018).

Moreover, protein-based nanomaterials are built from molecules that can self-assemble (Katyal et al. 2019). Finally, polysaccharides are natural compounds, the most used are chitosan, cellulose, and starch. The antimicrobial activity of chitosan has been demonstrated against many prokaryotes and eukaryotes (Barros and Casey 2020). Further, NPs are interesting since they induce defense mechanisms in plants. The proteins are involved in plant defense responses. When plant cells detect the presence of flagella or chitin in the cell walls of fungi, nearby cells are alerted by secreting chemical molecules to activate their defense system (Freeman and Beattie 2008).

Studies have reported the activation of antioxidant enzymes in the plant in response to exposure to nanomaterials (Figure 2) (Saha and Dutta Gupta 2018). Among these studies, the one on silicone NPs stands out, which increases the resistance of corn to fungal attacks by producing high concentrations of phenolic compounds (Rangaraj et al. 2014). Returning to the topic of chitosan, chitosan NPs activate genes related to defense, such as those that code for antioxidant enzymes, and those that increase the levels of total polyphenols (Picchi et al. 2021). Silicon dioxide nanoparticles (SiO_2NPs) increased antioxidant enzyme capacity, total phenols, and flavonoid content in hairy roots inoculated with *Agrobacterium rhizogenes* (Nourozi et al. 2019).

In plants, metal nanoparticles in some circumstances increase the activity of defense enzymes such as catalase (CAT) and peroxidase (POD), and reduce free radicals of reactive oxygen species (Gupta et al. 2018). Parveen and Siddiqui (2022), when spraying foliar Zirconium dioxide NP(ZrO_2NP) on tomato plants, found that ZrO_2NPs increased the activity of defense enzymes, i.e., superoxide dismutase (SOD), catalase (CAT), ascorbate peroxidase (APX) and phenylalanine ammonia-lyase (PAL) in the presence or absence of several pathogens. Nekrasova et al. (2011) reported that copper oxide NP also promotes increased levels of SOD, CAT, and lipid peroxidase. These are some studies in which NPs increased enzymatic activity in plants, with or without pathogens.

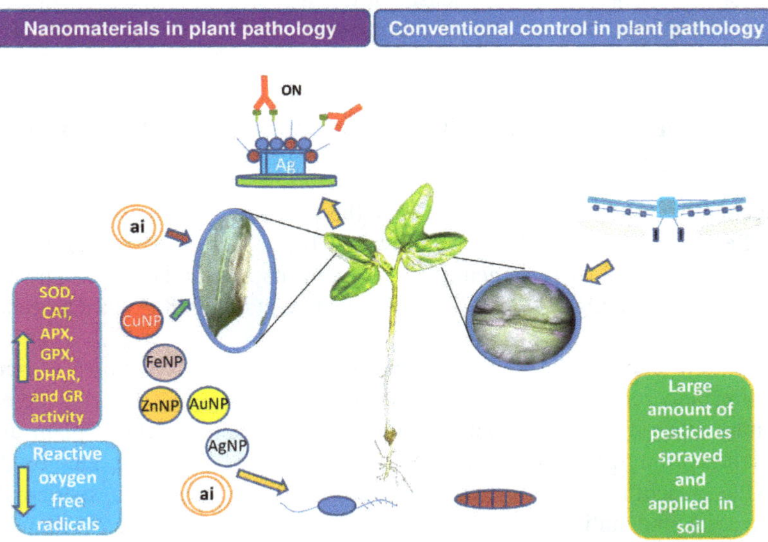

Figure 2: NPs can act as a preventive method before the pathogen establishes itself in the leaflets, or as a control for internal pathogens, including the inhabitants of the xylem and phloem. Nanosensors are useful for timely diagnosis, especially in the case of quarantined pathogens. Nanocarriers (micelles, liposomes or nanospheres) can carry the active ingredient or active molecules towards the pathogen and release them in a controlled way. Superoxide dismutase (SOD), catalase (CAT), ascorbate peroxidase (APX), guaiacol peroxidase (GPX), dehydroascorbate reductase (DHAR), and glutathione reductase (GR).

Moreover, foliar SiO_2NPs induce resistance in rice plants and protect them against *M. oryzae*. This systemic acquired resistance (SAR) depends on salicylic acid, which is one of the signaling compounds (Du et al. 2022).

Another advantage of nanomaterials is that pesticides and volatile compounds can be nanoencapsulated. The latter also stimulates the defense mechanisms of plants, and nanoencapsulated pesticides have advantages such as protection against photolysis, as well as volatilization and leaching is reduced (Nuruzzaman et al. 2016). This is how nanomaterials offer many advantages to plant pathology and their use will reduce environmental pollution, and will have a positive impact on food security and sovereignty.

Nanobiosensors

Nanosensors are just like ordinary sensors, albeit nanoscale in size. These small devices are based on biology, chemistry, and nanotechnology. They can detect gases, aromas, chemical contaminants, and pathogens. It binds to what it is desirable to detect, for example, a physicochemical signal (sensors) and a biological signal (biosensors), and exhibits a signal. They are very specific and selective towards their target compound like ions, proteins, toxins, etc. (Omanović-Mikličanina and Maksimović 2016).

The nanosensor is composed of a probe, a transducer, and a detector. When the analyte or biosubstance joins the sensor, it changes in the vibration frequency and finally emits a signal that warns of the presence of the target pathogen (Johnson et al. 2021).

Metallic nanoparticles for the management of phytopathogenic fungi on staple crops

Nanotechnology has been improving over the past few years in several important fields (Pérez-de-Luque 2017, Lira-Saldivar 2018). Among NPs, metal-based NPs are widely used against pathogenic microorganisms, particularly Pd, Zn, Ag, and Cu-based NP (Ulloa-Ogaz et al. 2017, Athié-García et al. 2018, Avila-Quezada et al. 2021).

This new application of nanotechnology can sustainably boost agricultural production and reduce the amount of chemical products used and generated in agriculture since it has the potential to optimize the management of plant diseases while safeguarding plant health (Lira-Saldivar 2018). For instance, zinc oxide NPs (ZnONPs) present unique properties such as optical, photocatalytic, semiconductor, antimicrobial and antifungal (Miri et al. 2019), and are a biosafe material for biological species, due to its use in the medical area (Mishra et al. 2017). In addition to this, they offer great potential to boost agricultural productivity, since it is the only metal found in six classes of enzymes: lyases, hydrolases, oxidoreductases, isomerases, transferases, and ligases (Singh et al. 2018).

The movement of nanomaterials within plants can be following the flow of the apoplast or symplast pathways, depending on the movement upstream or downstream. Several tissues must be crossed before reaching the vascular tissues; for the internalization of NPs within the cell, several mechanisms have been proposed: endocytosis, pore formation, through carrier proteins and through plasmodesmata (Pérez de Luque 2017).

Since ZnONPs are inside the plant, their antifungal activity may be due to several different processes: cell membrane disruption, protein damage, interruption of transmembrane electron transport, mitochondrial damage, cytoplasm leakage or production of reactive oxygen species (ROS) (Singh et al. 2018).

Phytopathogenic fungi affect the whole plants, roots, stems, leaves, flowers, and fruits, causing losses in millions (Agrios 2005); besides, the use and abuse of agricultural fungicides have generated resistance in phytopathogenic fungi (Sang and Lee 2020), environmental pollution (Goulson 2014) as well as public health problems because some of them are endocrine disruptors, genotoxic, carcinogenic, among other diseases (Mostafalou and Abdollahi 2013, Nicolopoulou-Stamati et al. 2016, Ordoñez-Beltran et al. 2019).

Legumes are one of the staple crops in Latin America and the rest of the world. Beans (*Phaseolus vulgaris* L.) are one of the main crops grown worldwide since it provides a large amount of fiber, protein, vitamins, and minerals to the diet (Chávez-Mendoza and Sánchez 2017). According to research conducted on beans and other legumes such as lentils, alfalfa, and soybeans, looking for the positive or negative effects of using ZnONPs as a nanofertilizer, nanocarrier or antifungal, some interesting aspects have been shown mostly *in vitro*, although there are few field experiments (Kutawa et al. 2021).

In a study carried out by Huang et al. (2014) in soybean, ZnONPs did significantly affect root length, creating a disruption in the rhizobium-legume symbiosis, delaying nodulation and therefore nitrogen fixation. Lakshmeesha et al. (2020) studied ZnONP biosynthetized from foliar extracts of *Melia azedarach* plants and found antifungal bioactivity against *Cladosporium cladosporioides* and *Fusarium oxysporum* in soybean. The ZnONP showed a decrease in the production of ergosterol and thus fungal biomass, as well as lipid peroxidation, reactive oxygen species, and loss of membrane integrity (Figure 3).

The fungal diseases that most affect beans such as anthracnose are caused by *Colletotrichum lindemuthianum* and root rot caused by *Fusarium oxysporum* f. sp. *phaseoli*, *Fusarium solani* f. sp. *Phaseoli*, and *Rhizoctonia solani* Khün (Anaya-López et al. 2021). Narendhran and Sivaraj (2016) tested ZnONP against three different phytopathogenic *Fusarium oxysporum* strains, finding a 20–25% decrease in mycelial growth *in vitro* at a concentration of 100 $\mu g\ mL^{-1}$.

Some phytopathogenic fungi can produce mycotoxins, hence the importance of determining whether the nanoparticles have the ability to reduce fungal activity and therefore their ability to produce toxins; Yehia and Ahmed (2013) treated mycotoxigenic fungi including *Fusarium oxysporum* with 12 mg L^{-1} of ZnONPs, obtaining a growth inhibition of 77% as well as a decrease in the production of fusaric acid from 39.0 to 0.2 mg g^{-1}.

Besides, Arciniegas-Grijalba et al. (2019) used ZnONPs in a nanobiohybrid composition *in vitro* against *Colletotrichum* sp., observing an inhibition percentage above 93%. Through TEM analysis, it was possible to observe irreversible physical

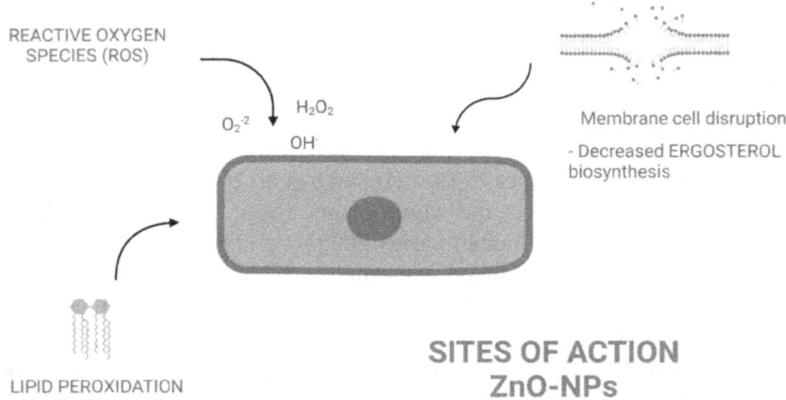

Figure 3: ZnONPs action sites in a fungal cell.

damage in the micro and ultrastructure of the fungus. The hyphae became thicker and with an appreciable reduction in conidia production.

Unfortunately, there is not enough *in vivo* or *in situ* evidence to assure that ZnONPs are effective in the field against phytopathogenic fungi, as observed in studies conducted *in vitro* (Kutawa et al. 2021). There is enough information on the biosynthesis of ZnONPs based on plants, bacteria, and fungi, but not enough information regarding the effect of these nanoparticles on phytopathogenic fungi in legumes (Singh et al. 2018).

Nanoparticles for the management of viruses and bacteria

Compared to bulk materials, the uniqueness of nanometric materials arises from their high surface area to volume ratio (Hernández-Díaz et al. 2021). NPs operate on the same size scale as cells; therefore, they are ideal for navigating systemic and cellular traffic.

Plant viruses represent a serious threat to global food security causing high losses in crop production (Bernardo et al. 2018), with an estimated annual economic impact of more than $30 million worldwide (Sastry and Zitter 2014). Cultivated plant viruses make up about 50% of known plant pathogens. The transmission of viruses in the environment, through insect vectors, plays an important role in transmitting more than 70% of the total number of viruses that infect plants (Hogenhout et al. 2008). Although plant viruses possess simple ribonucleic acid (RNA) or deoxyribonucleic acid (DNA)-based genomes, factors such as dynamic genome diversity, transmission by a variety of vectors, rapid evolution, and wide adaptive potential to several environmental conditions make them very difficult to control. Therefore, sustainable management strategies are required to control plant pathogenic viruses.

Different types of silver, aluminum oxide, zinc oxide, and titanium dioxide NPs have been used to control rice weevil and silkworm (Bombyx mori) grass disease caused by *Sitophilus oryzae* and the baculovirus BmNPV (Bombyx mori nuclear polyhedrosis virus), respectively. In the experiment of Goswami et al. (2010), metal NP formulations were applied to rice plants and kept in a plastic box with 20 adults of *S. oryzae*. Hydrophilic AgNPs were more effective on the first day. On the second day, more than 90% mortality was obtained with AgNPs and AlNPs. After seven days of exposure, mortality of 95 and 86% was observed with hydrophilic and hydrophobic AgNPs and almost 70% of the insects died when rice was treated with lipophilic AgNPs.

In the same experiment, mortality of 100% was observed when using AlNPs. Similarly, in another bioassay conducted for grass disease in silkworm (*B. mori*), a significant decrease in viral load was reported when *B. mori* leaves were treated with an ethanolic suspension of hydrophobic aluminosilicate NP (Goswami et al. 2010).

To control pests of virus vector insects, silica nanoparticles (SiO_2NP) were tested at different concentrations (75–425 mg L^{-1}) in broad beans and soybeans in the field, where SiO_2NPs reduced the populations of the three insect pests tested in the experiment (Thabet et al. 2021). The mechanisms of action between nanoparticles and viruses, vectors, and plants are shown in Figure 4.

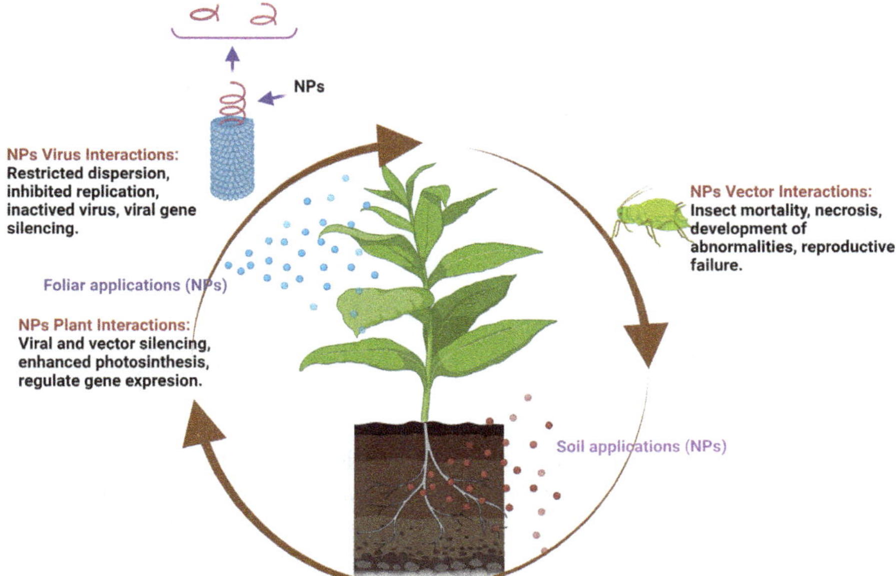

Figure 4: Mechanisms of action among nanoparticles, virus, vector and plant.

In recent studies, it has been found that CNT nanoparticles inhibit the replication and movement of Tobacoo Mosaic Virus (TMV) (Adeel et al. 2021). Besides, Fe_3O_4 induced the activation of antioxidants; thus, the resistance of the plant against TMV was enhanced through the upregulation of SA genes (Cai et al. 2020). Also, ZnONP promoted TMV deactivation and regulation of plant immunity (Cai et al. 2019), NiONP reduced virus titer and disease severity caused by Cucumber Mosaic Virus (CMV) (Derbalah and Elsharkawy 2019), and AgNP reduced Tomato Spotted Wilt Virus (TSWV) infection and inhibition of local lesions (Shafie et al. 2018).

Moreover, bacteria are particularly difficult to control in plants and this difficulty is even increased by the restriction in those countries where the use of antibiotics is prohibited, as in the European community. This makes the search for alternative control methods such as nanobiotechnology a promising tool. Metal NPs involve physical and chemical targeting to kill bacterial cells, while conventional antibiotics or biocides manifest biochemical destruction of cells resulting in the development of resistance (Javed and Nadhman 2020).

Avila-Quezada and Espino-Solis (2019) found that metallic NPs in pathogenic bacteria inhibits cell wall synthesis, binds to the cell membrane surface, damages DNA, affects the thiol group of enzymes, and inhibits biofilm formation. In experiments, the results obtained with metallic NPs are encouraging, e.g., biosynthesized ZnONPs (25 $\mu g\ mL^{-1}$) produced high suppression of the pathogenic bacteria *Pseudomonas aeruginosa* (Jayaseelan et al. 2012).

In a field study, TiO_2NPs of size 10 to 50 nm created an adhesive layer on the plant surface that prevented infection by bacteria. The treatment reduced the cucumber infection by *Pseudomonas syringae* pv. *lachrymans* by 69% (and 91% against the oomycetes *Pseudoperonospora cubensis*), and also increases photosynthetic activity by 30% (Cui et al. 2009). Moreover, colloidal AgNPs were evaluated by Javed and Nadhman (2020) and found promising antibacterial activity against the phytopathogenic bacteria *Xanthomonas vesicatoria*, *Pectobacterium carotovorum*, *Xanthomonas oryzae*, and *Ralstonia solanacearum*.

In another study, Fortunati et al. (2016) tested biopolymeric NPs based on poly(DL-lactide-co-glycolide) copolymers (PLGA) and cellulose nanocrystals (CNC) against the bacterial speck disease caused by *P. syringae* pv. *tomato*. They demonstrated that N was able to cover the surface, through a uniform distribution, and the tomato plants' surface had no damage, thus allowing regular development.

Nano-sized microfabricated xylem vessels have been shown to be very useful in understanding the mechanisms and kinetics of bacterial colonization in xylem vessels. This information generated is very useful because new disease control strategies can be developed (Zaini et al. 2009). Likewise, Hernández-Montelongo et al. (2016) showed the application of hyaluronan/chitosan nanofilms, assembled layer by layer as a potential antimicrobial nanomaterial for the quarantine bacterium *Xylella fastidiosa*. The best antibacterial effect of 3.0 pH/0.10 IS can be explained by its larger number of nitrogenated groups exposed on the surface, which included a higher concentration of free protonated ammonium groups (NH_3^+), and the antibacterial killer agent, which was mainly generated by the synthesis pH.

Field studies indicated that TiO_2 sol NPs (neutral, viscous colloid of 10–50 nm) inhibit pathogenic infection of cucumber by *P. syringae* pv. *lachrymans* in 68.6% and *P. cubensis* in 90.6% (Cui et al. 2009). Importantly, the authors also reported significant increases in photosynthetic activity (30%) compared to control plants. In another study, Huang et al. (2005) used MgONPs as an effective bactericide due to its strong interactions with negatively charged bacteria and spores, inhibiting *Staphylococcus aureus* by 99.9% and *Bacillus subtilis* by up to 97.3%.

Metallic nanoparticles as inducers of damage in the pathogen structure

Metal NPs are an emerging option for the management of plant pathogens because of the high surface-volume ratio which favors the interaction with the pathogen cell membrane (Hernández-Díaz et al. 2021). Two mechanisms have been proposed for the interaction of NPs with the cell membrane: in the first, the NPs of size 10 nm or less, that is, less than the diameter of the phospholipid bilayer, can form irreversible pores and translocate; and the second mechanism involves the larger NPs which enter through endocytosis, being able to deform the membrane (Linklater et al. 2020).

The interaction of Metal NPs with phytopathogens produce an increase in oxidative stress due to the ability of NPs to produce damaging radicals, participate in a redox reaction, and interact with DNA, lipids, and proteins which causes cell membrane damage, protein degradation, and mutagenesis in the pathogen (Tee et al. 2016, Al-khattaf 2021). Besides, electrostatic interaction affects membrane integrity, function, and polarization, resulting in the outflow of proteins, DNA, RNA, metabolites, and enzymes (Kumar et al. 2021), and also disruption of ATP by damaging mitochondrial components, inhibition of DNA synthesis and enzyme activity; these mechanisms can occur individually or in combination and depend on the physical characteristics (Hernández-Díaz et al. 2021).

It has been reported that AgNPs produce membrane despolarization, destabilization and increase permeability (Vazquez-Muñoz et al. 2019). In the study of Haroon et al. (2019), AgNPs increased membrane permeability in *Ralstonia solanacearum* allowing leakage of the nucleic acids and cell death. Moreover, it has been verified that AgNPs cause ROS, membrane damage, and the death of *Meloidogyne incognita* (Fouda et al. 2020). ROS had been found in larval tissues after exposure to AgNPs (Raj et al. 2017, Mao et al. 2018). Also, AgNPs increase the number of autolysosomes that aggravate autophagy, which suggests that the ROS accumulation by AgNPs activate autophagy (Mao et al. 2016).

Antiviral activity of AgNPs and metal oxide NPs link ZnO was evaluated and the direct interaction of NPs with the virus surface was investigated (Vargas-Hernandez et al. 2020). In an experiment on tomato plants, it was reported that AgNPs bind to Potato Virus Y and Tomato mosaic virus coat protein (El-Dougdoug et al. 2018).

For fungal control strategies, ZnONPs have more selectivity, better durability, and heat resistance than the other inorganic NPs. The main mechanism of these NPs is the surface production of ROS, superoxide anions and hydroxyl radicals cannot penetrate the membrane but cause the loss of cellular integrity, and hydrogen peroxide is able to cross the membrane and can cause the destructions of cellular components such as DNA, proteins, and lipids (da Silva et al. 2019).

CuNPs act as a fungicide against *Phytophthora capsici* (Pham et al. 2019), *Fusarium solani*, and *F. oxysporum*. Since the fungal cell wall is negatively charged due to glycoproteins (Chen et al. 2020), it is suggested that CuNPs promote disruption of the membrane surface and cell wall damage, and consequently, the leakage of cellular components (Pariona et al. 2019). It has also been reported that MgONPs cause the damage of *Thielaviopsis basicola* cell membrane (Chen et al. 2020). In conclusion, electrostatic interaction between metal NPs with cell membrane causes destabilization, loss of integrity, and ultimately leakage of the cytoplasm. These mechanisms of action depend on size, shape, and type of NP.

Nanoparticles for the control of root pathogens. Do these damage the soil microbiota?

Several studies on nanomaterials have focused their attention on evaluating their effect as fungicides since they alter the mechanisms in the fungal membrane, and therefore affect the integrity of the microorganisms (AbdelRahim et al. 2017). The fungicidal effect of metal and chitosan NPs has been demonstrated in many studies, finding that they inhibit the growth and germination of spores of root pathogens such as *Phytophthora* spp., *Thielaviopsis basicola*, *Fusarium* spp., among others (Giannousi et al. 2013, Velmurugan et al. 2015, Xue et al. 2016, Le et al. 2019, Chen et al. 2020).

The success of the control depends on several factors. Akpinar et al. (2021) found that 3 nm AgNPs at a dose of 100 ppm are capable of preventing the development of *Fusarium oxysporum* f. sp. *radicis-lycopersici*.

Other researchers have evaluated the effectiveness of different metallic NPs in various microorganisms to observe the effect of each interaction, finding that AgNPs have greater antifungal activity than CuNP in *Pilidium concavum* and *Pestalotia* sp. (Bayat et al. 2021).

Many more examples of the effectiveness of nanomaterials to control root pathogens can be found in recent literature. However, concerns about the ecological risk and the effect on the beneficial microbial community have been raised; thus, it is important to detect any adverse effects in a timely manner (Simonin and Richaume 2015). Recent research is aimed at evaluating biological interactions, including the symbiosis of mycorrhizal fungi that, among its many benefits, give the plant resistance to root pathogens (Song et al. 2015).

The use of nanocomposites could pose a threat to the soil microbiome because the NPs hamper the physicochemical properties of the soil along with the microbial metabolic activities in the rhizospheric soil (Khanna et al. 2021). However, recent studies show that the application of NPs can have either positive, negative, or unaltered effects on mycorrhizal and rhizobial symbioses (Tian et al. 2019). Once the microorganisms come into contact with the NPs, their toxicity can be determined as generated oxidative stress, membrane alterations, lipid peroxidation, and oxidation (Abbas et al. 2020).

Sarabia-Castillo et al. (2022) demonstrated that Fe_2O_3NPs at 180 mg per experimental unit in wheat roots did not interfere with the establishment and development of arbuscular mycorrhizal fungi. However, the dose of 3.2 mg kg^{-1}

of Fe_2O_3NP significantly decreased the percentage of colonization of *Glomus caledonium* in roots of *Trifolium repens* (Feng et al. 2013). In the study of Priyanka et al. (2020) they reported that TiO_2NPs decreased the formation of vesicles and arbuscules of arbuscular mycorrhizal fungi in *Oryza sativa* plants. Moreover, Wang et al. (2018) found that the dose of 500 mg kg^{-1} ZnONPs reduced the colonization rate of arbuscular mycorrhizal fungi in the root of *Zea mays* under different phosphorus concentrations.

In relation to the enzymatic activity in the rhizospheric soil, Zhao et al. (2020) reported that copper oxide NPs exhibited changes such as inhibition of nitrate reductase and nitric oxide reductase. Likewise, it has been reported that ZnNPs at low concentrations increase catalase and urease and cause a reduction in high doses. Also, CuNPs produce a decrease in urease and polyphenol oxidase (Galaktionova et al. 2019). Besides, Eivazi et al. (2018) reported a temporary effect of reduction in the activity of acid phosphatase, β-glucosaminidase, β-glucosidase, and arylsulfatase due to the effect of AgNPs.

Based on these studies, it can be concluded that the possible adverse effect of metallic NPs on the soil microbiota will depend largely on the dose of the NPs. The influence of nanomaterials on the soil ecosystem is not easy to understand, since plant responses are different in combination with the soil microbiota. Therefore, among the most relevant factors to measure the effect of metal NPs on the soil microbial community, in addition to concentration, are the exposure time and soil texture (Grün et al. 2019). Indeed, it is necessary to consider the long-term exposure of soil microorganisms to NPs to carry out the impact assessment.

There are still many interactions and factors that must be studied before using metallic NPs, which is essential to continue with future research involving various crops, associated microorganisms, and the different types of nanomaterials.

Conclusions

Nanophytopathology is a new and emerging field with a great potential for managing diseases of wild and cultivated plants and ensuring food for humanity.

New nanotechnologies offer a variety of applications such as early detection of diseases by using nanosensors, direct control with NPs or as carriers for the active ingredient or volatile compound to reduce populations of phytopathogens. There are several studies on the toxicity of nanoparticles to the aquatic and soil ecosystem. However, to understand the real problems generated by the nanoparticles used, more thorough experimental trials are needed. In the future, the plant-microbe/pathogen interaction will open up new avenues in nanophytopathology.

Acknowledgement

Mahendra Rai thankfully acknowledges the financial support rendered by the Polish National Agency for Academic Exchange (NAWA) (Project No. PPN/ULM/2019/1/00117/U/00001).

References

Abbas, Q., Yousaf, B., Ullah, H., Ali, M.U., Ok, Y.S. and Rinklebe, J. 2020. Environmental transformation and nano-toxicity of engineered nano-particles (ENPs) in aquatic and terrestrial organisms. Crit. Rev. Environ. Sci. Technol. 50(23): 2523–2581. https://doi.org/10.1080/10643389.2019.1705721.

AbdelRahim, K., Mahmoud, S.Y., Ali, A.M., Almaary, K.S., Mustafa, A.E.Z.M. and Husseiny, S.M. 2017. Extracellular biosynthesis of silver nanoparticles using *Rhizopus stolonifer*. Saudi J. Biol. Sc. 24(1): 208–216. https://doi.org/10.1016/j.sjbs.2016.02.025.

Adeel, M., Farooq, T., White, J.C., Hao, Y., He, Z. and Rui, Y. 2021. Carbon-based nanomaterials suppress Tobacco Mosaic Virus (TMV) infection and induce resistance in *Nicotiana benthamiana*. J. Hazard. Mater. 404: 124167. https://doi.org/10.1016/j.jhazmat.2020.124167.

Agrios, G.N. 2005. Plant Pathology. Fifth Edition. Elsevier Academic Press. USA, 948 pp.

Akpinar, I., Unal, M. and Sar, T. 2021. Potential antifungal effects of silver nanoparticles (AgNPs) of different sizes against phythopathogenic (FORL) strains. SN Appl. Sci. 3: 506. https://doi.org/10.1007/s424552-021-04524-5.

Al-khattaf, F.S. 2021. Gold and silver nanoparticles: Green synthesis, microbes, mechanism, factors, plant disease management and environmental risks. Saudi J. Biol. Sci. 28(6): 3624–3631. https://doi.org/10.1016/j.sjbs.2021.03.078.

Anaya-López, J.L., Ibarra-Pérez, F.J., Rodríguez-Cota, F.G., Ortega-Murrieta, P.F., Acosta-Gallegos, J.A. and Chiquito-Almanza, E. 2021. Grain legumes in Mexico: Improved varieties of bean and chickpea developed by INIFAP. Rev. Mexicana Cienc. Agríc. 25: 63–75. https://doi.org/10.29312/remexca.v12i25.2827.

Arciniegas-Grijalba, P.A., Patiño-Portela, M.C., Mosquera-Sánchez, L.P., Sierra, B.G., Muñoz-Florez, J.E., Erazo-Castillo, L.A. et al. 2019. ZnO-based nanofungicides: Synthesis, characterization and their effect on the coffee fungi *Mycena citricolor* and *Colletotrichum* sp. Mater. Sci. Eng. C 98: 808–825. https://doi.org/10.1016/j.msec.2019.01.031.

Athié-García, M.S., Piñón-Castillo, H.A., Muñoz-Castellanos, L.N., Ulloa-Ogaz, A.L., Martínez-Varela, P.I., Quintero-Ramos, A. et al. 2018. Cell wall damage and oxidative stress in *Candida albicans* ATCC10231 and *Aspergillus niger* caused by palladium nanoparticles. Toxicol. *In Vitro* 48: 111–120. https://doi.org/10.1016/j.tiv.2018.01.006.

Avila-Quezada, G., Silva-Rojas, H., Sanchez-Chávez, E., Leyva-Mir, G., Martínez-Bolaños, L., Guerrero-Prieto, V. et al. 2016. Seguridad alimentaria: la continua lucha contra las enfermedades de los cultivos. TECNOCIENCIA Chihuahua 10(3): 133–142. https://vocero.uach.mx/index.php/tecnociencia/article/view/176.

Avila-Quezada, G.D. and Espino-Solis, G.P. 2019. Silver nanoparticles offer effective control of pathogenic bacteria in a wide range of food products. *In*: Pathogenic Bacteria. IntechOpen, pp. 203–213. http://dx.doi.org/10.5772/interchopen.89403.

Avila-Quezada, G.D., Esquivel, J.F., Silva-Rojas, H.V., Leyva-Mir, S.G., Garcia-Avila, C. de J., Noriega-Orozco, L. et al. 2018. Emerging plant diseases under a changing climate scenario: Threats to our global food supply. Emir. J. Food Agricul. 30(6) https://doi.org/10.9755/ejfa.2018.v30.i6.1715.

Avila-Quezada, G.D., Golinska, P. and Rai, M. 2021. Engineered nanomaterials in plant diseases: Can we combat phytopathogens? Appl. Microbiol. Biotechnol. 106: 117–129. https://doi.org/10.1007/s00253-022-11765-w.

Avila-Quezada, G.D., Ingle, A.P., Golińska, P. and Rai, M. 2022. Strategic applications of nano-fertilizers for sustainable agriculture: Benefits and bottlenecks. Nanotechnol. Rev. 11: 2123–2140. https://doi.org/10.1515/ntrev-2022-0126.

Barhoum, A., García-Betancourt, M.L., Jeevanandam, J., Hussien, E.A., Mekkawy, S.A., Mostafa, M. et al. 2022. Review on natural, incidental, bioinspired, and engineered nanomaterials: History, definitions, classifications, synthesis, properties, market, toxicities, risks, and regulations. Nanomater. 12(2): 177. https://doi.org/10.3390/nano12020177.

Barros, C.H. and Casey, E. 2020. A review of nanomaterials and technologies for enhancing the antibiofilm activity of natural products and phytochemicals. ACS Appl. Nano Mater. 3(9): 8537–8556. https://doi.org/10.1021/acsanm.0c01586.

Bayat, M., Zargar, M., Chudinova, E., Astarkhanova, T. and Pakina, E. 2021. *In vitro* evaluation of antibacterial and antifungal activity of biogenic silver and copper nanoparticles: The first report of applying biogenic nanoparticles against *Pilidium concavum* and *Pestalotia* sp. fungi. Molecules 26: 5402. https://doi.org/10.3390/molecules26175402.

Behera, N., Arakha, M., Priyadarshinee, M., Pattanayak, B.S., Soren, S., Jha, S. et al. 2019. Oxidative stress generated at nickel oxide nanoparticle interface results in bacterial membrane damage leading to cell death. RSC Adv. 9(43): 24888–24894. https://doi.org/10.1039/c9ra02082a.

Bernardo, P., Charles-Dominique, T., Barakat, M., Ortet, P., Fernandez, E., Filloux, D. et al. 2018. Geometagenomics illuminates the impact of agriculture on the distribution and prevalence of plant viruses at the ecosystem scale. ISME J. 12(1): 173–184. https://doi.org/10.1038/ismej.2017.155.

Bové, J.M. 2012. Huanglongbing and the future of citrus in Sao Paulo, State, Brazil. J. Plant Pathol. 94(3): 465–467. https://www.jstor.org/stable/45156274.

Cai, L., Cai, L., Jia, H., Liu, C., Wang, D. and Sun, X. 2020. Foliar exposure of Fe_3O_4 nanoparticles on *Nicotiana benthamiana*: Evidence for nanoparticles uptake, plant growth promoter and defense response elicitor against plant virus. J. Hazard. Mater. 393: 122415. https://doi.org/10.1016/j.jhazmat.2020.122415.

Cai, L., Liu, C., Fan, G., Liu, C. and Sun, X. 2019. Preventing viral disease by ZnONPs through directly deactivating TMV and activating plant immunity in *Nicotiana benthamiana*. Environ. Sci. Nano 6(12): 3653−3669. https://doi.org/10.1039/C9EN00850K.

Chávez-Mendoza, C. and Sánchez, E. 2017. Bioactive compounds from mexican varieties of the common bean (*Phaseolus vulgaris*): Implications for health. Molecules 22: 1360. http://dx.doi.org/10.3390/molecules22081360.

Chen, J., Wu, L., Lu, S., Li, Z. and Ding, W. 2020. Comparative study on the fungicidal activity of metallic MgO nanoparticles and macroscale MgO against soilborne fungal phytopathogens. Front. Microbiol. 11: 365. https://doi.org/10.3389/fmicb.2020.00365.

Clark, S.L., Schlarbaum, S.E., Saxton, A.M. and Baird, R. 2019. Eight-year blight (*Cryphonectria parasitica*) resistance of backcross-generation American chestnuts (*Castanea dentata*) planted in the southeastern United States. For. Ecol. Manag. 433: 153–161. https://doi.org/10.1016/j.foreco.2018.10.060.

Cooke, D.E.L., Cano, L.M., Raffaele, S., Bain, R.A., Cooke, L.R., Etherington, G.J. et al. 2012. Genome analyses of an aggressive and invasive lineage of the Irish potato famine pathogen. PLoS Pathog. 8(10): e1002940. https://doi.org/10.1371/journal.ppat.1002940.

Cui, H., Zhang, P., Gu, W. and Jiang, J. 2009. Application of anatase TiO sol derived from peroxotitannic acid in crop diseases control and growth regulation. Nanotech. 2: 286–289.

da Silva, B.L., Abuçafy, M.P., Manaia, E.B., Junior, J.A.O., Chiari-Andréo, B.G., Pietro, R.C.R. et al. 2019. Relationship between structure and antimicrobial activity of zinc oxide nanoparticles: An overview. Int. J. Nanomed. 14: 9395. https://doi.org/10.2147/IJN.S216204.

Derbalah, A.S.H. and Elsharkawy, M.M. 2019. A new strategy to control Cucumber mosaic virus using fabricated NiO-Nanostructures. J. Biotechnol. 306: 134–141. https://doi.org/10.1016/j.jbiotec.2019.10.003.

Du, J., Liu, B., Zhao, T., Xu, X., Lin, H., Ji, Y. et al. 2022. Silica nanoparticles protect rice against biotic and abiotic stresses. J. Nanobiotechnol. 20(1): 1–18. https://doi.org/10.1186/s12951-022-01420-x.

Eivazi, F., Afrasiabi, Z. and Elizabeth, J.O.S.E. 2018. Effects of silver nanoparticles on the activities of soil enzymes involved in carbon and nutrient cycling. Pedosphere 28(2): 209–214. https://doi.org/10.1016/S1002-0160(18)60019-0.

El-Dougdoug, N.K., Bondok, A.M. and El-Dougdoug, K.A. 2018. Evaluation of silver nanoparticles as antiviral agent against ToMV and PVY in tomato plants. Middle East J. Appl. Sci. 8(1): 100–111. https://www.curresweb.com/mejas/mejas/2018/100-111.pdf.

Esfanjani, A.F., Assadpour, E. and Jafari, S.M. 2018. Improving the bioavailability of phenolic compounds by loading them within lipid-based nanocarriers. Trends Food Sci. Technol. 76: 56–66. https://doi.org/10.1016/j.tifs.2018.04.002.

FAO. 2019. New standards to curb the global spread of plant pests and diseases. https://www.fao.org/news/story/en/item/1187738/icode/.

FAO. 2021. Climate change fans spread of pests and threatens plants and crops, new FAO study. https://www.fao.org/news/story/es/item/1402920/icode/.

FAO, IFAD, UNICEF, WFP and WHO. 2019. The State of Food Security and Nutrition in the World 2019. Safeguarding against economics low downs and downturns. Rome, FAO. https://docs.wfp.

org/api/documents/WFP-0000106760/download/?_ga=2.34432033.1420231576.1654830500-2024328212.1654830500.

Farooq, T., Adeel, M., He, Z., Umar, M., Shakoor, N., da Silva, W. et al. 2021. Nanotechnology and plant viruses: An emerging disease management approach for resistant pathogens. ACS Nano 15(4): 6030–6037. https://doi.org/ 10.1021/acsnano.0c10910.

Feng, Y., Cui, X., He, S., Dong, G., Chen, M., Wang, J. and Lin, X. 2013. The role of metal nanoparticles in influencing arbuscular mycorrhizal fungi effects on plant growth. Environ. Sci. Technol. 47(16): 9496–9504. https://doi.org/10.1021/es402109n.

Fortunati, E., Rescignano, N. and Botticella, E. 2016. Effect of poly (DL-lactide-coglycolide) nanoparticles or cellulose nanocrystals based formulations on *Pseudomonas syringae* pv. *tomato* (Pst) and tomato plant development. J. Plant Dis. Prot. https://doi.org/10.1007/s41348016-0036-x.

Fouda, M.M., Abdelsalam, N.R., Gohar, I.M.A., Hanfy, A.E., Othman, S.I., Zaitoun, A.F. et al. 2020. Utilization of high throughput microcrystalline cellulose decorated silver nanoparticles as an eco-nematicide on root-knot nematodes. Colloids Surf. B 188: 110805. https://doi.org/10.1016/j.colsurfb.2020.110805.

Freeman, B.C. and Beattie, G.A. 2008. An overview of plant defenses against pathogens and herbivores. Plant Health Instr. https://doi.org/10.1094/PHI-I-2008-0226-01.

Galaktionova, L., Gavrish, I. and Lebedev, S. 2019. Bioeffects of Zn and Cu nanoparticles in soil systems. Toxicol. Environ. Health Sci. 11(4): 259–270. https://doi.org/10.1007/s13530-019-0413-5.

Gettens, R.J. 1962. Maya Blue: An unsolved problem in ancient pigments. Am. Antiq. 27: 557–564. https://doi.org/10.2307/277679.

Giannousi, K., Avramidis, I. and Dendrinou-Samara, C. 2013. Synthesis, characterization and evaluation of copper based nanoparticles as agrochemicals against *Phytophthora infestans*. RSC Adv. 3(44): 21743–21752. https://doi.org/10.1039/C3RA42118J.

Goswami, A., Roy, I., Sengupta, S. and Debnath, N. 2010. Novel applications of solid and liquid formulations of nanoparticles against insect pests and pathogens. Thin Solid Films 519(3): 1252–1257. https://doi.org/10.1016/j.tsf.2010.08.079.

Goulson, D. 2014. Pesticides linked to bird declines. Nature 511: 295–296. https://doi.org/10.1038/nature13642.

Grün, A.L., Manz, W., Kohl, Y.L., Meier, F., Straskraba, S., Jost, C. et al. 2019. Impact of silver nanoparticles (AgNP) on soil microbial community depending on functionalization, concentration, exposure time, and soil texture. Environ. Sci. Eur. 31(1): 1–22. https://doi.org/10.1186/s12302-019-0196-y.

Gupta, S.D., Agarwal, A. and Pradhan, S. 2018. Phytostimulatory effect of silver nanoparticles (AgNPs) on rice seedling growth: An insight from antioxidative enzyme activities and gene expression patterns. Ecotoxicol. Environ. Saf. 161: 624–633. https://doi.org/10.1016/j.ecoenv.2018.06.023.

Haroon, M., Zaidi, A., Ahmed, B., Rizvi, A., Khan, M.S. and Musarrat, J. 2019. Effective inhibition of phytopathogenic microbes by eco-friendly leaf extract mediated silver nanoparticles (AgNPs). Indian J. Microbiol. 59(3): 273–287. https://doi.org/10.1007/s12088-019-00801-5.

Hernández-Díaz, J.A., Garza-García, J.J.O., Zamudio-Ojeda, A., León-Morales, J.M., López-Velázquez, J.C., García-Morales, S. 2021. Plant-mediated synthesis of nanoparticles and their antimicrobial activity against phytopathogens. J. Sci. Food Agricul. 101(4): 1270–1287. https://doi.org/10.1002/jsfa.10767.

Hernández-Montelongo, J., Nascimento, V.F., Murillo, D., Taketa, T.B., Sahoo, P., de Souza, A.A. et al. 2016. Nanofilms of hyaluronan/chitosan assembled layer-bylayer: an antibacterial surface for *Xylella fastidiosa*. Carbohydr. Polym. 136: 1–11. https://doi.org/10.1016/j.carbpol.2015.08.076.

Hogenhout, S.A., Ammar, E.D., Whitfield, A.E. and Redinbaugh, M.G. 2008. Insect vector interactions with persistently transmitted viruses. Annu. Rev. Phytopathol. 46(1): 327–359. https://doi.org/10.1146/annurev.phyto.022508.092135.

Huang, B., Chen, F., Shen, Y., Qian, K., Wang, Y., Sun, C. et al. 2018. Advances in targeted pesticides with environmentally responsive controlled release by nanotechnology. Nanomater. 8(2): 102. https://doi.org/10.3390/nano8020102.

Huang, L., Li, D.Q., Lin, Y.J., Wei, M., Evans, D.G. and Duan, X. 2005. Controllable preparation of nano-MgO and investigation of its bactericidal properties. J. Inorg. Biochem. 99(5): 986–993. https://doi.org/10.1016/j.jinorgbio.2004.12.022.

Huang, Y.C., Fan, R., Grusak, M.A., Sherrier, J.D. and Huang, C.P. 2014. Effects of nano-ZnO on the agronomically relevant Rhizobium-legume symbiosis. Sci. Total Environ. 497: 78–90. http://dx.doi.org/10.1016/j.scitotenv.2014.07.100.

Javed, B. and Nadhman, A. 2020. Optimization, characterization and antimicrobial activity of silver nanoparticles against plant bacterial pathogens phyto-synthesized by *Mentha longifolia*. Mater. Res. Express 7(8): 085406. https://doi.org/10.1088/2053-1591/abak19.

Jayaseelan, C., Rahuman, A.A. and Kirthi, A.V. 2012. Novel microbial route to synthesize ZnO nanoparticles using *Aeromonas hydrophila* and their activity against pathogenic bacteria and fungi. Spectrochim. Acta A Mol. Biomol. Spectrosc. 90: 78–84. https://doi.org/10.1016/j.saa.2012.01.006.

Johnson, M.S., Sajeev, S. and Nair, R.S. 2021. Role of Nanosensors in agriculture. *In*: 2021 International Conference on Computational Intelligence and Knowledge Economy (ICCIKE), pp. 58–63. IEEE. https://doi.org/10.1109/ICCIKE51210.2021.9410709.

Kalia, A., Sharma, S.P., Kaur, H. and Kaur, H. 2020. Novel nanocomposite-based controlled-release fertilizer and pesticide formulations: Prospects and challenges. *In*: Multifunctional Hybrid Nanomaterials for Sustainable Agri-food and Ecosystems. Elsevier, pp. 99–134. https://doi.org/10.1016/B978-0-12-821354-4.00005-4.

Katyal, P., Meleties, M. and Montclare, J.K. 2019. Self-assembled protein-and peptide-based nanomaterials. ACS Biomater. Sci. Eng. 5(9): 4132–4147. https://doi.org/10.1021/acsbiomaterials.9b00408.

Kaushal, M. and Wani, S.P. 2017. Nanosensors: Frontiers in precision agriculture. *In*: Nanotechnology. Springer, Singapore, pp. 279–291. https://doi.org/10.1007/978-981-10-4573-8_13.

Khanna, K., Kohli, S.K., Handa, N., Kaur, H., Ohri, P., Bhardwaj, R. et al. 2021. Enthralling the impact of engineered nanoparticles on soil microbiome: A concentric approach towards environmental risks and cogitation. Ecotoxicol. Environ. Saf. 222: 112459. https://doi.org/10.1016/j.ecoenv.2021.112459.

Kowalczuk, A., Trzcinska, R., Trzebicka, B., Müller, A.H., Dworak, A. and Tsvetanov, C.B. 2014. Loading of polymer nanocarriers: Factors, mechanisms and applications. Prog. Polym. Sci. 39(1): 43–86. https://doi.org/10.1016/j.progpolymsci.2013.10.004.

Kumar, P., Pandhi, S., Mahato, D.K., Kamle, M. and Mishra, A. 2021. Bacillus-based nano-bioformulations for phytopathogens and insect–pest management. Egypt. J. Biol. Pest Control 31(1): 1–11. https://doi.org/10.1186/s41938-021-00475-6.

Kumar, R., Duhan, J.S., Manuja, A., Kaur, P., Kumar, B. and Sadh, P.K. 2022. Toxicity assessment and control of early blight and stem rot of *Solanum tuberosum* L. by mancozeb-loaded chitosan–gum Acacia nanocomposites. J. Xenobiot. 12(2): 74–90. https://doi.org/10.3390/jox12020008.

Kutawa, A.B., Ahmad, K., Ali, A., Hussein, M.Z., AbdulWahab, M.A., Adamu, A. et al. 2021. Trends in nanotechnology and its potentialities to control plant pathogenic fungi: A review. Biology 10: 881. https://doi.org/10.3390/biology10090881.

Lakshmeesha, T.R., Murali, M., Ansari, M.A., Udayashankar, A.C., Alzohairy, M.A., Almatroudi, A. et al. 2020. Biofabrication of zinc oxide nanoparticles from *Melia azedarach* and its potential in controlling soybean seed-borne phytopathogenic fungi. Saudi J. Biol. Sci. 27(8): 1923–1930. https://doi.org/10.1016/j.sjbs.2020.06.013.

Le, V.T., Bach, L.G., Pham, T.T., Le, N.T.T., Phan Ngoc, U.T., Nguyen Tran, D.H. et al. 2019. Synthesis and antifungal activity of chitosan-silver nanocomposite synergize fungicide against *Phytophthora capsici*. J. Macromol. Sci. A 56(6): 522–528. https://dx.doi.org/10.1080/10601325.2019.1586439.

Linklater, D.P., Baulin, V.A., Le Guével, X., Fleury, J.B., Hanssen, E., Nguyen, T.H.P. et al. 2020. Antibacterial action of nanoparticles by lethal stretching of bacterial cell membranes. Adv. Mater. 32(52): 2005679. https://doi.org/10.1002/adma.202005679.

Lira-Saldivar, R.H., Méndez-Argüello, B., Vera-Reyes, I. and de los Santos Villarreal, G. 2018. Agronanotechnology: A new tool for modern agriculture. Rev. Fac. Ciencias Agrar. 50(2): 95–411. https://www.cabdirect.org/cabdirect/abstract/20193435227.

Liu, T., Wan, A., Liu, D. and Chen, X. 2017. Changes of races and virulence genes in *Puccinia striiformis* f. sp. *tritici*, the wheat stripe rust pathogen, in the United States from 1968 to 2009. Plant Dis. 101: 1522–1532. https://doi.org/10.1094/PDIS-12-16-1786-RE.

Mao, B.H., Chen, Z.Y., Wang, Y.J. and Yan, S.J. 2018. Silver nanoparticles have lethal and sublethal adverse effects on development and longevity by inducing ROS-mediated stress responses. Sci. Rep. 8(1): 1–16. https://doi.org/10.1038/s41598-018-20728-z.

Mao, B.H., Tsai, J.C., Chen, C.W., Yan, S.J. and Wang, Y.J. 2016. Mechanisms of silver nanoparticle-induced toxicity and important role of autophagy. Nanotoxicol. 10(8): 1021–1040. https://doi.org/10.1080/17435390.2016.1189614.

Miri, A., Mahdinejad, N., Ebrahimy, O., Khatami, M. and Sarani, M. 2019. Zinc oxide nanoparticles: Biosynthesis, characterization, antifungal and cytotoxic activity. Mater. Sci. Eng. C 104: 109981. https://doi.org/10.1016/j.msec.2019.109981.

Mishra, P.K., Mishra, H., Ekielski, A., Talegaonkar, S. and Vaidya, B. 2017. Zinc oxide nanoparticles: A promising nanomaterial for biomedical applications. Drug Discov. Today 22(12): 1825–1834. http://dx.doi.org/10.1016/j.drudis.2017.08.006.

Moreno, P., Ambros, S., Albiach-Martí, M.R., Guerri, J. and Peña, L. 2008. Citrus tristeza virus: A pathogen that changed the course of the citrus industry. Mol. Plant Pathol. https://doi.org/10.1111/J.1364-3703.2007.00455.X.

Mostafalou, S. and Abdollahi, M. 2013. Pesticides and human chronic diseases: Evidences, mechanisms, and perspectives. Toxicol. Appl. Pharmacol. 268(2): 157–177. https://doi.org/10.1016/j.taap.2013.01.025.

Narendhran, S. and Sivaraj, R. 2016. Biogenic ZnO nanoparticles synthesized using *L. aculeata* leaf extract and their antifungal activity against plant fungal pathogens. Bull. Mater. Sci. 39(1): 1–5. https://doi.org/10.1007/s12034-015-1136-0.

Nekrasova, G.F., Ushakova, O.S., Ermakov, A.E., Uimin, M.A. and Byzov, I.V. 2011. Effects of copper (II) ions and copper oxide nanoparticles on *Elodea densa* Planch. Russ. J. Ecol. 42: 458–463. https://doi.org/10.1134/S1067413611060117.

Nicolopoulou-Stamati, P., Maipas, S., Kotampasi, C., Stamatis, P. and Hens, L. 2016. Chemical pesticides and human health: The urgent need for a new concept in agriculture. Front. Public Health 4: 148. https://doi.org/10.3389/fpubh.2016.00148.

Nourozi, E., Hosseini, B., Maleki, R. and Mandoulakani, B.A. 2019. Pharmaceutical important phenolic compounds overproduction and gene expression analysis in *Dracocephalum kotschyi* hairy roots elicited by SiO_2 nanoparticles. Ind. Crops Prod. 133: 435–446. https://doi.org/10.1016/j.indcrop.2019.03.053.

Nurfatihah, Z. and Siddiquee, S. 2019. Nanotechnology: Recent trends in food safety, quality and market analysis. pp. 283–293. *In*: Siddiquee, S., Melvin, G. and Rahman, M. (eds.). Nanotechnology: Applications in Energy, Drug and Food. Springer, Cham. https://doi.org/10.1007/978-3-319-99602-8_14.

Nuruzzaman, M.D., Rahman, M.M., Liu, Y. and Naidu, R. 2016. Nanoencapsulation, nano-guard for pesticides: A new window for safe application. J. Agric. Food Chem. 64(7): 1447–1483. https://doi.org/10.1021/acs.jafc.5b05214.

Omanović-Mikličanina, E. and Maksimović, M. 2016. Nanosensors applications in agriculture and food industry. Bull. Chem. Technol. Bosnia Herzegovina 47: 59–70. http://hemija.pmf.unsa.ba/glasnik/files/Issue%2047/5-59-70-Omanovic.pdf.

Ordoñez-Beltrán, V., Frías-Moreno, M.N., Parra-Acosta, H. and Martínez-Tapia, M.E. 2019. Estudio sobre el uso de plaguicidas y su posible relación con daños a la salud. Revista Toxicol. 36(2): 148–153. https://www.redalyc.org/articulo.oa?id=91967023011.

Pariona, N., Mtz-Enriquez, A.I., Sánchez-Rangel, D., Carrión, G., Paraguay-Delgado, F. and Rosas-Saito, G. 2019. Green-synthesized copper nanoparticles as a potential antifungal against plant pathogens. RSC Adv. 9(33): 18835–18843. https://doi.org/10.1039/c9ra03110c.

Parveen, A. and Siddiqui, Z.A. 2022. Zirconium dioxide nanoparticles affect growth, photosynthetic pigments, defense enzymes activities and mitigate severity of bacterial and fungal diseases of tomato. Gesunde Pflanzen, 1–14. https://doi.org/10.1007/s10343-022-00636-z.

Pérez-de-Luque, A. 2017. Interaction of nanomaterials with plants: What do we need for real applications in agriculture? Front. Environ. Sci. 5(12): 1–7. https://doi.org/10.3389/fenvs.2017.00012.

Pham, N.D., Duong, M.M., Le, M.V. and Hoang, H.A. 2019. Preparation and characterization of antifungal colloidal copper nanoparticles and their antifungal activity against *Fusarium oxysporum* and *Phytophthora capsici*. Comptes Rendus Chimie, 22(11-12): 786–793. https://doi.org/10.1016/j.crci.2019.10.007.

Picchi, V., Gobbi, S., Fattizzo, M., Zefelippo, M. and Faoro, F. 2021. Chitosan nanoparticles loaded with N-acetyl cysteine to mitigate ozone and other possible oxidative stresses in durum wheat. Plants 10(4): 691. https://doi.org/10.3390/plants10040691.

Priyanka, P., Kumar, D., Yadav, A. and Yadav, K. 2020. Nanobiotechnology and its application in agriculture and food production. *In*: Thangadurai, D., Sangeetha, J. and Prasad, R. (eds.). Nanotechnology for Food, Agriculture, and Environment. Nanotechnology in the Life Sciences. Springer, Cham. https://doi.org/10.1007/978-3-030-31938-0_6.

Rai, M. and Ingle, A. 2012. Role of nanotechnology in agriculture with special reference to management of insect pests. Appl. Microbiol. Biotechnol. 94(2): 287–293. https://doi.org/10.1007/s00253-012-3969-4.

Raj, A., Shah, P. and Agrawal, N. 2017. Dose-dependent effect of silver nanoparticles (AgNPs) on fertility and survival of *Drosophila*: An *in-vivo* study. PLoS One 12(5): e0178051. https://doi.org/10.1371/journal.pone.0178051.

Rangaraj, S.R., Gopalu, K., Muthusamy, P., Rathinam, Y., Venkatachalam, R. and Narayanasamy, K. 2014. Augmented biocontrol action of silica nanoparticles and *Pseudomonas fluorescens* bioformulant in maize (*Zea mays* L.). RSC Adv. 4: 8461–8465. https://doi.org/10.1039/C3RA46251J.

Rivas-Valencia, P., Rosales-Rivas, L.A., Ávila-Quezada, G.D. and Martínez-Martínez, T.O. 2021. Economy of the mexican agriculture sector in times of COVID-19. Mexican J. Phytopathol. 39(4): 176–190. https://doi.org/10.18781/R.MEX.FIT.2021-21.

Rizzo, D., Garbelotto, M. and Hansen, E.M. 2005. *Phythophora ramorum*: Integrative research and management of an emerging pathogen in California and Oregon forests. Annu. Rev. Phytopathol. 43: 309–335. https://doi.org/10.1146/annurev.phyto.42.040803.140418.

Saha, N. and Dutta Gupta, S. 2018. Promotion of shoot regeneration of *Swertia chirata* by biosynthesized silver nanoparticles and their involvement in ethylene interceptions and activation of antioxidant activity. PCTOC 134(2): 289–300. https://doi.org/10.1007/s11240-018-1423-8.

Sang, H. and Lee, H.B. 2020. Molecular mechanisms of succinate dehydrogenase inhibitor resistance in phytopathogenic fungi. Res. Plant Dis. 26(1): 1–7. https://doi.org/10.5423/RPD.2020.26.1.1.

Sarabia-Castillo, C.R., Torres-Gómez, A.P., Flores-Rentería, D.Y. and Fernández-Luqueño, F. 2022. Influencia de nanopartículas de Fe_2O_3 sobre la colonización de hongos micorrízicos arbusculares en raíces de trigo. 45 Congreso Ciencia del suelo, 446–457. https://www.researchgate.net/publication/359152567_Influencia_de_nanoparticulas_de_Fe_2_O_3_sobre_la_colonizacion_de_hongos_micorrizicos_arbusculares_en_raices_de_trigo?enrichId=rgreq-50f319a676d27156d6a8e9b4c12ef16c-XXX&enrichSource=Y292ZXJQYWdlOzM1OTE1MjU2NztBUzoxMTMyMjUzNDEzNDE2OTYwQDE2NDY5NjE2NTgxMTk%3D&el=1_x_2&_esc=publicationCoverPdf.

Sastry, K.S. and Zitter, T.A. 2014. Management of virus and viroid diseases of crops in the tropics. pp. 149–480. *In*: Sastry, K.S. and Zitter, T.A. (eds.) Plant Virus and Viroid Diseases in the Tropics: Epidemiology and Management. Springer: Dordrecht, Netherlands. https://doi.org/10.1007/978-94-007-7820-7_2.

Saunders, D.G.O., Pretorius, Z.A. and Hovmøller, M.S. 2019. Tackling the re-emergence of wheat stem rust in Western Europe. Communic. Biol. 2: 51. https://doi.org/10.1038/s42003-019-0294-9.

Savary, S., Willocquet, L., Pethybridge, S.J., Esker, P., McRoberts, N. and Nelson, A. 2019. The global burden of pathogens and pests on major food crops. Nat. Ecol. Evol. 3: 430–439. https//doi.org/10.1038/s41559-018-0793-y.

Schneider, K., van der Werf, W., Cendoya, M., Mourits, M., Navas-Cortes J.A., Vicent, A. et al. 2020. Impact of *Xylella fastidiosa* subspecies *pauca* in European olives. PNAS 117: 9250–9259. https://doi.org/10.1073/pnas.1912206117.

Schumann, G.L. 1991. Plant Diseases: Their Biology and Social Impact. St Paul, USA, APS Press.

Shafie, R.M., Salama, A.M. and Farroh, K. 2018. Silver nanoparticles activity against Tomato spotted wilt virus. Middle East J. Appl. Sci. 7: 1251–1267. https://www.curresweb.com/mejar/mejar/2018/1251-1267.pdf.

Simonin, M. and Richaume, A. 2015. Impact of engineered nanoparticles on the activity, abundance, and diversity of soil microbial communities: A review. Environ. Sci. Pollut. Res. 22(18): 13710–13723. https://doi.org/10.1007/s11356-015-4171-x.

Singh, A., Singh, N.Á., Afzal, S., Singh, T. and Hussain, I. 2018. Zinc oxide nanoparticles: A review of their biological synthesis, antimicrobial activity, uptake, translocation and biotransformation in plants. J. Mater. Sci. 53(1): 185–201. https://doi.org/10.1007/s10853-017-1544-1.

Singh, R.P., Hodson, D.P., Huerta-Espino, J., Jin, Y., Bhavani, S., Njau, P. et al. 2011. The emergence of Ug99 races of the stem rust fungus is a threat to world wheat production. Annu. Rev. Phytopathol. 49: 465–481. https://doi.org/10.1146/annurev-phyto-072910-095423.

Slavin, Y.N., Asnis, J., Häfeli, U.O. and Bach, H. 2017. Metal nanoparticles: Understanding the mechanisms behind antibacterial activity. J. Nanobiotechnol. 15(1): 1–20. https://doi.org/10.1186/s12951-017-0308-z.

Song, Y., Chen, D., Lu, K., Sun, Z. and Zeng, R. 2015. Enhanced tomato disease resistance primed by arbuscular mycorrhizal fungus. Front. Plant Sci. 6(786): 1–13. https://doi.org/10.3389/fpls.2015.00786.

Tee, J.K., Ong, C.N., Bay, B.H., Ho, H.K. and Leong, D.T. 2016. Oxidative stress by inorganic nanoparticles. Wiley Interdisciplinary Reviews: Nanomed. Nanobiotechnol. 8(3): 414–438. https://doi.org/10.1002/wnan.1374.

Thabet, A.F., Boraei, H.A., Galal, O.A., El-Samahy, M.F., Mousa, K.M., Zhang, Y.Z. et al. 2021. Silica nanoparticles as pesticide against insects of different feeding types and their non-target attraction of predators. Sci. Rep. 11(1): 1–13. https://doi.org/10.1038/s41598-021-93518-9.

Tian, H., Kah, M. and Kariman, K. 2019. Are nanoparticles a threat to mycorrhizal and rhizobial symbioses? A critical review. Front. Microbiol. 10: 1660. https://doi.org/10.3389/fmicb.2019.01660.

Ulloa-Ogaz, A.L., Piñón-Castillo, H.A., Muñoz-Castellanos, L.N., Athie-García, M.S., Ballinas-Casarrubias, M.D.L., Murillo-Ramirez, J.G. et al. 2017. Oxidative damage to *Pseudomonas aeruginosa* ATCC 27833 and *Staphylococcus aureus* ATCC 24213 induced by CuO-NPs. Environ. Sci. Pollut. Res. 24(27): 22048–22060. https://doi.org/10.1007/s11356-017-9718-6.

UN. 2019. United Nations, Department of Economic and Social Affairs, Population Division. Growing at a slower pace, world population is expected to reach 9.7 billion in 2050 and could peak at nearly 11 billion around 2100. https://www.un.org/development/desa/en/news/population/world-population-prospects-2019.html.

UNEP. 2016. Food Systems and Natural Resources: A Report of the Working Group on Food Systems of the International Resource Panel. Westhoek, H., Ingram, J., Van Berkum, S., Özay, L. and Hajer, M. (eds.). United Nations Environment Programme. https://www.resourcepanel.org/sites/default/files/documents/document/media/food_systems_summary_report_english.pdf.

Vargas-Hernandez, M., Macias-Bobadilla, I., Guevara-Gonzalez, R.G., Rico-Garcia, E., Ocampo-Velazquez, R.V., Avila-Juarez, L. et al. 2020. Nanoparticles as potential antivirals in agriculture. Agriculture 10(10): 444. https://doi.org/10.3390/agriculture10100444.

Vazquez-Muñoz, R., Meza-Villezcas, A., Fournier, P.G.J., Soria-Castro, E., Juarez-Moreno, K., Gallego-Hernández, A.L. et al. 2019. Enhancement of antibiotics antimicrobial activity due to the silver nanoparticles impact on the cell membrane. PloS One 14(11): e0224904. https://doi.org/10.1371/journal.pone.0224904.

Velmurugan, P., Sivakumar, S., Young-Chae, S., Seong-Ho, J., Pyoung-In, Y., Jeong-Min, S. et al. 2015. Synthesis and characterization comparison of peanut shell extract silver nanoparticles with commercial silver nanoparticles and their antifungal activity. J. Ind. Eng. Chem. 31: 51–54. https://doi.org/10.1016/j.jiec.2015.06.031.

Wang, F., Jing, X., Adams, C.A., Shi, Z. and Sun, Y. 2018. Decreased ZnO nanoparticle phytotoxicity to maize by arbuscular mycorrhizal fungus and organic phosphorus. Environ. Sci. Pollut. Res. 25: 23736–23747. https://doi.org/10.1007/s11356-018-2452-x.

Xue, B., He, D., Gao, S., Wang, D., Yokoyama, K. and Wang, L. 2016. Biosynthesis of silver nanoparticles by the fungus *Arthroderma fulvum* and its antifungal activity against genera of *Candida, Aspergillus* and *Fusarium*. Int. J. Nanomed. 11: 1899–1906. https://doi.org/10.2147/IJN.S98339.

Yehia, R.S. and Ahmed, O.F. 2013. *In vitro* study of the antifungal efficacy of zinc oxide nanoparticles against *Fusarium oxysporum* and *Penicilium expansum*. African J. Microbiol. Res. 7(19): 1917–1923. https://doi.org/10.5897/AJMR2013.5668.

Zaini, P.A., De La Fuente, L., Hoch, H.C. and Burr, T.J. 2009. Grapevine xylem sap enhances biofilm development by *Xylella fastidiosa*. FEMS Microbiol. Lett. 295(1): 129–134. https://doi.org/10.1111/j.1574-6968.2009.01597.x.

Zhao, S., Su, X., Wang, Y., Yang, X., Bi, M., He, Q. et al. 2020. Copper oxide nanoparticles inhibited denitrifying enzymes and electron transport system activities to influence soil denitrification and N_2O emission. Chemosphere 245: 125394. https://doi.org/10.1016/j.chemosphere.2019.125394.

CHAPTER 2

The Perspective of Nanotechnology in Plant Protection

Oluwatoyin Adenike Fabiyi

Introduction

The study of extraordinarily small substances is termed nanotechnology. These small substances are engaged in all science disciplines like engineering, chemistry, physics, agriculture, biology and medicine. The dimensions of nanoparticles are of the order of billions of meters (10^{-9} m, i.e., nm) (Fabiyi et al. 2020a). Nanoparticles are tailored to deliver a new and refined outlook to man, animal and plant disease determination. Nanoscale gadgets and appliances act on molecules of biological origin on both sides, that is, inside and on the surface of diseased cells. Nanotechnology has the huge potential to transform human lives, from the use of cheap, lightweight solar plastics, which makes solar energy widely available to cleaning of toxic chemical spills, as well as air-borne pollutants (Khan and Rizvi 2014). Nanotechnology has helped humans to make electrical lines, solar cells, and biofuels more efficient, and made nuclear reactors safer. This has led to immense progress in health care by facilitating the detection and treatment of complex diseases. Similarly, bottlenecks in crop production such as pest infestation, soil pollution, climate variability, low crop yield and substandard irrigation have been allayed.

The general application of nanotechnology is represented in Figure 1. This situation brought about an increase in application of inputs such as fertilizers and pesticides to boost soil nutrients and reduce the risk of pests and disease attacks. Although these exertions are vital to crop growth and development, over-reliance and unrestricted use of these materials have ruined the natural functioning of the soil, and the ecosystem is vitiated and deficient in important nutrients and minerals as a fallout of modification of the soil's chemical properties from extreme use of pesticides and

Department of Crop Protection, Faculty of Agriculture, University of Ilorin, Ilorin, Nigeria.
Email: fabiyitoyinike@hotmail.com

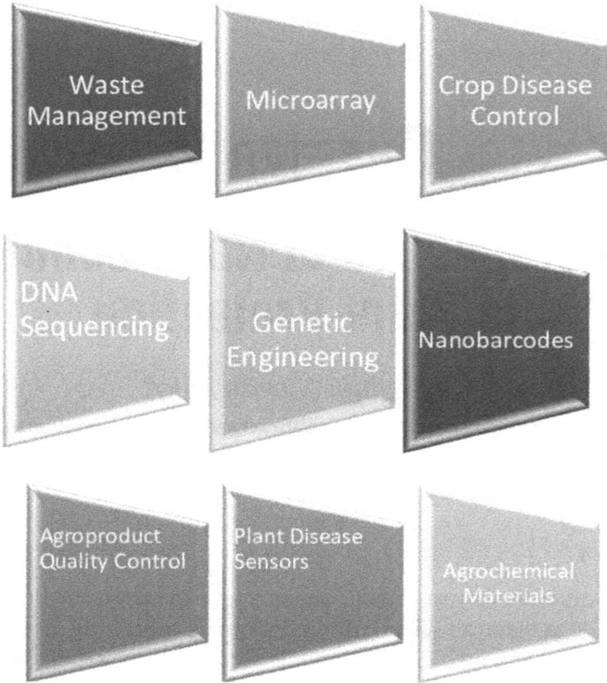

Figure 1: General application of nanotechnology.

fertilizers (Atolani et al. 2014a, b, Fabiyi et al. 2020b). To address challenges of crop production occasioned by food insecurity, diverse alternatives to pest control have been proposed (Fabiyi et al. 2021a, Fabiyi and Olatunji 2021a, b, Fabiyi et al. 2022, Fabiyi 2022a). Nevertheless, technically high agricultural farming systems supported by nanotechnology could be engaged in the bid or an attempt to revitalize the soil. The pursuit of nanotechnology for dependable agricultural practices would imply rock bottom use of mineral fertilizers and pesticides. It is expected that the farmlands will be preserved and soil beneficial micro-organisms will be prevented from eradication. Primarily, crop productivity is augmented by the use of nanomaterials through intensification of structured inputs for plant protection in agriculture. Dependable onsite conveyance of pesticides and nutrients is expected to reduce the conventional and general practice of application. Distinctly, it is projected that with nanotechnology, crops would adapt to factors associated with climate change and intense weather conditions on the field. In addition, nanotechnology will enhance in-depth control and monitoring of soil quality, plant health, pesticide delivery, disease occurrence and severity. Besides the aforementioned, agro-nanotechnology is expected to mend problems linked to conventional agriculture and reduce uncertainties. This chapter abridges agricultural activity with advancements in nanotechnology innovations for reasonable agricultural farming methods. Possible nanotools for a way forward in crop protection are equally examined.

Nanotechnology in agriculture

In agriculture, nanotechnology is employed in various areas such as in the production of chemical fertilizers, pesticides and other nanomaterials for improved plant health, growth and yield. This is achieved by producing and using nanomaterials which help to increase efficiency of output and also minimize relevant losses. More so, nanomaterials act as unique carriers of pesticides and fertilizers to facilitate intent on-site supervised conveyance of nutrients coupled with increase in crop development and management. Through their head-on premeditated and accurate application, nanotools like nano biosensors are employed in the operation and successful use of agricultural inputs in the form of various pesticides and fertilizers, thus bracing the growth of high-tech agricultural farms. The amalgamation of nanotechnology and biology into nano sensors has significantly reinforced the probable ability of the instrument to detect, perceive and recognize impairments in the environment (Verma et al. 2018). Furthermore, nanotechnology in agriculture includes the use of nanomaterials in diverse areas of agriculture, viz., crop growth, crop improvement, crop protection, precision farming, tolerance to stress and soil boosting. To improve or enhance plant growth, nano fertilizers are used, while nanotechnology for the cultivation of the transgenic plant, nano pesticides, nanosensors, nanoparticles and nanomaterials are substances in the nanoscale used for crop improvement, protection, precision farming, tolerance to stress and soil enhancement, respectively (Ghormade et al. 2011, Zaytsera et al. 2016, Loh et al. 2016, Kumar et al. 2019, McShane and Sunahara 2015, Khodakovskaya et al. 2009, Srinivasan and Saraswathi 2010). Farming systems and all areas of agriculture have been appreciably boosted through the use of nanotechnology. Crop production standard has improved with enhancement of all areas of farming (Figure 2).

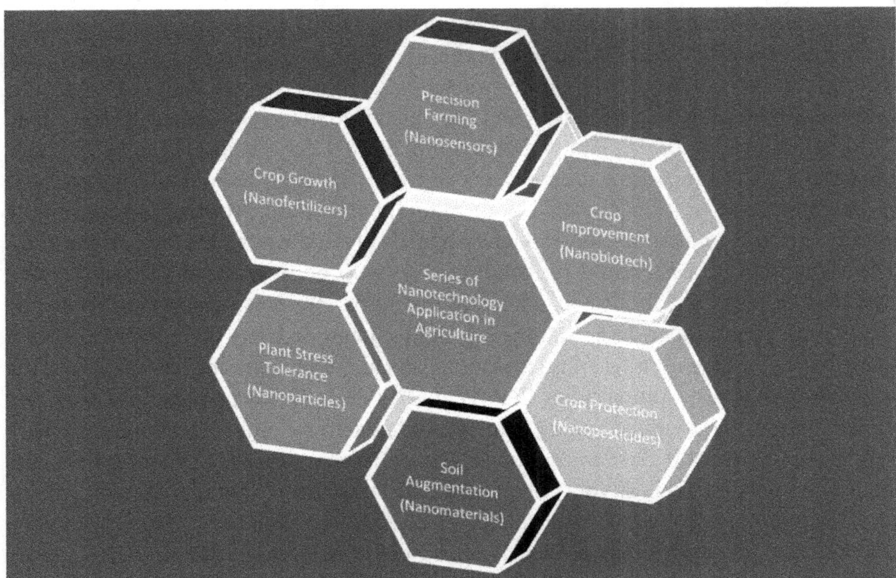

Figure 2: Application and significance of nanotechnology in agriculture.

World agriculture has been greatly revolutionized via the advent of engineered nanomaterials. Sustainable agriculture and global food demand have witnessed expansive development, thus guaranteeing the global food demand projection. Environmental protection, most especially elimination of pollution, is crucial in sustainable agriculture. A pollutant-free environment is pivotal for international trade and agriculture. The likelihood of environmental pollution is minimized by the application of nanotechnology in agriculture. It emboldens green agriculture with minimal danger to the environment; this is realized through nanomaterials which furnishes good management of plant protection inputs (Verma et al. 2018).

Nanotechnology and plant development

The use of agrochemicals in the nanoscale for the improvement of plant growth in recent years cannot be overemphasized. This practice is done mainly to boost food production using the best and healthy agricultural practices. Nano fertilizers (NF), Nano pesticides (NPst), nano organic and inorganic Nanoparticles (NPs), and nanosensors are some of the various nano substances that have been used to successfully boost plant production and also improve plant health (Galraith 2007). The influence of nanomaterials (NMs) on the growth and sprouting of plants with the main objective of promoting their use in agricultural applications (Torley et al. 2007) is important. The properties of NPs bring about countless changes in the physiological and physical traits and attributes of the plant (Akamwai et al. 2005), while the success of the NPs depends on the reactivity of the metal used, concentration, size and surface area after formation (Gopinath et al. 2014). Metal oxide NPs, metal NPs and carbon NMs (Torney et al. 2007) are the classes of nano particles, which are customarily used in agriculture. The NPs are capable of penetrating plant cells and, in some cases, transporting chemicals and DNA into the plant system (Akamwami et al. 2005, Gopinath et al. 2014). The most recent method for the synthesis of NPs is the green synthesis, where plant extracts are mostly used. This method of synthesis from proportionate metal ions is low-priced, chemical free, low-risk for biological applications and well-disposed to the environment (Ajey et al. 2014, Kitching et al. 2015, Fabiyi and Olatunji 2018, Fabiyi et al. 2021b). Generally, the methods of nanoparticle preparation are mainly through biological, physical and chemical synthesis. NPs obtained with minimal flaws are processed from minute entities through oxidation and reduction which are known as biological methods. The biological process could be conducted *in vivo* or *in vitro* and it requires the presence of plant metabolites phenolics, terpenoids, proteins, sugars, enzymes and flavonoids which function as stabilizing and reducing agents (Iravani 2011, Fabiyi et al. 2018). A large number of micro-organisms and plants are incorporated into the processing of green nanoparticles (Fabiyi et al. 2020c). Bacteria, algae and fungi are now a part of the synthesis. Gold nanoparticles are sometimes compounded with *Medicago sativa* and *Sesbania* species, while nanomaterials of inorganic nature such as copper, nickel, silver, cobalt and zinc have been synthesized with plants like *Brassica juncea*, *Medicago sativa*, *Helianthus annus*, *Eucalyptus officinalis*, *Cnidoscolus aconitifolius* and *Ficus mucoso* (Jampilek and Kral'ova 2015, Dubey and Mailapalli 2016, Joga et al. 2016, Fabiyi and Olatunj 2018, Fabiyi et al. 2020c,

Fabiyi 2021a). Sulphide nanoparticles like cadmium, silicon, zinc and gold are produced with microorganisms, and the ones of significance in the process are *Clostridium thermoaceticum*, *Desulfovibrio desulfuricans*, *Pseudomonas stuzeri*, *Klebsiella aerogens*, *Aspergillus flavus* and *Aspergillus fumigatus* (Gogas et al. 2012). Several other microorganisms have been integrated into nanoparticle formulations, predominantly in silver nano particle formulation (Table 1).

Table 1: Micro-organisms and nanoparticles' production.

Microorganisms	NPs	Size	Shape	References
Bacillus krulwichiae M2.5 and *Bacillus cellulosilyticus* M4.1	Silver	25.88 10.49 nm	Spherical, irregular	Galvez et al. (2019)
Bacillus licheniformis strain GPI-2	Gold	20–45 nm	Triangular, irregular, hexagonal	Thakur et al. (2018)
Pleurotus ostreatus	Silver	<40nm	Spherical	Al-Bahrani et al. (2017)
Trichoderma viride (MTCC 5661)	Silver	2–5 nm	Spherical	Kumari et al. (2017)
		40–65 nm	Rectangular	
		50–100 nm	Pentagonal and hexagonal	
Penicillium nalgiovense AJ12	Silver	25.2 ± 2.8 nm	Spherical	Maliszewska et al. (2014)
Macrophomina phaseolina	Silver	5–40 nm	Spherical	Chowdhury et al. (2014)
Humicola sp.	Silver	5–25 nm	Spherical	Syed et al. (2013)
Aspergillus terreus	Silver	1–20 nm	Spherical	Li et al. (2012)
Trichoderma Reesei	Silver	5–50 nm	–	Vahabi et al. (2011)
Penicillium brevicompactum	Silver	23–105 nm	–	Shaligram et al. (2009)
Penicillium fellutanum	Silver	5–25 nm	Spherical	Kathiresan et al. (2009)
Phoma glomerata	Silver	60–80 nm	Spherical	Birla et al. (2009)
Cladosporium cladosporioides	Silver	35 nm	Polydisperse and spherical	Balaji et al. (2009)
Coriolus versicolor	Silver	25–75, 444–491 nm	Spherical	Sanghi and Verma (2009)
Aspergillus flavus	Silver	8.92 nm	Spherical	Vigneshwaran et al. (2007)
Pediococcus pentosaceus	Silver	28.2–122 nm	–	Shahverdi et al. (2007)
Phanerochaete chrysosporium	Silver	5–200 nm	Pyramidal	Vigneshwaran et al. (2006)
Aspergillus fumigatus	Silver	5–25 nm	Spherical, triangular	Bhainsa and D'Souza (2006)
Fusarium oxysporum	Silver	20–50 nm	Spherical	Durán et al. (2005)

The most fascinating thing about nanotechnology is that it has helped to improve crop management and also reduced wastes in the customary agricultural operation, and therefore provides feasible progress (Nair et al. 2010, Yang et al. 2016). Traditionally, the spraying method is used in the application of agrochemicals, and in cases of granular formulations, broadcasting may be employed. For this reason, a minimal quantity of pesticides actually reached the desired goal. In this circumstance, the vital concentration and requisite quantity necessary for either plant growth or disease control are not realized. More so, since the appropriate distribution of agrochemicals into plant cells is statutory in precision farming, it is of great importance to ensure proper delivery to plants and soil (Panjatte et al. 2016). Several factors such as evaporation, microbial degradation, hydrolysis, chelation, leaching, photolysis and runoff minimize the amount of fertilizer and agrochemicals. The emergence of nanotechnology has corrected all these flaws in agriculture (Li et al. 2018, Shajaei et al. 2019). Extra attention has been redirected to nanotechnology in agriculture. Pesticide application mode such as controlled release has provided solutions to some of the problems encountered with pesticide application in customary farming systems (Dijk and Meijerink 2016, Fabiyi et al. 2020b). Loss and harmful effects of pesticides are greatly minimized through the technique of controlled delivery. Nanotechnology has advanced sufficiently from minor *in vitro* experimental trials to field applications where a sufficient amount of obligatory quantities of pesticides are delivered to plants and soil over time for attainment of biological capability (Kale and Gawade 2016). Delivery has been achieved through incorporation of nanomaterials into agrochemical carriage systems. Effective distribution of pesticides and fertilizers for improved plant health could be ascribed to some significant attributes of nanoparticles, which includes an easy attachment, large surface area and a quick mass transfer (Li et al. 2014); this subsequently leads to enhanced crop productivity. Nanomaterials also help to improve the strength and balance of pesticides and with this, they are safeguarded from degradation; ultimately, their effectiveness is increased while the toxicity of these pesticides is reduced in the environment. Some examples of nanoparticles that have been synthesized over the years to improve plant growth and health include silver nanoparticles (Ag-NPs), hydroxy fullerenes, iron (II) oxide (FeO) copper oxide nanoparticles (CuNPs), zinc oxide (ZnO), titanium dioxide (TiO_x), silicon dioxide (SiO_2), gold nanoparticles (AuNPs), Fe/SiO_2 and Multiwall Carbon Nanotubes (MWNCNTs).

Nanotechnology in crop production

Pests and disease are the bane of yield loss in crop production; diverse soil-dwelling micro-organisms such as fungi, bacteria and plant parasitic nematodes equally contribute significantly to a minimal yield of diverse crops (Atolani and Fabiyi 2020, Fabiyi et al. 2020d, Fabiyi 2020, Fabiyi 2021b, c, d, 2022b). Ameliorating soil fertility and crop output through the use of supplementary materials such as nanoparticles (Li et al. 2014) is germane to agriculturists, because food security is a global basic human right that is under consequential problem across the world. The limitation of obtainable natural resources is a threat to food security. Food security is intimidated comparatively due to the constrain of accessible natural means. It is expected that there will be an upsurge in the present human population worldwide, with an estimated value of about nine billion by the year 2050 (Kitching et al. 2015). An increase in intensive farming is necessary in order to meet the expected upsurge in food clamour which is equally expected to rise up to about 90% (Solanki et al. 2015). Intensive farming which in due course steers into an aggressive state of depletion of soil fertility and decrease in agricultural produce is being practiced and there are indications that roughly about 40% of world agricultural land has gravely degenerated, progressing into a drastic loss in soil fertility owing to exhaustive farming operation (Sabir et al. 2014). Hence, enormous quantity of fertilizers is used to upgrade soil fertility and crop yield (Godfray et al. 2010). Similarly, farm yield is associated with fertilizer application and efficient use of farm inputs. Again, the concentration of fertilizers getting to plants is dependent on nutrient use efficiency (Shukla et al. 2019). Researchers have reported that a large part of the sprayed or applied pesticide is lost due to several natural factors in the field (Lateef et al. 2016). The implication of this is that frequent use of excessive fertilizer turns out unfavourably with negative effects on nutrient equilibrium which is inherent in the soil. Apart from this, the unnecessary and excessive use of pesticides and fertilizers has impaired the environment, and water bodies are contaminated through leaching; thus, affected rivers become useless and unsafe owing to dissolved toxic substances (Millan et al. 2008). In order to curb the heavy contamination of the environment, which might consequently affect both plant and animal lives, it is important to produce engineered materials, which will help reduce the risk of contamination. The materials are then produced by the use of nanotechnology. The diverse use of nanomaterials in crop protection is presented in Table 2. Application of nanomaterials in agriculture has demonstrated absolutely a different technique for agricultural development, which is tailored to surmount the accustomed problems associated with the present traditional practice (Abdel-Aziz et al. 2016). The dawn of green nanotechnology has excitingly transformed the international agricultural structure and nanomaterials in form of nano-fertilizers, guaranteeing the expectation of worldwide food requests and dependable agricultural practices. In an effort to attenuate the macro and micronutrients inadequacy by improving nutrient use capability and to surmount the unwelcome situation of bioaccumulation, nano fertilizers can be a superlative relief (Dwivedi et al. 2016). Based on their activities, nano fertilizer can be categorized into different classes depending on their mode of action. There are a number of nanocomposite fertilizers that are combined as a single device; these include slow-release fertilizers and magnetic fertilizers, which are designed to supply micro and macronutrients to improve agriculture (Shalaby et al. 2016). The main strategy of production is through encapsulation of vital nutrients with nanomaterials. The process of production involves chemical and physical techniques which are termed bottom-up and top-down techniques, respectively; in due course, nutrients are then encapsulated into nanoporous materials. In some cases, the nanomaterials may be coated with a polymer film of thin nature, emulsions or particles of nanoscale dimension may also be used for delivery. Adequate and balanced fertilizer management with quality seeds combined with irrigation is known

Table 2: Diverse use of nanomaterials in crop protection.

Nanoparticles	**Crop pathogens and diseases**	**References**
Silver nanoparticles	*Alternaria alternata*	Krishnaraj et al. (2012), Bryaskova et al. (2014)
	Sclerotinia sclerotiorum	
	Macrophomina phaseolina	
	Rhizoctonia solani	
	Botrytis cinerea	
	Curvularia lunata	
	Sunn-hemp rosette virus	Jain and Kothari (2014)
	Bean yellow mosaic virus	Elbeshehy et al. (2015)
	Heterodera saccharri	Fabiyi et al. (2018a), Fabiyi et al. (2018b), Fabiyi et al. (2020)
	Meloidogyne incognita	Fabiyi and Olatunji (2018), Fabiyi (2020), Fabiyi et al. (2021a)
Titanium dioxide nanoparticles	Bacteria and inactivation of viruses	Sadeghi et al. (2017)
Poly-dispersed gold nanoparticles	Barley yellow mosaic virus	Alkubaisi et al. (2015)
Chitosan	Mosaic virus of alfalfa	Kochkina et al. (1994), Pospieszny et al. (1991), Chirkov (2002)
Fusarium sp., *Botrytis* sp. Bean mild mosaic virus *Pyricularia grisea*		Kashyap et al. (2015)
Tobacco mosaic virus Tobacco necrosis virus		Malerba and Raffaella (2018)
Oleander aphid (*Aphis nerii*) Cotton leafworm (*Spodoptera littoralis*) Root-knot nematode (*Meloidogyne javanica*) Nymphs of the pear psylla (*Cacopsylla pyricola*)		
Zinc nanoparticles	*Pseudomonas aeruginosa*, *Aspergillus flavus*	Jayaseelan et al. (2012), Rajput et al. (2018)
Ag NPs/PVP (hybrid mate rials based on polyvinylpyrrolidone with silver nanoparticles)	*Staphylococcus aureus* (gram positive bacteria), *E. coli* (gram negative bacteria), *P. aeruginosa* (gram-negative bacteria), as well as spores of *Bacillus subtilis*	Azam et al. (2012)
Candida albicans, C. krusei, C. tropicalis, C. glabrata, and *Aspergillus brasiliensis*		Bryaskova et al. (2011)
CuO NPs	*S. aureus, Bacillus subtilis, P. aeruginosa*, and *Escherichia coli*	Bryaskova et al. (2011), Azam et al. (2012)
Manganese and zinc	Damping off and charcoal rot diseases in sunflower	Abd El-Hai et al. (2009)

to increase output in agriculture to about 35–40% (Li et al. 2009). Chitosan, zeolites, and clay applied through porous nanomaterials minimises nitrogen loss in the field, by regulating release on a demand basis and boosting uptake for plants (Rico et al. 2011, Monreal et al. 2016). The solubility of phosphate minerals is aided by ammonium-charged zeolites. The charged zeolites also intensify the accessibility of phosphorus for plant use (Rico et al. 2011). More so, the loss of nutrients by leaching could be controlled by extending time of a function to lengthen potassium nitrate delivery with graphene oxide films, a carbon-containing nanomaterial (Jampilek and Kral'ova 2015). Research revealed that application of nano calcite ($CaCO_3$– 40%) coupled with nano SiO_2 (4%), MgO (1%) and Fe_2O_3 (1%) has significantly enhanced the uptake of phosphorus and other nutrients like Fe, Mg, Ca, Zn and Mn (Millan et al. 2008). Grain yields have been augmented with carbon nanoparticles coupled with fertilizer. Examples of such yield enhancement have been recorded in crops such as maize with 10.93%, soybean had 16.74%, 28.81% was noted for wheat, 10.29% for rice while 19.76% yield increment was observed in vegetables (Li et al. 2009, Jampilek and Kral'ova 2015). The porosity of root and leaf surfaces is at nanoscale, so the use of nanomaterials amplifies intake of nutrients by plants, thus improving the plant growth and biology (Monreal et al. 2016). Yield variables such as crop index, harvest index and mobilization index remarkably increased with the use of NPK-Chitosan fertilizer as against what was noted in control plants (Liu et al. 2009, Swaminathan et al. 2013). The horizon of nanotechnology is greatly widened in agriculture with the amalgamation of the nano in micro and macronutrients delivery. The addition of nano formulas and encapsulation of micro and macronutrients for use as controlled and slow release has conveniently solved the problem of nutrient deficiency in plants. By extension, reduction of macronutrients in plants equally influence humans when they consume unhealthy nutrient deficient foods. Some of the fallouts in health of humans include general health problems, anaemia, growth retardation and low cognition (Ali et al. 2019, Peteu et al. 2010). However, supplementing plants with nano formulated nutrients by way of slow release mechanism has enhanced the use of vital nutrients by plants, thus improving plant health, development and yield of crops (Khodakovskaya et al. 2009). This will invariably get to humans and any anomaly will be corrected by consuming healthy and nutritious food. The use of nano zinc oxide at low dosage has been reported to augment growth of roots, shoots and physiological reaction of plants grown in zinc deficient soil (Gao et al. 2011, Joshi et al. 2018). Furthermore, it has also been shown that utilization of zinc oxide nanoparticles, coupled with some different types of fertilizer in soil lacking in zinc, promoted nutrient use efficiency with a striking improvement in productivity of plants such as barley by 91% compared to the $ZnSO_4$ used as control, which increases productivity by 31% (Godfray et al. 2010). More so, researchers have shown that the implementation of control loss fertilizer in wheat boosted yield by 5.5%, there was about 9.8% rise in extra soil mineral nitrogen, while there was 24.5 and 21.6% drop in leaching and nitrogen loss from the soil, respectively, with nanomaterial fertilizers in comparison to customary fertilizer (Kole et al. 2013). The application of Multi-Walled Carbon Nanotubes (MWCNTs) has also been demonstrated to positively influence the sprouting of various crop varieties such as garlic, peanut, barley, soybean, tomato, maize and wheat (Davarpanah et al. 2016, Yousefzadah et al. 2016). Application of zeolite, nano silicon dioxide (SiO_2) and titanium dioxide (TiO_2) emphatically pushed up seed sprouting in crops. Fe/SiO_2 nano formations are equally known to have huge effects on seed sprouting. In addition, vegetable species and grain crops which comprise wheat, potato, onion, tomato, peanut, spinach, mustard, mung bean and soybean are described to have brought forth excellent fruit and leaves with the use of hydroxyl-fullerenes, MWCNTs, zinc oxide (ZnO), TiO_2, ZnFeCu-oxide and iron (II) oxide (FeO) nano formulations (Swaminathan et al. 2013). OH-functionalized fullerenes (fullerol, a carbon nano material) have regularly brought into play favourable influence on plant development as demonstrated by Gao et al. (2011); they documented that cell division were activated by fullerenes and there was a great improvement in hypocotyl growth in the thale cress, commonly known as Arabidopsis. Similarly, the levels of active principles and bioactive compounds like inulin, charantin, lycopene and cucurbitacin-B were elevated in *Momordica charantia*, while there was a remarkable increase in yield to the tune of 128%, and the number and size of fruits were equally raised (Lu et al. 2002, Banarjee et al. 2016). Essential oil constituents in *Dracocephalum moldavica* were intensified with nano iron fertilizer; this was coupled with improvement in vegetative growth and agronomic characters of the plant (Yousefzadeh and Sabaghnia 2016). The use of zinc and boron nano fertilizers did not have any negative effect on the characteristics of *Punica granatum*. Fruit yield and quality were upgraded with an increase in the TSS value of the fruit up to 7.6% (Davarpanah et al. 2016). The ability of nanomaterials to infiltrate seed coats has prompted the capability of plants to absorb and use water; this boosts the enzymatic system and refines the process of sprouting with accelerated seedling growth (Liu et al. 2016, Yousefzadah et al. 2016). The reports above exhibited the outlook and prospect of nanomaterials in agricultural science. There is affirmation that fertilizers incorporated into nanomaterials ensure long time accessibility of nutrients to the growing crops, which is not the case when the customary mode of application of fertilizers is employed. In all, the implementation of nanotechnology in all fields of agriculture is encouraged, owing to the potentiality of nanomaterials to revamp agriculture. Degenerated lands could equally be restored with the mastery of nano formulations, which promise a reassuring and encouraging result. Routinely, the concentration, size, nature, and mode of application and dosage of nanomaterials, as well as the general features of plant samples to be used for the research is a very important aspect to be studied in boosting plant health and growth for improved productivity in agriculture.

Conclusion

Nanotechnology is the most advanced research in all fields of scientific endeavour for the production of sophisticated and advanced materials for human use or consumption. In agriculture, the development of nanotechnologies such as nano-sensors, nano-pesticides, nano-fertilizers, and nanoparticles have helped in diverse areas such as precision farming, improved crop yield, and growth as well as enhanced plant health. All of these have contributed to the production of an abundance of different types of healthy crops for animal use and human consumption without posing a threat of toxicity in the environment. This has also reduced the cost and duration of crop production and growth, thus fostering sustainable agriculture with consistent improvement in production. In pursuance of disease management in agriculture, particularly in foremost crops, metal predicated pesticide residue analysis, nanosensors and biosensors could be employed in disease identification and forecasting. Consequently, colossal yield loss in principal crops would be curtailed. Furthermore, the wastage of resources by the smallholder farmers under the unregistered informal sector would be lessened.

References

Abd El-Hai, K.M., El-Metwally, M.A., El-Baz, S.M. and Zeid, A.M. 2009. The use of antioxidants and microelements for controlling damping-off caused by *Rhizoctonia solani* and charcoal rot caused by *Macrophomina phasoliana* on sunflower. Plant Pathol. J. 8(3): 79–89.

Abdel-Aziz, H.M.M., Hasaneen, M.N.A. and Omer, A.M. 2016. Nano chitosan-NPK fertilizer enhances the growth and productivity of wheat plants grown in sandy soil. Span. J. Agric. Res. 20 14: 17.

Al-Bahrani, R., Raman, J., Lakshmanan, H., Hassan, A.A. and Sabaratnam, V. 2017. Green synthesis of silver nanoparticles using tree oyster mushroom *Pleurotus ostreatus* and its inhibitory activity against pathogenic bacteria. Mater. Lett. 186: 21–25.

Ajey Singha, N., Singha, B., Imtiyaz Hussaina, Himani Singha and Singhb, S.C. 2014. Plant-nanoparticle interaction: An approach to improve agricultural practices and plant productivity. Intl. J. Pharm. Sci. Invention 4: 25–40.

Ali, S., Rizwan, M., Noureen, S., Anwar, S., Ali, B., Naveed, M., Abd_Allah, E.F., Alqarawi, A.A. and Ahmad, P. 2019. Combined use of biochar and zinc oxide nanoparticle foliar spray improved the plant growth and decreased the cadmium accumulation in rice (*Oryza sativa* L.) Plant. Environ. Sci. Pollut. Res. 26: 11288–11299.

Alkubaisi, N.A.O., Aref, N.M.M.A. and Hendi, A.A. 2015. Method of inhibiting plant virus using gold nanoparticles. US Patents US9198434B11.

Ankamwar, B., Chaudhary, M. and Sastry, M. 2005. Gold nanotriangles biologically synthesized using tamarind leaf extract and potential application in vapor sensing. Synthesis and Reactivity in Inorganic Metal-Organic and Nano Metal Chemistry35: 19–26. http://dx.doi.org/10.1081/SIM-20004752.

Atolani, O., Fabiyi, O.A. and Olatunji, G.A. 2014a. Isovitexin from *Kigelia pinnata*, a potential eco-friendly nematicidal agent. Trop. Agric. 91(2): 67–74.

Atolani, O., Fabiyi, O.A. and Olatunji, G.A. 2014b. Nematicidal isochromane glycoside from *Kigelia pinnata* leaves. Acta Agric. Slov. 104(1): 25–31.

Atolani, O. and Fabiyi, O.A. 2020. Plant parasitic nematodes management through natural products: Current progress and challenges. pp. 297–315. *In*: Ansari, R.A., Rizvi, R. and Mahmood, I. (eds.). Management of Phytonematodes: Recent Advances and Future Challenges. Singapore.

Azam, A.S., Ahmed, M., Oves, M.S., Khan, S.S. and Memic, H.A. 2012. Antimicrobial activity of metal oxide nanoparticles against gram-positive and gram-negative bacteria: A comparative study. Int. J. Nanomedicine 7: 6003–6009.

Banerjee, J. and Kole, C. 2016. Plant nanotechnology: An overview on concepts, strategies, and tools. pp. 1–14. *In*: Kole, C., Kumar, D. and Khodakovskaya, M. (eds.). Plant Nanotechnology. Springer, Cham, Switzerland.

Bryaskova, R., Pencheva, D., Nikolov, S. and Kantardjiev, T. 2014. Synthesis and comparative study on the antimicrobial activity of hybrid materials based on silver nanoparticles (AgNPs) stabilized by polyvinylpyrrolidone (PVP). J. Chem. Biol. 4: 185–191.

Chirkov, S. 2002. The antiviral activity of chitosan (review). Appl. Biochem. Microbiol. 38: 1–8.

Chowdhury, S., Basu, A. and Kundu, S. 2014. Green synthesis of protein capped silver nanoparticles from phytopathogenic fungus *Macrophomina phaseolina* (Tassi) Goid with antimicrobial properties against multidrug-resistant bacteria. Nano Res. Lett. 9: 365.

Davarpanah, S., Tehranifar, A., Davarynejad, G., Abadia, J. and Khorasani, R. 2016. Effects of foliar applications of zinc and boron nano-fertilizers on pomegranate (*Punica granatum* cv. Ardestani) fruit yield and quality. Sci. Hortic. 210: 57–64.

Dijk, V.M. and Meijerink, G.W. 2014. A review of food security scenario studies: Gaps and ways forward. pp. 30–32. *In*: Achterbosch, T.J., Dorp, M., van Driel, W.F., van Groot, J.J., Lee, J., van der Verhagen, A. and Bezlepkina, I. (eds.). The Food Puzzle: Pathways to Securing Food for All. Wageningen UR, Wageningen, Netherlands.

Dubey, A. and Mailapalli, D.R. 2016. Nanofertilisers, nanopesticides, nanosensors of pest and nanotoxicity in agriculture. pp. 307–330. *In*: Lichtfouse, E. (ed.). Sustainable Agriculture Reviews. Springer, Cham, Switzerland, 19.

Dwivedi, S., Saquib, Q., Al-Khedhairy, A.A. and Musarrat, J. 2016. Understanding the role of nanomaterials in agriculture. pp. 271–288. *In*: Singh, D.P., Singh, H.B. and Prabha, R. (eds.). Microbial Inoculants in Sustainable Agricultural Productivity. Springer: New Delhi, India, 2.

Elbeshehy, E.K.F., Elazzazy, A.M. and Aggelis, G. 2015. Silver nanoparticles synthesis mediated by new isolates of Bacillus sp. nanoparticle characterization and their activity against bean yellow mosaic virus and human pathogens. Front. Microbiol. 6: 453.

Fabiyi, O.A. and Olatunji, G.A. 2018. Application of green synthesis in nano particles preparation: *Ficus mucoso* extracts in the management of *Meloidogyne incognita* parasitizing groundnut *Arachis hypogea.* Indian J. Nematol. 48(1): 13–17.

Fabiyi, O.A., Olatunji, G.A. and Saadu, A.O. 2018a. Suppression of agricultural waste silver nano particles. The Journal of Solid Waste Management 44(2): 87–91.

Fabiyi, O.A., Olatunji, G.A., Osunlola, O.S. and Umar, K.A. 2018b. Efficacy of agricultural wastes in the control of rice cyst nematode (*Heterodera sacchari*). Agric. conspec. Sci. 83(4): 329–334.

Fabiyi, O.A. 2020. Growth and yield response of groundnut *Arachis hypogaea* (Linn.) under *Meloidogyne incognita* infection to furfural synthesised from agro-cellulosic materials. J. Trop. Agric. 58(2): 241–245.

Fabiyi, O.A., Alabi, R.O. and Ansari, R.A. 2020a. Nanoparticles' Synthesis and their application in the management of phytonematodes: An overview. pp. 125–140. *In*: Ansari, R.A., Rizvi, R. and Mahmood, I. (eds.). Management of Phytonematodes: Recent Advances and Future Challenges. Singapore.

Fabiyi, O.A., Atolani, O. and Olatunji, G.A. 2020d. Toxicity Effect of *Eucalyptus globulus* to *Pratylenchus* spp of *Zea mays.* Sarhad J. Agric. 36(4): 1244–1253.

Fabiyi, O.A., Olatunji, G.A., Atolani, O. and Olawuyi, R.O. 2020c. Preparation of bio-nematicidal nanoparticles of *Eucalyptus officinalis* for the control of cyst nematode (*Heterodera sacchari*). J. Anim. Plant. Sci. 30(5): 1172–1177.

Fabiyi, O.A., Saliu, O.D., Claudius-Cole, A.O., Olaniyi, I.O., Oguntebi, O.V and Olatunji, G.A. 2020b. Porous starch citrate biopolymer for controlled release of carbofuran in the management of root knot nematode *Meloidogyne incognita.* Biotechnol. Rep. 25(e00428).

Fabiyi, O.A. 2021a. Sustainable management of *Meloidogyne incognita* infecting carrot: Green synthesis of silver nanoparticles with *Cnidoscolus aconitifolius*: (*Daucus carota*). Vegetos 34(2): 277–285.

Fabiyi, O.A. 2021b. Evaluation of plant materials as root-knot nematode (*Meloidogyne incognita*) suppressant in okro (*Abelmuscous esculentus*). Agric. Conspec. Sci. 86(1): 51–56.

Fabiyi, O.A. 2021c. Evaluation of nematicidal activity of *Terminalia glaucescens* fractions against *Meloidogyne incognita* on *Capsicum chinense.* Hort. Res. 29(1): 67–74.

Fabiyi, O.A. 2021d. Application of furfural in sugarcane nematode pest management. Pak. J. Nematol. 39(2): 151–155. DOI | https://dx.doi.org/10.17582/journal.pjn/2021.39.2.151.155.

Fabiyi, O.A. and Olatunji, G.A. 2021a. Environmental sustainability: Bioactivity of *Leucaena leucocephala* leaves and pesticide residue analysis in tomato fruits. Acta Univ. Agric. et Silvic. Mendelianae Brun. 69(4): 473–480.

Fabiyi, O.A. and Olatunji, G.A. 2021b. Toxicity of derivatized citrulline and extracts of water melon rind (*Citrullus lanatus*) on root-knot nematode (*Meloidogyne incognita*). J. Trop. Agric. 98(4): 347–355.

Fabiyi, O.A., Claudius-Cole, A.O. and Olatunji, G.A. 2021a. *In vitro* assessment of n-phenyl imides in the management of *Meloidogyne incognita.* Sci. Agric. Bohem. 52(3): 60–65.

Fabiyi, O.A., Claudius-Cole, A.O., Olatunji, G.A., Abubakar, D.O. and Adejumo, O.A. 2021b. Evaluation of the *in vitro* response of *Meloidogyne incognita* to silver nano particle liquid from agricultural wastes. Agrivita J. Agricultural Sci. 43(3): 524–534.

Fabiyi, O.A. 2022a. Fractions from *Mangifera indica* as an alternative in *Meloidogyne incognita* management. Pak. J. Nematol. 40(1): 65–74.

Fabiyi, O.A. 2022b. Evaluation of weeds against root-knot nematode (*Meloidogyne incognita*) in vegetables. Sarhad J. Agric. 38(4): 1289–1299.

Fabiyi, O.A., Baker, M.T., Claudius-Cole, A.O., Alabi, R.O. and Olatunji, G.A. 2022. Application of cyclic imides in the management of root-knot nematode (*Meloidogyne incognita*) on cabbage. Indian Phytopathol. 75(2): 457–466. https://doi.org/10.1007/s42360-022-00480-1.

Galbraith, D.W. 2007. Nanobiotechnology: Silica breaks through in plants. Nat. Nanotechnol. 2(5): 272–273.

Galvez, A.M., Ramos, K.M., Alexis, J.T. and Baculi, R. 2019. Bacterial exopolysaccharide-mediated synthesis of silver nanoparticles and their application on bacterial biofilms. J. Microbiol. Biotechnol. Food Sci. 8(4): 970.

Gao, J., Wang, Y., Folta, K.M., Krishna, V., Bai, W., Indeglia, P., Georgieva, A., Nakamura, H., Koopman, B. and Moudgil, B. 2011. Polyhydroxy fullerenes (fullerols or fullerenols): Beneficial effects on growth and lifespan in diverse biological models. PLoS ONE 6: e19976. Doi:10.1371/ journal. pone.0019976.

Ghormade, V., Deshpande, M.V. and Paknikar, K.M. 2011. Perspectives for nano-biotechnology enabled protection and nutrition of plants. Biotechnol. Adv. 29: 792–803.

Godfray, H.C.J., Beddington, J.R., Crute, I.R., Haddad, L., Lawrence, D., Muir, J.F., Pretty, J., Robinson, S., Thomas, S.M. and Toulmin, C. 2010. Food security: The challenge of feeding 9 billion people. Science 327: 812–818.

Gogos, A., Knauer, K. and Bucheli, T.D. 2012. Nanomaterials in plant protection and fertilization: Current state, foreseen applications, and research priorities. J. Agric. Food Chem. 60: 9781–9792.

Gopinath, K., Shanmugam, V.K., Gowri Senthilkumar, V., Kumaresan, S. and Arumugam, A. 2014. Antibacterial activity of ruthenium nanoparticles synthesized using *Gloriosa superba* L. leaf extract. J. Nanostruct. Chem. 24: 83.

Iravani, S. 2011. Green synthesis of metal nanoparticles using plants. Green Chem. 13: 2638–2650.

Jain, D. and Kothari, S. 2014. Green synthesis of silver nanoparticles and their application in plant virus inhibition. J. Mycol. Plant Pathol. 44: 21.

Jampilek, J. and Kral'ova, K. 2015. Application of nanotechnology in agriculture and food industry, its prospects and risks. Ecol. Chem. Eng. S-Chemia I Inzynieria Ekologiczna S 22: 321–361.

Jayaseelan, C., Rahuman, A.A., Kirthi, A.V., Marimuthu, S., Santhoshkumar, T., Bagavan, A., Gaurav, K., Karthik, L. and Rao, K.V.B. 2012. Novel microbial route to synthesize ZnO nanoparticles using Aeromonas hydrophila and their activity against pathogenic bacteria and fungi. Spectrochim Acta A Mol. Biomol. Spectrosc. 90: 78–84.

Joga, M.R., Zotti, M.J., Smagghe, G. and Christiaens, O. 2016. RNAi efficiency, systemic properties, and novel delivery methods for pest insect control: What we know so far. Front. Physiol. 7: 553.

Joshi, A., Kaur, S., Dharamvir, K., Nayyar, H. and Verma, G. 2018. Multi-walled carbon nanotubes applied through seedpriming influence early germination, root hair, growth and yield of bread wheat (*Triticum aestivum* L.). J. Sci. Food Agri. 98: 3148–3160.

Kale, A.P. and Gawade, S.N. 2016. Studies on nanoparticle induced nutrient use efficiency of fertilizer and crop productivity. Green Chem. Technol. Lett. 2: 88–92.

Kashyap, P.L., Xiang, X. and Heiden, P. 2015. Chitosan nanoparticle based delivery systems for sustainable agriculture. Int. J. Boil Macromol. 77: 36–51.

Kathiresan, K., Manivannan, S., Nabeal, M.A. and Dhivya, B. 2009. Studies on silver nanoparticles synthesized by a marine fungus, *Pencillium fellutanum* isolated from coastal mangrove sediment. Colloids Surface B Biointerfaces 71: 133–137.

Khan, M.R. and Rizvi, T.F. 2014. Nano technology scope and application in plant disease management. Plant Pathol. J. 13: 214–231.

Khodakovskaya, M., Dervishi, E., Mahmood, M., Xu, Y., Li, Z., Watanabe, F. and Biris, A.S. 2009. Carbon nanotubes are able to penetrate plant seed coat and dramatically affect seed germination and plant growth. ACS Nano 4: 3221–3227.

Khodakovskaya, M.V. and Biris, A.S. 2010. Method of using carbon nanotubes to affect seed germination and plant growth. U.S. Patent Application. 13/509,487.

Kitching, M., Ramani, M. and Marsili, E. 2015. Fungal biosynthesis of gold nanoparticles: Mechanism and scale up. Microb. Biotechnol. 8: 904–917.

Kochkina, Z., Pospeshny, G. and Chirkov, S. 1994. Inhibition by chitosan of productive infection of T-series bacteriophages in the *Escherichia coli* culture. Mikrobiologia 64: 211–215.

Kole, C., Kole, P., Randunu, K.M., Choudhary, P., Podila, R., Ke, P.C., Rao, A.M. and Marcus, R.K. 2013. Nanobiotechnology can boost crop production and quality: First evidence from increased plant biomass, fruit yield and phytomedicine content in bitter melon (*Momordica charantia*). BMC Biotechnol. 13: 37.

Kottegoda, N., Munaweera, I., Madusanka, N. and Karunaratne, V.A. 2011. Green slow-release fertilizer composition based on urea-modified hydroxyapatite nanoparticles encapsulated wood. Curr. Sci. 101: 73–78.

Krishnaraj, C., Ramachandran, R., Mohan, K. and Kalaichelvan, P. 2012. Optimization for rapid synthesis of silver nanoparticles and its effect on phytopathogenic fungi. Spectrochim Acta Part A Mol. Biomol. Spectrosc. 93: 95–99.

Kumar, S., Nehra, M., Dilbaghi, N., Marrazza, G., Hassan, A.A. and Kim, K.-H. 2019. Nano-based smart pesticide formulations: Emerging opportunities for agriculture. Journal of Controlled Release 294: 131–153.

Kumari, M., Pandey, S., Giri, V.P., Bhattacharya, A., Shukla, R., Mishra, A. and Nautiyal, C.S. 2017. Tailoring shape and size of biogenic silver nanoparticles to enhance antimicrobial efficacy against MDR bacteria. Microb. Pathol. 105: 346–355.

Lateef, A., Nazir, R., Jamil, N., Alam, S., Shah, R., Khan, M.N. and Saleem, M. 2016. Synthesis and characterization of zeolite based nano–composite: An environment friendly slow release fertilizer. Microporous Microporous Mater. 232: 174–183.

Li, C., Li, Y., Li, Y. and Fu, G. 2018. Cultivation techniques and nutrient management strategies to improve productivity of rain-fed maize in semi-arid regions. Agric. Water Manag. 210: 149–157.

Li, G., He, D., Qian, Y., Guan, B., Gao, S., Cui, Y., Yokoyama, K. and Wang, L. 2012. Fungus-mediated green synthesis of silver nanoparticles using *Aspergillus terreus*. Int. J. Mol. Sci. 13: 466–476.

Li, S.X., Wang, Z.H., Miao, Y.F. and Li, S.Q. 2014. Soil organic nitrogen and its contribution to crop production. J. Integr. Agric. 13: 2061–2080.

Liu, J., Zhang, Y. and Zhang, Z. 2009. The application research of nano-biotechnology to promote increasing of vegetable production. Hubei Agric. Sci. 1: 041.

Liu, R.H., Kang, Y.H., Pei, L., Wan, S.Q., Liu, S.P. and Liu, S.H. 2016. Use of a new controlled-loss-fertilizer to reduce nitrogen losses during winter wheat cultivation in the Danjiangkou reservoir area of China. Commun. Soil Sci. Plant Anal. 47: 1137–1147.

Loh, X.J., Lee, T.-C., Dou, Q. and Deen, G.R. 2016. Utilising inorganic nano carriers for gene delivery. Biomater. Sci. 4: 70–86.

Lu, C., Zhang, C., Wen, J., Wu, G. and Tao, M.X. 2002. Research of the effect of nanometer materials on germination and growth enhancement of Glycine max and its mechanism. Soybean Sci. 21: 168–171.

Malerba, M. and Raffaella, C. 2018. Recent advances of chitosan applications in plants. Polymers 10: 118. https://doi.org/10.3390/polym10020118.

Maliszewska, I., Juraszek, A. and Bielska, K. 2014. Green synthesis and characterization of silver nanoparticles using ascomycota fungi *Penicillium nalgiovense* AJ12. J. Clust. Sci. 25: 989–1004.

McShane, H.V. and Sunahara, G.I. 2015. Environmental perspectives. In Nano Engineering; Elsevier: Amsterdam, Netherlands, 257–283.

Millan, G., Agosto, F., Vazquez, M., Botto, L., Lombardi, L. and Juan, L. 2008. Use of clinoptilolite as a carrier for nitrogen fertilizers in soils of the Pampean regions of Argentina. Cienc. Investig. Agrar. 35: 293–302.

Monreal, C.M., DeRosa, M., Mallubhotla, S.C., Bindraban, P.S. and Dimkpa, C. 2016. Nanotechnologies for increasing the crop use efficiency of fertilizer-micronutrients. Biol. Fertil. Soils 52: 423–437.

Nair, R., Varghese, S.H., Nair, B.G., Maekawa, T., Yoshida, Y. and Kumar, D.S. 2010. Nanoparticulate material delivery to plants. Plant Sci. 179: 154–163.

Panpatte, D.G., Jhala, Y.K., Shelat, H.N. and Vyas, R.V. 2016. Nanoparticles: The next generation technology for sustainable agriculture. *In*: Microbial Inoculants in Sustainable Agricultural Productivity; Springer: New Delhi, 289–300.

Peteu, S.F., Oancea, F., Sicuia, O.A., Constantinescu, F. and Dinu, S. 2010. Responsive polymers for crop protection. Polymers 2: 229–251.

Pospieszny, H., Chirkov, S. and Atabekov, J. 1991. Induction of antiviral resistance in plants by chitosan. Plant Sci. 79: 63–68.

Rico, C.M., Majumdar, S., Duarte-Gardea, M., Peralta-Videa, J.R. and Gardea-Torresdey, J.L. 2011. Interaction of nanoparticles with edible plants and their possible implications in the food chain. J. Agric. Food Chem. 59: 3485–3498.

Sabir, A., Yazar, K., Sabir, F., Kara, Z., Yazici, M.A. and Goksu, N. 2014. Vine growth, yield, berry quality attributes and leaf nutrient content of grapevines as influenced by seaweed extract (*Ascophyllum nodosum*) and nanosize fertilizer pulverizations. Sci. Hortic. 175: 1–8.

Sadeghi, R., Rodriguez, R.J., Yao, Y. and Kokini, J.L. 2017. Advances in nanotechnology as they pertain to food and agriculture: Benefits and risks. Annu. Rev. Food Sci. Technol. 8: 467–492.

Shalaby, T.A., Bayoumi, Y., Abdalla, N., Taha, H., Alshaal, T., Shehata, S., Amer, M., Domokos-Szabolcsy, É. and El Ramady, H. 2016. Nanoparticles, soils, plants and sustainable agriculture. pp. 283–312. *In*: Shivendu, R., Nandita, D. and Eric, L. (eds.). Nanoscience in Food and Agriculture 1. Springer: Cham, Switzerland.

Shaligram, N.S., Bule, A., Bhambure, R., Singhal, R.S., Singh, S.K., Szakacs, G. and Pandey, A. 2009. Biosynthesis of silver nanoparticles using aqueous extract from the compacting producing fungal strain. Process Biochem. 44: 939–943.

Shang, Y., Hasan, M.K., Ahammed, G.J., Li, M., Yin, H. and Zhou, J. 2019. Applications of nanotechnology in plant growth and crop protection: A review. Molecules 24(14): 2558. Doi: 10.3390/molecules24142558. PMID: 31337070; PMCID: PMC6680665.

Shojaei, T.R., Salleh, M.A.M., Tabatabaei, M., Mobli, H., Aghbashlo, M., Rashid, S.A. and Tan, T. 2019. Applications of nanotechnology and carbon nanoparticles in agriculture. pp. 247–277. *In*: Suraya, A.R., Raja, N.I.R.O. and Mohd, Z.H. (eds.). Synthesis, Technology and Applications of Carbon Nanomaterials; Elsevier, Amsterdam, Netherlands.

Shukla, P., Chaurasia, P., Younis, K., Qadri, O.S., Faridi, S.A. and Srivastava, G. 2019. Nanotechnology in sustainable agriculture: Studies from seed priming to post-harvest management. Nanotechnol. Environ. Eng. 4: 11.

Solanki, P., Bhargava, A., Chhipa, H., Jain, N. and Panwar, J. 2015. Nano-fertilizers and their smart delivery system. pp. 81–101. *In*: Rai, M., Ribeiro, C., Mattoso, L. and Duran, N. (eds.). Nanotechnologies in Food and Agriculture. Springer: Cham, Switzerland.

Srinivasan, C. and Saraswathi, R. 2010. Nano-agriculture-carbon nanotubes enhance tomato seed germination and plant. Current Science 99(3): 274–275.

Swaminathan, S., Edward, B.S. and Kurpad, A.V. 2013. Micronutrient deficiency and cognitive and physical performance in Indian children. Eur. J. Clin. Nutr. 67: 467–474.

Syed, A., Saraswati, S., Kundu, G.C. and Ahmad, A. 2013. Biological synthesis of silver nanoparticles using the fungus Humicola sp.. and evaluation of their cytotoxicity using normal and cancer cell lines. Spectrochim Acta A Mol. Biomol. Spectros. 114: 144–147.

Thakur, R.K., Shirkot, P. and Dhirta, B. 2018. Studies on effect of gold nanoparticles on *Meloidogyne incognita* and tomato plants growth and development. BioRxiv: 428144. https://doi.org/10.1101/428144.

Torney, F., Trewyn, B.G., Lin, V.S. and Wang, K. 2007. Mesoporous silica nanoparticles deliver DNA and chemicals into plants. Nat. Nanotechnol. 2(5): 295–300.

Vahabi, K., Mansoori, G.A. and Karimi, S. 2011. Biosynthesis of silver nanoparticles by fungus *Trichoderma reesei* (a route for large-scale production of AgNPs). Inter. Sci. J. 1: 65–79.

Verma, S.K., Das, A.K., Patel, M.K., Shah, A., Kumar, V. and Gantait, S. 2018. Engineered nanomaterials for plant growth and development: A perspective analysis. Sci. Total Environ. 630: 1413–1435.

Worrall, E., Hamid, A., Mody, K., Mitter, N. and Pappu, H. 2018. Nanotechnology for plant disease management. Agronomy 8: 285. Doi:10.3390/agronomy8120285.

Yang, H., Xu, M., Koide, R.T., Liu, Q., Dai, Y., Liu, L. and Bian, X. 2016. Effects of ditch-buried straw return on water percolation, nitrogen leaching and crop yields in a rice-wheat rotation system. J. Sci. Food Agric. 96(4): 1141–1149. Doi: 10.1002/jsfa.7196. Epub 2015 Apr 24. PMID: 25847361.

Yousefzadeh, S. and Sabaghnia, N. 2016. Nano-iron fertilizer effects on some plant traits of dragonhead (*Dracocephalum moldavica* L.) under different sowing densities. Acta Agric. Slov. 107: 429–437.

Zaytseva, O. and Neumann, G. 2016. Carbon nanomaterials: Production, impact on plant development, agricultural and environmental applications. Chem. Biol. Technol. Agric. 3: 17. Doi: 10.1186/s40538-016-0070-8.

Section II

Detection and Management of Plant Diseases and Pests

Chapter 3

Nanobiosensors for Detection of Plant Pathogens

Yulia Plekhanova, *Sergei Tarasov* and *Anatoly Reshetilov**

Introduction

The identification of plant pathogens is related both to the productivity of agricultural structures and food safety issues and thus to the life and health of people. About 15% of economically important crops in the world are damaged by plant diseases, and 30% of plant diseases are induced by plant viruses, which leads to substantial economic losses in agriculture (Jablonski et al. 2021). The main sources of plant infectious diseases are the microorganisms shown in Figure 1.

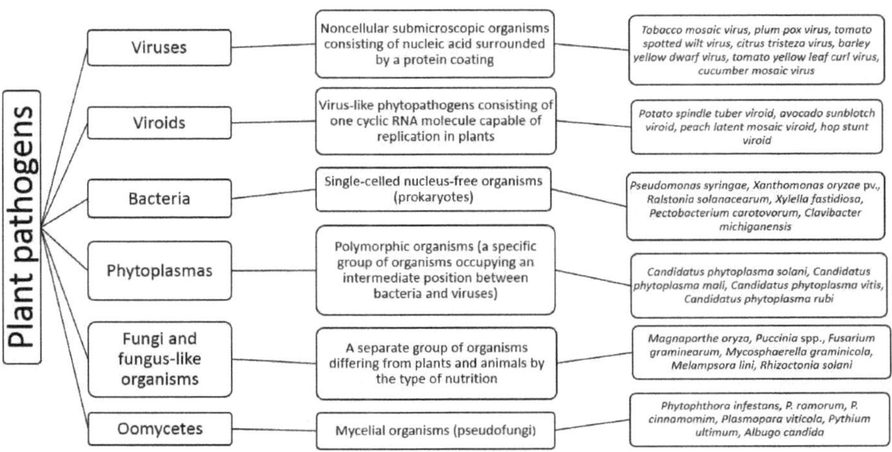

Figure 1: The main sources of infectious plant diseases.

Laboratory of Biosensors, G.K. Skryabin Institute of Biochemistry and Physiology of Microorganisms, FRC Pushchino Centre for Biological Research, Russian Academy of Sciences, 142290 Pushchino, Russian Federation.

* Corresponding author: anatol@ibpm.pushchino.ru

In addition, factors causing plant diseases can be not only living organisms, but also chemical compounds synthesized by humans, or other abiotic factors. However, in this chapter, we focus on infectious plant diseases caused by three main pathogens: viruses, bacteria and fungi. The Food and Agriculture Organization of the United Nations estimates that plant diseases cost the global economy around US$ 220 billion per year, with 20–40% of crop production lost to pests (Nature 2021). The first step in protecting plants from pathogenic diseases is the detection of a pathogen. A large number of different methods are used to detect various plant diseases. First of all, these are visual observations that make it possible to identify affected plants by external signs. However, in most cases, external signs manifest themselves when the plant has already been long infected with pathogens. Therefore, the efforts of researchers in recent years have been aimed at early detection of plant pathogens. Early pathogen detection is important not only to assess the health of plants but also to minimize the risks of spreading infections in plants of other areas. For these tasks, methods based on the analysis of nucleic acids and proteins are often used, such as polymerase chain reaction (PCR), enzyme-linked immunosorbent assay (ELISA), immunofluorescence, gas chromatography and others. It should be noted that all methods of pathogen detection can be divided into direct and indirect. In direct methods, the presence of a pathogen causing an infectious disease of a plant, or a biomolecular marker isolated from the tissues of the affected plant, is detected. Indirect methods are based on the detection of changes in the properties of plants when they are affected by plant pathogens. In recent years, biosensors have proven themselves as alternative methods for conducting highly specific and accurate analysis in such areas as clinical diagnostics, control of biotechnological processes in production, environmental pollution analysis, etc. Biosensor technologies are a promising area of research in the field of plant pathogens' detection. The use of various bioreceptors in them, such as DNA probes, antibodies, phages, etc., makes it possible to obtain highly specific and sensitive devices for fieldwork. The aim of the present chapter is to focus on the role of nanobiosensors in early detection and management of plant pathogens. Special attention is paid to the use of various nanomaterials in the composition of nanobiosensors.

Biosensors for detection of plant pathogens

A biosensor is an analytical device containing a bioreceptor (biological recognition element) and a transducer used to detect an analyte. The biorecognizing component interacts with the analyte, and changes in its physico-chemical properties are recorded by the converter. An antibody, DNA, enzyme, tissue, whole cell, etc., can act as a bioreceptor. Depending on the type of transducer, biosensors are divided into electrochemical, optical, acoustic, colorimetric and piezoelectric. In this chapter, we look at which bioreceptors and transducers are used to detect specific plant pathogens. Conditionally, there are two ways to identify pathogens – determining the source of infection (viruses or bacteria that cause the disease) as well as indirect methods based on the analysis of volatile organic compounds (VOCs) that are released by plants as a protective mechanism against pathogens.

Direct methods of analysis

Direct methods of detecting pathogens using biosensors are based, in most cases, on the use of antibodies, DNA probes or phages as bio-identifying elements.

DNA sensors

The main principle of DNA-based biosensors relies on hybridization or hydrogen bonding between a target DNA sequence and a DNA probe sequence that is immobilized on a sensing platform (Dyussembayev et al. 2021). The advantages of DNA biosensors include the fact that the detection of pathogenic microorganisms is possible even in cases of infection of samples with single pathogens. The presence of DNA sequence as a detection probe provides high specificity to biosensors. DNA-based biosensors, thanks to their ability to detect at the molecular level, make it possible to detect diseases at the early stages before any visual symptoms appear (Vatankhah et al. 2022). Biosensors based on electrochemical signal converters have become the most widely used for determining plant pathogens. Methods such as voltammetry (Malecka et al. 2014, Wongkaew and Poosittisak 2014), chronoamperometry (Umer et al. 2021) and electrochemical impedance spectroscopy (Mohd Said et al. 2018) are used. For example, Franco et al. (2019) described an electrochemical nanobiosensor for detecting *Phytophthora palmivora*, a known cocoa pathogen that causes serious crop losses. The authors have developed sandwich hybrids between two oligonucleotide probes and the genomic DNA of *P. palmivora*. The detection limit is 0.30 ng DNA μL^{-1}. The biosensor is used to analyze real samples of cocoa pods, which opens prospects for using this method in the analysis of field samples.

In later works, the surface of the electrodes of electrochemical biosensors began to be modified with various nanomaterials designed to stabilize bioreceptors or to enhance the biosensor signal. Nanomaterials have a high ratio of surface area to volume (Hwang et al. 2020), which leads to more efficient use of the analyzer surface. Nanomaterials are of particular importance for electrochemical devices since most nanomaterials have high electrical conductivity (Cho et al. 2020). The use of nanomaterials in biosensor devices in combination with DNA makes it possible to track biological processes at the physiological level with much greater accuracy. Table 1 shows examples of the use of nanomaterials in electrochemical DNA sensors for the determination of plant pathogens.

Most often, gold nanoparticles serve as a nanomaterial facilitating the immobilization of the DNA sample. Gold nanoparticles have a high surface-to-volume ratio, good stability and high conductivity, thanks to which they effectively hold bioreceptors on the electrode surface and increase the efficiency and sensitivity of the biosensor without changing its structure. The surface of the working electrodes is modified with gold nanoparticles, and then DNA molecules are bound via sulfide bonds. In addition, gold nanoparticles increase the ability of DNA to immobilize and hybridize, and support the biological activity of DNA (Khater et al. 2019). For example, Lau et al. (2017) developed a nanoparticle-based electrochemical biosensor for the detection of plant-pathogen DNA on disposable screen-printed carbon

Table 1: Examples of the use of nanomaterials in electrochemical DNA sensors.

Nanomaterial	Crop	Pathogen	Detection limit	References
Reduced graphene oxide and gold nanoparticles	*Rosa hybrida* L.	*Agrobacterium tumefaciens*	0.87×10^{-13} M	Vatankhah et al. 2022
Electrically active magnetic (EAM) nanoparticles	*Theobroma cacao* L.	*Phytophthora palmivora*	0.30 ng DNA μL^{-1}	Franco et al. 2019
Gold nanoparticles	*Arabidopsis thaliana*	*Pseudomonas syringae*	214 pM	Lau et al. 2017
Graphene	Rice	*Ustilaginoidea virens*	10 fM	Rana et al. 2021
Gold nanoparticle	Citrus	*Citrus tristeza*	100 nM	Khater et al. 2019
Gold nanoparticles and anodic aluminium oxide	*Phalaenopsis*	*Odontoglossum* ringspot virus	0.345 ng/mL	Jian et al. 2018
Copper(II) particles	Over 100 species	*Xylella fastidiosa*	–	Machini and Oliveira-Brett 2021
Three-dimensional copper nanostructures reinforced graphene nanohybrid (Cu Ns@GO)	Cotton	Cotton leaf curl Khokhran virus strain Burewala (CLCuKoV-Bur)	600 pM	Rafiq et al. 2021
Carbon nanotubes, copper nanoparticles	Cotton	Geminiviruses	0.01 ng μL^{-1}	Tahir et al. 2018

electrodes. The authors used differential pulse voltammetry for rapid and high-precision detection of *Pseudomonas syringae*-infected plant samples; moreover, this biosensor was shown to be effective even at the stages before the visual manifestation of infection on the plant. This method turned out to be more sensitive than standard PCR/gel electrophoresis methods by more than 10,000 times. Khater et al. (2019) developed a label-free impedimetric biosensor for detecting the nucleic acid of the citrus tristeza virus (CTV). This virus is by far the most dangerous of all viruses capable of infecting citrus crops. The sensor platform based on a screen-printed carbon electrode was modified with electrodeposited gold nanoparticles, which made it possible to effectively immobilize the probes of thiolated single-stranded DNA, as well as to increase the conductivity of the electrode. The biosensor was able to selectively detect CTV nucleic acids in the presence of other non-specific DNA with the detection limit of 100 nM.

Gold nanoparticles can also be used in composites with other nanomaterials. In particular, Vatankhah et al. (2022) developed an electrochemical DNA-based biosensor for the detection of tms2 gene of *Agrobacterium tumefaciens*. This bacterium causes crown gall infection in *Rosa hybrida* L. The development of galls leads to overall weakness, a decrease in foliage, water stress and loss of vigour in infected plants. Vatankhah and coauthors modified the graphite electrode with a nanocomposite comprising reduced graphene oxide and gold nanoparticles.

The frequent use of graphite nanomaterials is due to their biocompatibility and high conductivity, as well as the possibility of creating three-dimensional electrodes (Omar et al. 2021). Tahir et al. (2018) have proposed a 3D biosensor, in which carbon nanotubes are used as a conductive frame and copper nanoparticles, to detect agroviruses. The biosensor was used for early detection of infection of cotton with geminiviruses, which are a major threat to cotton plants in many countries. The authors used differential pulse voltammetry with methylene blue as a redox indicator. The biosensor was characterized by high stability, losing only 4.3% of the initial signal value after four weeks. The detection limit was found to be 0.01 ng/μL.

Quantitative analysis using electrochemical DNA sensors is based on the registration of changes in the electrochemical characteristics (conductivity, current) of the electrode during the DNA hybridization reaction on its surface. If this reaction is accompanied by changes in optical properties, then it can also be registered visually. For example, Zhan et al. (2018) developed a gold nanoparticle-based lateral flow biosensor to detect *Phytophthora infestans*, which infects potatoes and tomatoes. The biosensor had a high specificity and a detection limit of 0.1 pg mL^{-1}. Besides using optical DNA methods, it is possible to qualitatively determine the presence of certain bacterial infections (Zhao et al. 2011, Vaseghi et al. 2013). Application of complementary metal–oxide–semiconductor (CMOS) technologies also enables creating quantitative optical sensors. In particular, Zamir et al. (2020) described the concept of a novel CMOS-based biosensor for the detection of *Colletotrichum gloeosporioides* fungus in harvested crops. The sensitivity of the biosensor was 10 nM RNA; it is worth noting that detecting quiescent fungi by any other method is practically impossible.

Biosensors based on acoustic detection of pathogen DNA are also known. In particular, Papadakis et al. (2015) combined a commercial acoustic biosensor system with a multiplex PCR reaction for the simultaneous detection of three of the most economically/scientifically important plant bacterial pathogens: *Ralstonia solanacearum*, *Pseudomonas syringae* pv. *tomato* and *Xanthomonas campestris* pv. *vesicatoria*. DNA binding affects the characteristics of the acoustic waves propagating on the sensor surface, i.e., velocity and energy, which in turn are monitored as changes in frequency and dissipation. Changes in the oscillation frequency recorded by the sensor are directly related to the amount of DNA present in the sample. The detection limit was found to be 10^1–10^4 CFU depending on the measurement conditions.

Thus, DNA sensors for detecting pathogens can determine their presence in the early stages of infection, when plant infections are asymptomatic. Nucleic acid-based methods are sufficiently accurate and specific for detecting a single target pathogen in a mixture containing more than one analyte and are highly effective for detecting multiple targets. However, there are some limitations when using DNA sensors to detect pathogens with low titres in materials such as seeds and insect vectors, or at the early stages of infection. In the case of even minimal contamination of the sample, the probability of a false-positive result increases dramatically. That is to say, the method has very high requirements for sterility at all stages of analysis.

Immunosensors

Immunoassay technologies offer high specificity and capabilities for detecting plant viruses *in situ*; however, they also have some limitations regarding low virus concentrations and disposable sensors. They are usually used to confirm plant diseases after the appearance of visual symptoms (Jablonski et al. 2021). The principle of immunosensors' functioning is the binding of a specific antibody with a detectable antigen on the transducer, as a result of which the signal of the transducer changes. Most often, electrochemical signal converters are used to create such biosensors, but other types of converters can also be used – surface plasmon resonance (change in refractive index), microweights on a quartz crystal (change in mass per unit area of a crystal by measuring the frequency change of a quartz crystal resonator), and cantilever-based sensors (measurement of changes in resonant frequency).

Table 2 shows examples of using different types of immunosensors to determine plant pathogens.

The simplest method for the end-user is immunochromatographic analysis. The final result is presented in the form of a test strip, which changes colour depending on the presence or absence of pathogens, as well as their amount. Figure 2 presents a scheme for the formation of this type of biosensor. Conjugates of gold nanoparticles and antibodies were used as a bioidentifying element, and the sensor itself makes use of the sandwich immunoassay format. The testing results can be obtained in 10 min and detected visually. Besides visual detection, photoluminescence methods are used. Changes in the photoluminescence signal of nanostructured material are used for the detection of analyte and the determination of its amount by the characterization of the photoluminescence spectra. For example, in Tereshchenko et al. (2017), a zinc oxide film is used as a photoluminescent material. The intensity of its photoluminescence changes when an antibody immobilized on its surface interacts with proteins of grapevine virus A. The sensitivity of this label-free biosensor towards the GVA-antigens was determined in the range from 1 pg/mL to 10 ng/mL.

Another method of optical detection in immunosensors is based on the phenomenon of fluorescence. To obtain the results, special fluorescent probes are used, which are excited at a certain wavelength. When the labelled antibody interacts with the analyzed antigen, the signal intensity changes, which allows quantitative analysis using this type of sensors. At the time of writing, the known biosensors are those for the early diagnosis of Asian soybean rust caused by the fungus *Phakopsora pachyrhizi* (Miranda et al. 2013), rice bacterial leaf blight disease caused by *Xanthomonas oryzae* pv. *oryzae* (Awaludin et al. 2020) and tristeza disease caused by citrus tristeza virus (CTV) (Shojaei et al. 2016). Due to the low titre or uneven distribution of CTV in field samples, its detection using conventional detection methods may be difficult. For this reason, Shojaei et al. (2016) used a nanomaterial, cadmium-telluride quantum dots. Quantum dots were conjugated with specific antibodies against the CTV shell protein, which was immobilized on the surface of gold nanoparticles. The developed system showed a higher sensitivity and specificity compared to solid-phase enzyme immunoassay with the detection limit of 0.13 μg/L and sensitivity and specificity of 93% and 94%, respectively. The advantage of fluorescent devices is that they can be used to detect infections of almost any type;

Table 2: Types of immunosensors for plant pathogens' determination.

Principle	Method	Nanomaterial	Crop	Pathogen	Parameter (detection limit)	References
Electrochemical	Square wave voltammetry	Gold nanoparticles	Tree sap	*Candidatus phytoplasma aurantifolia*	1.5 ng mL^{-1}	Ebrahimi et al. 2019
	Differential pulse voltammetry	Graphene oxide	Tomato (*Solanum lycopersicum*), cowpea (*Vigna unguiculata*) and benth *Nicotiana benthamiana*	Groundnut bud necrosis orthotospovirus	5.7 ± 0.7 ng mL^{-1}	Chaudhary et al. 2021
	Amperometry		Walnuts	*Xanthomonas arboricola*	1.5×10^2 cfu mL^{-1}	Regiart et al. 2017
	Differential pulse voltammetry	Colloidal gold nanoparticles	Orchids	Simultaneous detection of *Phalaenopsis* virus, *Odontoglossum* ringspot virus (ORSV) and *Cymbidium* mosaic virus (CymMV)	0.5 ng/mL	Lin et al. 2022
	Chrono-amperometry	Gold nanoparticles	Cucumbers	Cucumber mosaic virus	0.1 mg/mL	Rafidah et al. 2016
	Amperometry	Gold nanoparticles	Citrus trees	Citrus tristeza virus	0.3 fg mL^{-1}	Freitas et al. 2019
	Electrolyte-gated organic field-effect transistor	–	Stone fruit trees	Plum pox virus	sub ng/mL	Berto et al. 2019
	Linear sweep voltammetry	Gold nanoparticles	Maize	*Pantoea stewartii* subsp. *stewartia* NCPPB 449	7.8×10^3 cfu/mL	Zhao et al. 2014
Electrochemistry+ optics	Cyclic voltammetry, surface plasmon resonance (SPR)	–	Wheat	Karnal bunt caused by fungus *Tilletia indica*	0.625 ng/μL	Singh et al. 2014

Table 2 contd. ...

...Table 2 contd.

Principle	Method	Nanomaterial	Crop	Pathogen	Parameter (detection limit)	References
Optical	Lateral flow immunoassay	Gold nanoparticles	Grapes	Grapevine leafroll-associated virus 3	–	Byzova et al. 2018
	Photo-luminescence		Wheat	Karnal bunt caused by fungus *Tilletia indica*	10 pg	Mishra et al. 2020
	Fluorescence resonance energy transfer (FRET)	Cadmium-telluride quantum dots, gold nanoparticles	Citrus trees	Citrus tristeza virus	0.13 μg mL^{-1}	Shojaei et al. 2016
	Non-FRET nanobiosensor	Quantum dots	Citrus trees	Citrus tristeza virus	246 ng/mL	Safarnejad et al. 2017
	Fluorescence	Graphene quantum dots and gold nanoparticles	Rice	*Xanthomonas oryzae* pv. *oryzae*	22 cfu mL^{-1}	Awaludin et al. 2020
	Photoluminiscence	ZnO films	Grapes	Grapevine virus A	1 pg/mL	Tereshchenko et al. 2017
	Surface plasmon resonance	–	Bananas	*Pseudocercospora fijiensis*	11.7 μg mL^{-1}	Luna-Moreno et al. 2019
	Fluorescence	–	Soybeans	*Phakopsora pachyrhizi*	2.2 ng/mL (8–12 spores/mL)	Miranda et al. 2013
	Lateral flow assay (colorimetry)	Gold nanoparticles	Ginseng	*Alternaria panax*	0.01 pg	Wei et al. 2018

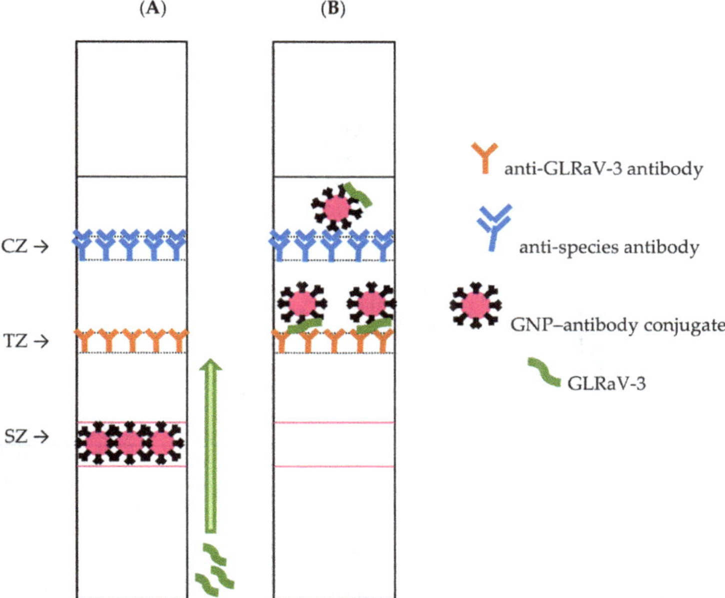

Figure 2: Scheme of the test strip for sandwich immunochromatographic assay of GLRaV-3 before (A) and after (B) analysis. Reprinted from Byzova et al. (2018) © MDPI.

however, expensive signal transducers and specific probes are needed to create such devices.

If antibodies are immobilized on a specialized substrate, it becomes possible to register the effect of surface plasmon resonance (SPR) when they interact with antigens. The SPR method allows direct registration of almost any intermolecular interactions in real-time without the use of any labels or associated processes (Nguyen et al. 2015). The scheme of formation of such biosensors is presented in Luna-Moreno et al. (2019) (see Figure 3 here), which developed a biosensor for the detection of *Pseudocercospora fijiensis* pathogen. The polyclonal antibody anti-HF1 was immobilized on a preliminarily bifunctionalized gold surface of the chip, and measurements were based on a direct immunoassay where the immobilized antibodies anti-HF1 bind to the antigen HF1 (GPI protein isolated from the *P. fijiensis* cell wall) in the sample. The developed biosensor was successfully tested on banana leaves' extracts, and the sensitivity of the analysis was 0.0021 reflectance units per ng mL^{-1}.

Electrochemical immunoassay methods are used in biosensors to determine plant pathogens along with optical ones. The most commonly used techniques are differential pulse and cyclic voltammetry, as well as amperometry. Besides, there are devices based on organic field-effect transistors. The advantages of electrochemical assays may include the low-cost production of electrodes; also, detection by electrochemical methods is independent of the turbidity of the sample and its other optical properties. Recently, electrochemical analysis methods have appeared that allow measuring the concentration of pathogens without using any specialized labels. In particular, to determine *Candidatus phytoplasma aurantifolia* causing witches

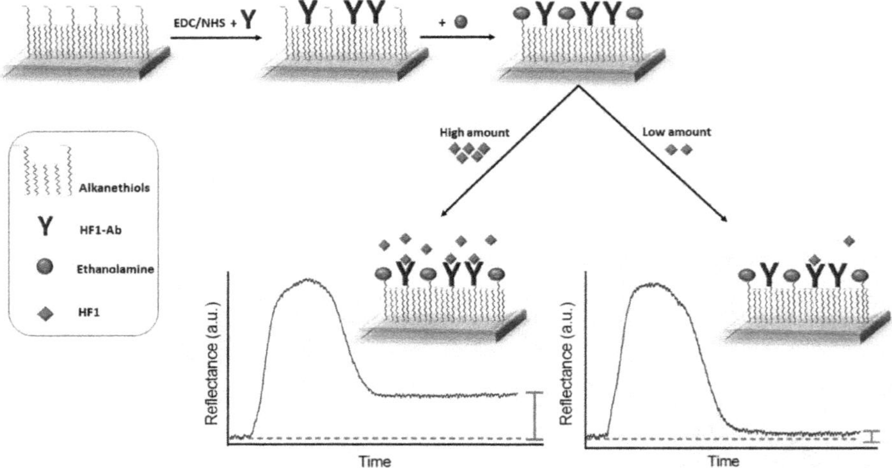

Figure 3: Schematic of the SPR immunoassay for detecting HF1 and illustrative representation of sensograms. Reprinted from Luna-Moreno et al. (2019) © MDPI.

broom diseases, Ebrahimi et al. (2019) used the method of square wave voltammetry, where antibodies to this pathogen were immobilized on a thiol terminated surface of a gold electrode modified by gold nanoparticles. Chauhary et al. (2021) used differential pulse voltammetry for the detection of groundnut bud necrosis orthotospovirus. The response measurements of the immunoelectrodes revealed a sensitivity of 221 ± 1 μA μg^{-1} mL^{-1} and the detection limit of 5.7 ± 0.7 ng mL^{-1} for the GBNV antigen.

Organic electronic materials, which provide low production costs in environmentally friendly conditions and on flexible substrates, were used to create an electrolyte-gated organic field-effect transistor of an immunosensor to detect the plum pox virus. This pathogen is responsible for sharka, a highly infectious disease affecting stone fruit trees and causing severe economic damage (Berto et al. 2019). The advantage of such devices is a tremendously low detection limit and no need to use labels. The principle of operation of the device consists in measuring transconductance changes upon binding of the plum pox virus to the surface-immobilized antibodies as a function of plum pox virus concentration. The authors estimated the theoretical detection limit of their biosensor as 180 pg/mL.

Immunoassay is most often used to identify viruses since in the case of fungi and bacteria, it is quite difficult to obtain antibodies with the required specificity. The resulting antibodies may be specific both exclusively to individual life forms of microorganisms and to a group of species. Nevertheless, in recent years there have been more and more works aimed at the label-free rapid detection of various types of pathogens.

Phage-based sensors

Phages are small viruses discovered in the early 20th century (Twort 1915, D'Herelle 1917) that lack their own metabolic machinery; instead, they use their host bacterial cells for propagation. Phages or bacteriophages provide natural affinity to their host bacterial cells and can serve as the recognition element for electrochemical sensors. Such sensors possess high specificity, sensitivity and simplicity (Xu et al. 2019a). A rather large number of bacteriophages have been discovered, which leads to the creation of a large number of biosensors based on them for the detection of plant diseases. A technology based on bioluminescent phages was developed to determine the presence of *Pseudomonas cannabina* pv. *alisalensis* affecting vegetables of the cruciferous family (Schofield et al. 2012, 2013). The main advantage of this technology is that nucleic acids of only viable bacterial cells are detected, as a result of which no false-positive results can be obtained. However, expression of the reporter phage may be suppressed by the presence of certain chemical compounds in the tested leaves, such as glucosinolate and isothiocyanate thioesters.

Such sensors have been designed mainly to detect the presence of bacteria dangerous to human health and their toxins in food and environmental matrices (Quintela and Wu 2020). However, they are also promising bioreceptors for creating biosensors that detect diseases caused by bacterial plant pathogens (Vu and Oh 2020). The main limitation of biosensors based on bacterial phages is that they can only be used to detect bacteria, but not fungi and viruses.

Indirect methods

It is possible to determine not only the pathogen that causes a plant disease but also to detect an infection by indirect signs. Such detection methods are called indirect methods. Their principle is based on the fact that plants generally respond to pathogen infection through its defence reactions and various physiological processes such as respiration, photosynthesis, nutrient translocation, transpiration, growth and development, primary carbon metabolism, etc. (Kumar and Arora 2020). Microbial infection often leads to the triggering of protective mechanisms inside plants, which can lead to local accumulation of phytoalexins, increased activity of some enzymes, the appearance of reactive oxygen species (Lebeda et al. 2001), and the release of certain chemicals into the environment (Zeilinger et al. 2016). All these effects can be used to detect plant pathogenic infections. In particular, indirect methods based on the analysis of VOCs, which plants secrete as a protective mechanism against the attack of pathogens, are very common (Scala et al. 2013). Besides, indirect methods of analysis include the determination of toxins or other chemicals released by pathogens themselves during their vital activity.

It is known that many pathogenic fungi produce mycotoxins, which are dangerous not only for plants, but also for humans. For example, zearalenone, an estrogenic metabolite secreted by *Fusarium* species when infecting corn, wheat, barley and rice, is dangerous for humans and livestock, because it can infect food samples, causing infertility, abortion or other breeding problems, especially in swine. In Xu et al. (2019a), a lateral flow immunosensor with gold nanoparticles was developed, allowing the detection of this toxin in corn samples with high sensitivity (7.4 pg/mL). In Hossain and Maragos (2018), a method was proposed for the determination of three mycotoxins isolated by *Fusarium* species – deoxynivalenol, zearalenone and T-2 toxin in wheat samples. The method is based on imaging surface plasmon resonance; a competitive immunoassay scheme is used to determine mycotoxins. The absence of mycotoxins is the most important indicator of food quality, so many works are devoted to the creation of biosensors to assess the presence of various types of toxins in food (Oliveira et al. 2019, Agriopoulou et al. 2020, Yan et al. 2020).

Worldwide, post-harvest losses are estimated at 40–50% of the harvested crop, mainly due to rot caused by microorganisms (Chalupowicz et al. 2020). To reduce these losses, there is a growing demand for effective early warning and real-time detection tools to track the development of spoilage of fruits and vegetables after harvest and throughout the supply chain. For example, in the process of fungal infection in plants, the appearance of certain chemical compounds is detected, the number of which can determine the presence and stage of the disease. A common indicator of plant infection by pathogenic microorganisms is their release of VOCs. VOCs are biomolecules and metabolites with high vapour pressure, low boiling point and low molecular weight. Plants secrete various VOCs into their immediate environment to provide many different functions of vital activity. These VOCs can originate from protective compounds, plant hormones, insect pheromones, digestive or metabolic byproducts, and compounds derived from cell damage (Cardoso et al. 2022). The VOC profiles of healthy crops may change after infection with insect pests or infection with pathogenic microorganisms (Yadav and Chhibbar 2018). Such changes can then be used as disease markers in fresh agricultural products at the early stages of infection and disease development. Enzyme biosensors can be used to determine VOCs, since many VOCs released by infected cultures are alcohols and aldehydes, and can be oxidized by enzymes (Kulagina et al. 1999, Thomas et al. 1999).

For example, *p*-ethylguaiacol is formed during leather rot in strawberries, caused by *Phytophthora cactorum*. Therefore, detecting the production of *p*-ethylguaiacol by strawberry plants will be a useful indicator for the confirmation of leather rot disease. Fang et al. (2014) have proposed an electrochemical sensor based on screen-printed electrodes, which allows for determining ultra-low concentrations of *p*-ethylguaiacol with a sensitivity of 174–188 mA cm^{-2} mM^{-1}. To enhance the biosensor signal, the electrode surface was modified with SnO_2 or TiO_2 nanoparticles. *p*-Ethylguaiacol was also determined using an amperometric biosensor based on the enzyme tyrosinase (Fang and Ramasamy 2016), which catalyzes the oxidation of phenols. A nanomaterial, carbon nanotubes, was also used to amplify the signal; the biosensor demonstrated a high sensitivity of 4.0 ± 0.5 μA cm^{-2} μM^{-1} with a measurement range of 0–100 μM. At the same time, some compounds present in plants to protect against pathogens can cause enzyme inhibition. This effect can be used to create biosensors for the determination of these substances. For example, citral (a mixture of neral and geranial), which is present in the essential oils of many plants, has an inhibitory activity in relation to tyrosinase (Capetti et al. 2021).

In 2019, Wang and coworkers developed a gas biosensor for the detection of ethylhexanol, linalool, tetradecene and phenylacetaldehyde biomarkers released by Huanglongbing disease-infected citrus trees (Wang et al. 2019). Single-walled carbon nanotubes (SWCNTs) decorated with single-stranded DNA were immobilized on the working electrode of the biosensor; the principle of analysis was to register electrostatic interactions of the DNA and VOC molecules with SWCNTs. The use of neural networks enabled distinguishing four VOCs over a wide concentration range of 5 to 100%.

Penicillium digitatum is the main post-harvest citrus pathogen causing losses in all citrus-growing countries. A study by Chalupowicz et al. (2020) presented a new biosensor based on whole cells for detecting the presence of *Penicillium digitatum* in oranges. The approach was based on the luminescent reactions of bacteria to changes in the content of VOCs after infection of oranges with pathogenic microorganisms. The differences between VOC samples in infected and uninfected fruits were tracked using gas chromatography–mass spectrometry, and then using four different genetically modified bioluminescent bacterial strains. The strains used made it possible to detect the pathogen on the third day of infection, before the appearance of visible signs of fungal infection on the surface of the orange.

In recent years, a technology called electronic nose (e-nose) has been used to detect plant diseases and pathogens. The principle of operation of e-nose is to change the electrical conductivity of sensors depending on their chemical interaction with the surrounding gas phase (Cellini et al. 2017). Air samples containing different VOC compositions are described by unique e-nose profiles, which makes it possible to create highly selective devices for detecting a variety of pathogens. For example, in Hazarika et al. (2020) an electronic nose was used to detect citrus tristeza virus in Khasi mandarin orange plants; the results were compared with the data of a standard PCR test. The developed e-nose system had a small response time (3 min) and recovery time (10 min) and allowed achieving a measurement accuracy of 97.58%. This enables the use of the technology for rapid and mass screening of plantations and nurseries, which is otherwise a costly, elaborate and time-consuming affair using the traditional molecular techniques. Miskolczi et al. (2020) suggested using insect odorant receptors, some of which are extremely sensitive to certain biologically relevant VOCs, as e-nose bioreceptors.

The use of nanomaterials makes it possible to improve the sensitivity and accuracy of the analysis using e-nose. Wang et al. (2020) suggested using an e-nose modified with single-stranded DNA and single-walled carbon nanotube to detect *p*-ethylphenol, an important plant VOC released by *Phytophthora cactorum* in infected strawberries. The bioelectronic nose was used to measure the content of 4-ethyl phenol over a wide range of 0.25–100%, which was further diagnosed to verify whether *P. cactorum* infected strawberries or not. Modification of e-nose by graphene oxide and gold nanoparticles (Li et al. 2021) enables continuous VOC analysis of plants in their natural habitats. The authors developed wearable sensors for instant monitoring of plant host responses for early disease diagnosis and rapid stress identification of living plants. The size of the sensor pad containing the sensors was 12 mm × 30 mm, and the weight was approximately 0.7 g. The system was able to differentiate 13 individual plant VOCs with an accuracy of 97.6%, which will allow the commercialization of such systems.

Electronic noses are a universal tool that can theoretically solve any problem. But the extrapolation of the result from one set of samples to another, the sensitivity of the gas analyzer to humidity and the high cost limit their use. Natural variations of VOC profiles within plant species complicate their accurate detection since they can change not only under the action of plant pathogens but also due to other factors (stress, temperature, abiotic effects, etc.). Accordingly, these methods do not always allow for highly selective analysis. Thus, despite constant efforts, the use of electronic nose technology in agriculture still occurs only at the laboratory level and mainly for research purposes. Besides, it should be noted that most studies in this field of research are about detection of bacterial or fungal infections, while diseases with other agents have been less focused on (Mohammad-Razdari et al. 2022).

Conclusions

This chapter described the currently existing laboratory models of biosensors that can be used to detect a variety of plant pathogens. Such methods as PCR, FISH, ELISA and GC-MS are still widely used in agriculture, but their principles are also becoming the basis for the development of new types of devices – nanobiosensors. Standard methods are relatively difficult to operate, are time-consuming and require qualified technical personnel, and biosensors are devoid of most of these disadvantages. Besides using biosensors, it is possible to detect plant diseases at early stages and monitor them in real-time, which makes them suitable for on-field testing. A wide variety of modern nanomaterials with such properties as the high surface area to volume ratio, ability to incorporate bio-recognition molecules, and the possibility of tuning specific properties contributes to the improvement of the analytical characteristics of biosensors developed on their basis and leads to the expansion of their application in agriculture. Another advantage of using nanobiosensors is their affordability – the massive reduction in cost per sample as compared with PCR or GC-MS (Dyussembayev et al. 2021). Despite all the advantages of nanobiosensors over the conventional methods mentioned in this chapter, further research aimed at stabilizing and long-term activity of bioreceptors of these devices is needed for the successful commercialization of this technology. Affordable, fast, highly sensitive and specific nanobiosensors for detecting plant pathogens may become widespread in the very near future if this task is successfully solved.

References

Agriopoulou, S., Stamatelopoulou, E. and Varzakas, T. 2020. Advances in analysis and detection of major mycotoxins in foods. Foods 9(4): 518. Doi: 10.3390/foods9040518.

Awaludin, N., Abdullah, J., Salam, F., Ramachandran, K., Yusof, N.A. and Wasoh, H. 2020. Fluorescence-based immunoassay for the detection of *Xanthomonas oryzae pv. oryzae* in rice leaf. Anal. Biochem. 610: 113876. Doi: 10.1016/j.ab.2020.113876.

Berto, M., Vecchi, E., Baiamonte, L., Condò, C., Sensi, M., Di Lauro, M., Sola, M., De Stradis, A., Biscarini, F., Minafra, A. and Bortolotti, C.A. 2019. Label free detection of plant viruses with organic transistor biosensors. Sens. Actuators B Chem. 281: 150–156. Doi: 10.1016/j.snb.2018.10.080.

Byzova, N., Vinogradova, S., Porotikova, E., Terekhova, U., Zherdev, A. and Dzantiev, B. 2018. Lateral flow immunoassay for rapid detection of grapevine leafroll-associated virus. Biosensors 8(4): 111. Doi: 10.3390/bios8040111.

Capetti, F., Tacchini, M., Marengo, A., Cagliero, C., Bicchi, C., Rubiolo, P. and Sgorbini, B. 2021. Citral-containing essential oils as potential tyrosinase inhibitors: A bio-guided fractionation approach. Plants 10: 969. https://doi.org/10.3390/plants10050969.

Cardoso, R.M., Pereira, T.S., Facure, M.H.M., dos Santos, D.M., Mercante, L.A., Mattoso, L.H.C. and Correa, D.S. 2022. Current progress in plant pathogen detection enabled by nanomaterials-based (bio)sensors. Sens. Actuators Rep. 4: 100068. Doi.org/10.1016/j.snr.2021.100068.

Cellini, A., Blasioli, S., Biondi, E., Bertaccini, A., Braschi, I. and Spinelli, F. 2017. Potential applications and limitations of electronic nose devices for plant disease diagnosis. Sensors 17(11): 2596. https://doi.org/10.3390/s17112596.

Chalupowicz, D., Veltman, B., Droby, S. and Eltzov, E. 2020. Evaluating the use of biosensors for monitoring of *Penicillium digitatum* infection in citrus fruit. Sens. Actuators B Chem. 311: 127896. Doi: 10.1016/j.snb.2020.127896.

Chaudhary, M., Verma, S., Kumar, A., Basavaraj, Y.B., Tiwari, P., Singh, S., Chauhan S.K., Kumar, P. and Singh, S.P. 2021. Graphene oxide based electrochemical immunosensor for rapid detection of groundnut bud necrosis orthotospovirus in agricultural crops. Talanta 235: 122717. Doi: 10.1016/j.talanta.2021.122717.

Cho, I.H., Kim, D.H. and Park, S. 2020. Electrochemical biosensors: Perspective on functional nanomaterials for on-site analysis. Biomater. Res. 24: 6. https://doi.org/10.1186/s40824-019-0181-y.

D'Herelle, F. 1917. Sur un microbe invisible antagoniste des bacilles dysentériques. Comptes Rendus l'Académie Des Sci. D 165: 373–375.

Dyussembayev, K., Sambasivam, P., Bar, I., Brownlie, J.C., Shiddiky, M.J.A. and Ford, R. 2021. Biosensor technologies for early detection and quantification of plant pathogens. Front. Chem. 9: 636245. Doi: 10.3389/fchem.2021.636245.

Ebrahimi, M., Norouzi, P., Safarnejad, M.R., Tabaei, O. and Haji-Hashemi, H. 2019. Fabrication of a label-free electrochemical immunosensor for direct detection of Candidatus phytoplasma aurantifolia. J. Electroanal. Chem. 851: 113451. Doi: 10.1016/j.jelechem.2019.113451.

Fang, Y. and Ramasamy, R.P. 2016. Detection of p-ethylphenol, a major plant volatile organic compound, by tyrosinase-based electrochemical biosensor. ECS J. Solid State Sci. Technol. 5(8): M3054–M3059. Doi:10.1149/2.0101608jss.

Fang, Y., Umasankar, Y. and Ramasamy, R.P. 2014. Electrochemical detection of p-ethylguaiacol, a fungi infected fruit volatile using metal oxide nanoparticles. The Analyst 139(15): 3804–3810. Doi:10.1039/c4an00384e.

Franco, A.J.D., Merca, F.E., Rodriguez, M.S., Balidion, J.F., Migo, V.P., Amalin, D.M., Alocilja E.C. and Fernando, L.M. 2019. DNA-based electrochemical nanobiosensor for the detection of *Phytophthora palmivora* (Butler) Butler, causing black pod rot in cacao (*Theobroma cacao* L.) pods. Physiol. Mol. Plant Pathol. 107: 14–20. Doi: 10.1016/j.pmpp.2019.04.004.

Freitas, T.A., Proença, C.A., Baldo, T.A., Materón, E.M., Wong, A., Magnani, R.F. and Faria, R.C. 2019. Ultrasensitive immunoassay for detection of Citrus tristeza virus in citrus sample using disposable microfluidic electrochemical device. Talanta 205: 120110. Doi: 10.1016/j.talanta.2019.07.005.

Hazarika, S., Choudhury, R., Montazer, B., Medhi, S., Goswami, M.P. and Sarma, U. 2020. Detection of Citrus tristeza virus in mandarin orange using a custom-developed electronic nose system. IEEE Trans. Instrum. Meas. 69(11): 9010–9018. Doi: 10.1109/TIM.2020.2997064.

Hossain, M.Z. and Maragos, C.M. 2018. Gold nanoparticle-enhanced multiplexed imaging surface plasmon resonance (iSPR) detection of *Fusarium mycotoxins* in wheat. Biosens. Bioelectron. 101: 245–252. Doi: 10.1016/j.bios.2017.10.033.

Hwang, H.S., Jeong, J.W., Kim, Y.A. and Chang, M. 2020. Carbon nanomaterials as versatile platforms for biosensing applications. Micromachines 11(9): 814. https://doi.org/10.3390/mi11090814.

Jablonski, M., Poghossian, A., Keusgen, M., Wege, C. and Schöning, M.J. 2021. Detection of plant virus particles with a capacitive field-effect sensor. Anal. Bioanal. Chem. 413(22): 5669–5678. Doi: 10.1007/s00216-021-03448-8.

Jian, Y.-S., Lee, C.-H., Jan, F.-J. and Wang, G.-J. 2018. Detection of odontoglossum ringspot virus infected phalaenopsis using a nanostructured biosensor. J. Electrochem. Soc. 165(9): H449–H454.

Khater, M., Escosura-Muñiz, A., Quesada-González, D. and Merkoçi, A. 2019. Electrochemical detection of plant virus using gold nanoparticle modified electrodes. J. Anal. Chim. Acta 2670: 31102–31104. Doi: 10.1016/j.aca.2018.09.031.

Kulagina, N.V., Shankar, L. and Michael, A.C. 1999. Monitoring glutamate and ascorbate in the extracellular space of brain tissue with electrochemical microsensors. Anal. Chem. 71: 5093–5100.

Kumar, V. and Arora, K. 2020. Trends in nano-inspired biosensors for plants. Mater. Sci. Technol. 3: 255–273. Doi: 10.1016/j.mset.2019.10.004.

Lau, H.Y., Wu, H., Wee, E.J.H., Trau, M., Wang, Y. and Botella, J.R. 2017. Specific and sensitive isothermal electrochemical biosensor for plant pathogen DNA detection with colloidal gold nanoparticles as probes. Sci. Rep. 7(1): 38896. Doi: 10.1038/srep38896.

Lebeda, A., Luhova, L., Sedlarova, M. and Jancova, D. 2001. The role of enzymes in plant–fungal pathogens interactions. J. Plant Dis. Protect. 108: 89–111.

Li, Z., Liu, Y., Hossain, O., Paul, R., Yao, S., Wu, S., Ristaino, J.B., Zhu, Y. and Wei, Q. 2021. Real-time monitoring of plant stresses via chemiresistive profiling of leaf volatiles by a wearable sensor. Matter. 4: 2553–2570. 10.1016/j.matt.2021.06.009.

Lin, W.-P., Wang, W.-J., Lee, C.-H., Jan, F.-J. and Wang, G.-J. 2022. A two-in-one immunoassay biosensor for the simultaneous detection of Odontoglossum ringspot virus and Cymbidium mosaic virus. Sens. Actuators B Chem. 350: 130875.

Luna-Moreno, D., Sánchez-Álvarez, A., Islas-Flores, I., Canto-Canche, B., Carrillo-Pech, M., Villarreal-Chiu, J.F. and Rodríguez-Delgado, M. 2019. Early detection of the fungal banana black sigatoka pathogen *Pseudocercospora fijiensis* by an SPR immunosensor method. Sensors 19(3): 465. https://doi.org/10.3390/s19030465.

Machini, W.B.S. and Oliveira-Brett, A.M. 2021. *In situ* electrochemical investigation of the interaction between bacteria *Xylella fastidiosa* DNA and copper(II) using DNA-electrochemical biosensors. Electrochem. Commun. 125: 106975. https://doi.org/10.1016/j.elecom.2021.106975.

Malecka, K., Michalczuk, L., Radecka, H. and Radecki, J. 2014. Ion-channel genosensor for the detection of specific DNA sequences derived from plum pox virus in plant extracts. Sensors 14(10): 18611–18624. Doi: 10.3390/s141018611.

Miranda, B.S., Linares, E.M., Thalhammer, S. and Kubota, L.T. 2013. Development of a disposable and highly sensitive paper-based immunosensor for early diagnosis of Asian soybean rust. Biosens. Bioelectron. 45: 123–128. Doi: 10.1016/j.bios.2013.01.048.

Mishra, M., Singh, S.K., Bhardwaj, A., Kumar, L., Singh, M.K. and Sundaram, S. 2020. Development of a diatom-based photoluminescent immunosensor for the early detection of karnal bunt disease of wheat crop. ACS Omega 5(14): 8251–8257. Doi: 10.1021/acsomega.0c00551.

Miskolczi, T.D., Zboray, K., Keszőce, A., Quddoos, Z., Ambrózy, Z., Hamow, K.Á., Toth, A., Sagi, L., Szelenyi, M.O., Radvanyi, D., Foldi, M. and Lukács, P. 2020. Development of a smell biosensor system for early detection of plant diseases. Biophys. J. 118(3): 315a. Doi: 10.1016/j.bpj.2019.11.1774.

Mohammad-Razdari, A., Rousseau, D., Bakhshipour, A., Taylor, S., Poveda, J. and Kiani, H. 2022. Recent advances in E-monitoring of plant diseases. Biosens. Bioelectron. 201: 113953. https://doi.org/10.1016/j.bios.2021.113953.

Mohd Said, N.A., Abu Bakar, N. and Lau, H.Y. 2018. Development of DNA biosensor for impedimetric detection of *Erwinia mallotivora* (papaya dieback) as early warning system tool in plant disease management. National Conference on Agricultural and Food Mechanization 2018 (NCAFM 2018).

Nguyen, H.H., Park, J., Kang, S. and Kim, M. 2015. Surface plasmon resonance: A versatile technique for biosensor applications. Sensors 15: 10481–10510. https://doi.org/10.3390/s150510481.

Oliveira, S., da Silva Junior, A.G., de Andrade, C.A.S. and Lima Oliveira, M.D. 2019. Biosensors for early detection of fungi spoilage and toxigenic and mycotoxins in food. Curr. Opin. Food Sci. 29: 64–79. Doi: 10.1016/j.cofs.2019.08.004.

Omar, M.H., Razak, K.A., Ab Wahab, M.N. and Hamzah, H.H. 2021. Recent progress of conductive 3D-printed electrodes based upon polymers/carbon nanomaterials using a fused deposition modelling (FDM) method as emerging electrochemical sensing devices. RSC Advances 11(27): 16557–16571. Doi: 10.1039/d1ra01987b.

Papadakis, G., Skandalis, N., Dimopoulou, A., Glynos, P. and Gizeli, E. 2015. Bacteria murmur: Application of an acoustic biosensor for plant pathogen detection. PLOS ONE 10(7): e0132773. Doi: 10.1371/journal.pone.0132773.

Pathogens, precipitation and produce prices. 2021. Nat. Clim. Chang. 11: 635. https://doi.org/10.1038/s41558-021-01124-4.

Quintela, I.A. and Wu, V.C.H. 2020. A sandwich-type bacteriophage-based amperometric biosensor for the detection of shiga toxin-producing Escherichia coli serogroups in complex matrices. RSC Adv. 10: 35765. https://doi.org/10.1039/D0RA06223E.

Rafidah, A.R., Faridah, S., Shahrul, A.A., Mazidah, M. and Zamri, I. 2016. Chronoamperometry measurement for rapid cucumber mosaic virus detection in plants. Procedia Chem. 20: 25–28. Doi: 10.1016/j.proche.2016.07.003.

Rafiq, A., Taj, A., Haider, S., Tahir, M.A., Zia, R., Moschou, D. and Amin, I. 2021. Biologically prepared copper-graphene nanohybrid as the interface of microchips for sensitive detection of crop viruses. J. Mater. Res. Technol. 12: 727–738. Doi: 10.1016/j.jmrt.2021.03.005.

Rana, K., Mittal, J., Narang, J., Mishra, A. and Pudake, R.N. 2021. Graphene based electrochemical DNA biosensor for detection of false smut of rice (*Ustilaginoidea virens*). Plant Pathol. J. 37(3): 291–298. DOI: https://doi.org/10.5423/PPJ.OA.11.2020.0207.

Regiart, M., Rinaldi-Tosi, M., Aranda, P.R., Bertolino, F.A., Villarroel-Rocha, J., Sapag, K. and Fernández-Baldo, M.A. 2017. Development of a nanostructured immunosensor for early and *in situ* detection of *Xanthomonas arboricola* in agricultural food production. Talanta 175: 535–541. Doi: 10.1016/j.talanta.2017.07.086.

Safarnejad, M.R., Samiee, F., Tabatabie, M. and Mohsenifar, A. 2017. Development of quantum dot-based nanobiosensors against Citrus Tristeza Virus (CTV). Sens. Transducers 213(6): 54–60.

Scala, A., Allmann, S., Mirabella, R., Haring, M.A. and Schuurink, R.C. 2013. Green leaf volatiles: A plant's multifunctional weapon against herbivores and pathogens. Int. J. Mol. Sci. 14: 17781–17811.

Schofield, D.A., Bull, C.T., Rubio, I., Wechter, W.P., Westwater, C. and Molineux, I.J. 2012. Development of an engineered bioluminescent reporter phage for detection of bacterial blight of crucifers. Appl. Environ. Microbiol. 8: 3592–3598. https://doi.org/10.1128/AEM.00252-12.

Schofield, D., Bull, C.T., Rubio, I., Wechter, W.P., Westwater, C. and Molineux, I.J. 2013. Light-tagged bacteriophage as a diagnostic tool for the detection of phytopathogents. Bioengineered 4: 50–54.

Shojaei, T.R., Salleh, M.A.M., Sijam, K., Rahim, R.A., Mohsenifar, A., Safarnejad, R. and Tabatabaei, M. 2016. Detection of Citrus tristeza virus by using fluorescence resonance energy transfer-based biosensor. Spectrochim. Acta A Mol. Biomol. Spectrosc. 169: 216–222. Doi: 10.1016/j.saa.2016.06.052.

Singh, S., Gupta, A.K., Gupta, S., Gupta, S. and Kumar, A. 2014. Surface plasmon resonance (SPR) and cyclic voltammetry based immunosensor for determination of teliosporic antigen and diagnosis of karnal bunt of wheat using anti-teliosporic antibody. Sens. Actuators B Chem. 191: 866–873. Doi: 10.1016/j.snb.2013.10.049.

Tahir, M.A., Bajwa, S.Z., Mansoor, S., Briddon, R.W., Khan, W.S., Scheffler, B.E. and Amin, I. 2018. Evaluation of carbon nanotube based copper nanoparticle composite for the efficient detection of agroviruses. J. Hazard. Mater. 346: 27–35. Doi: 10.1016/j.jhazmat.2017.12.007.

Tereshchenko, A., Fedorenko, V., Smyntyna, V., Konup, I., Konup, A., Eriksson, M. and Bechelany, M. 2017. ZnO films formed by atomic layer deposition as an optical biosensor platform for the detection of grapevine virus A-type proteins. Biosens. Bioelectron. 92: 763–769. Doi: 10.1016/j.bios.2016.09.071.

Thomas, S.G., Phillips, A.L. and Hedden, P. 1999. Molecular cloning and functional expression of gibberellin 2-oxidases, multifunctional enzymes involved in gibberellin deactivation. Proc. Natl. Acad. Sci. USA 96: 4698–4703.

Twort, W. 1915. An investigation on the nature of ultra-microscopic viruses. Lancet II 186: 1241–1243.

Umer, M., Binte Aziz, N., Al Jabri, S., Bhuiyan, S.A. and Shiddiky, M.J.A. 2021. Naked eye evaluation and quantitative detection of the sugarcane leaf scald pathogen, *Xanthomonas albilineans*, in sugarcane xylem sap. Crop Pasture Sci. 72: 361–371. https://doi.org/10.1071/CP20416.

Vaseghi, A., Safaie, N., Bakhshinejad, B., Mohsenifar, A. and Sadeghizadeh, M. 2013. Detection of *Pseudomonas syringae* pathovars by thiol-linked DNAGold nanoparticle probes. Sens. Actuators B Chem. 181: 644–651. Doi: 10. 1016/j.snb.2013.02.018.

Vatankhah, A., Reezi, S., Izadi, Z., Ghasemi-Varnamkhasti, M. and Motamedi, A. 2022. Development of an ultrasensitive electrochemical biosensor for detection of *Agrobacterium tumefaciens* in *Rosa hybrida* L. Measurement 187: 110320. https://doi.org/10.1016/j.measurement.2021.110320.

Vu, N. and Oh, C. 2020. Bacteriophage usage for bacterial disease management and diagnosis in plants. Plant Pathol. J. 36(3): 204–217. DOI: 10.5423/PPJ.RW.04.2020.0074.

Wang, H., Ramnani, P., Pham, T., Villarreal, C.C., Yu, X., Liu, G. and Mulchandani, A. 2019. Gas biosensor arrays based on single-stranded DNA-functionalized single-walled carbon nanotubes for the detection of volatile organic compound biomarkers released by huanglongbing disease-infected citrus trees. Sensors 19: 4795. https://doi.org/10.3390/s19214795.

Wang, H., Wang, Y., Hou, X. and Xiong, B. 2020. Bioelectronic nose based on single-stranded DNA and single-walled carbon nanotube to identify a major plant volatile organic compound (p-ethylphenol) released by *Phytophthora cactorum* infected strawberries. Nanomaterials 10: 479. https://doi.org/10.3390/nano10030479.

Wei, S., Sun, Y., Xi, G., Zhang, H., Xiao, M. and Yin, R. 2018. Development of a single-tube nested PCR-lateral flow biosensor assay for rapid and accurate detection of *Alternaria panax* Whetz. PLOS ONE 13(11): e0206462. Doi: 10.1371/journal.pone.0206462.

Wongkaew, P. and Poosittisak, S. 2014. Diagnosis of sugarcane white leaf disease using the highly sensitive DNA based voltammetric electrochemical determination. Am. J. Plant Sci. 5: 2256–2268. Doi: 10.4236/ajps.2014.515240.

Xu, J., Chau, Y. and Lee, Y. 2019a. Phage-based electrochemical sensors: A review. Micromachines 10(12): 855. Doi: 10.3390/mi10120855.

Xu, S., Zhang, G., Fang, B., Xiong, Q., Duan, H. and Lai, W. 2019b. Lateral flow immunoassay based on polydopamine-coated gold nanoparticles for the sensitive detection of zearalenone in maize. ACS Appl. Mater. Interfaces 11(34): 31283–31290. Doi: 10.1021/acsami.9b08789.

Yadav, S. and Chhibbar, A.K. 2018. Plant–virus interactions. *In*: Singh, A. and Singh, I. (eds.). Molecular Aspects of Plant-Pathogen Interaction. Springer, Singapore. https://doi.org/10.1007/978-981-10-7371-7_3.

Yan, C., Wang, Q., Yang, Q. and Wu, W. 2020. Recent advances in aflatoxins detection based on nanomaterials. Nanomaterials 10(9): 1626. Doi: 10.3390/nano10091626.

Zamir, D., Galsarker, O., Alkan, N. and Eltzov, E. 2020. Detection of quiescent fungi in harvested fruit using CMOS biosensor: A proof of concept study. Talanta 217: 120994. Doi: 10.1016/j.talanta.2020.120994.

Zeilinger, S., Gupta, V., Dahms, T.E.S., Silva, R.N., Singh, H.B., Upadhyay, R.S., Gomes, E.V., Tsui, C.K.M. and Nayak, C. 2016. Friends or foes? Emerging insights from fungal interactions with plants. FEMS Microbiol. Rev. 40: 182–207.

Zhan, F., Wang, T., Iradukunda, L. and Zhan, J. 2018. A gold nanoparticle-based lateral flow biosensor for sensitive visual detection of the potato late blight pathogen, *Phytophthora infestans*. Anal. Chim. Acta. 1036: 153–161. Doi: 10.1016/j.aca.2018.06.083.

Zhao, W., Lu, J., Ma, W., Xu, C., Kuang, H. and Zhu, S. 2011. Rapid on-site detection of *Acidovorax avenae* subsp. *citrulli* by gold-labeled DNA strip sensor. Biosens. Bioelectron. 26(10): 4241–4244. Doi: 10.1016/j.bios.2011.04.004.

Zhao, Y., Liu, L., Kong, D., Kuang, H., Wang, L. and Xu, C. 2014. Dual amplified electrochemical immunosensor for highly sensitive detection of *Pantoea stewartii* subsp. *stewartii*. ACS Appl. Mater. Interfaces 6(23): 21178–21183. Doi: 10.1021/am506104r.

Chapter 4

Bioprospecting of Nanoparticles against Phytopathogens

Sougata Ghosh,[1,2,*] *Bishwarup Sarkar*[3] and *Sirikanjana Thongmee*[1]

Introduction

The population of the world is increasing at an alarming rate, which has compelled agriculturalists to focus on the increased demand for food (Elmer et al. 2018). While several extensive kinds of research are going on for the production of genetically modified crops and plants with higher productivity and nutritional yield, the majority of agriculturalists believe that alleviation of crop losses because of abiotic and biotic stresses can be a crucial solution to the food crisis (Kumar et al. 2022, Nitnavare et al. 2022a). Almost 25% of the plant loss throughout the world is contributed by phytopathogens, which inflict deleterious effects on plants resulting in diseased crops and plant products (Kumar et al. 2022, Ghosh and Sarkar 2022a–c). Although several chemical pesticides, insecticides, fungicides, and herbicides are available, they often contribute to large-scale environmental pollution due to their refractory nature (Ghosh et al. 2022a, b). Hence, biocontrol of these phytopathogens is considered as a major alternative to counter the detrimental effects of microbial infections and pest infestations. However, rising resistance against biocontrol as well as chemical agents emphasizes the growing need to develop promising complementary and alternative strategies for advanced crop protection.

Nanotechnology provides a promising facile and eco-friendly solution to this problem wherein the unique properties of nanomaterials can be applicable for targeted inhibition of the phytopathogens (Hernández-Díaz et al. 2021, Ghosh and Kikani 2022). Biogenic nanoparticles with attractive physico-chemical and

[1] Department of Physics, Faculty of Science, Kasetsart University, Bangkok, Thailand.
[2] Department of Microbiology, School of Science, RK. University, Rajkot, Gujarat, India.
[3] College of Science, Northeastern University, Boston, MA, USA.
* Corresponding author: ghoshsibb@gmail.com

optoelectronic properties can significantly inhibit both bacterial and fungal pathogens (Rai et al. 2009a). Biosynthesis of nanoparticles is an ecofriendly, rapid, and efficient approach, which can result in the fabrication of diverse types of nanoparticles with exotic shapes and industrial as well as biomedical applications (Rai et al. 2009b). Hence, this chapter reviews the pertinence of nanoparticles in the biocontrol of phytopathogens. Several studies are reported to biosynthesize nanoparticles such as copper, silver, selenium, gold, titanium oxide, and zinc oxide using microbial biomass and/or cell-free extracts as well as plant extracts which demonstrated considerable anti-phytopathogenic properties (Ranpariya et al. 2021, Singh et al. 2021). Some of the phytopathogens which are targeted by these nanoparticles include *Alternaria solani*, *Bipolaris sorokiniana*, *Fusarium oxysporum*, *Phytophthora infestans*, *Poria hypolateritia*, *Rhizoctonia solani*, *Xanthomonas oryzae* pv. *oryzae*, and gold barley yellow dwarf virus-PAV which are summarized in Table 1. These potential pathogens are known causative agents of early blight, root rot, spot blotch, late blight, leaf blight, and others. Many of these nanoparticles are able to induce systemic resistance in plants against these phytopathogens while some facilitate intracellular leakage of the phytopathogen through alteration of the cell membrane which results in cell death whereas certain nanoparticles also inhibit the growth of the pathogen through DNA replication interference. Hence, further evaluation and regulation of the toxicity, efficacy, and *in situ* applicability could provide useful insights for the bioprospecting of these nanoparticles against phytopathogens.

Engineered nanoparticles

Copper

Various studies have reported the biomedical applications and potential antagonistic effect of copper nanoparticles (CuNPs) against pathogens, which can be beneficial for plant disease management as well (Bhagwat et al. 2018, Ghosh 2018). Hassan et al. (2018) reported the biosynthesis of CuNPs that displayed antimicrobial activity and biocontrol properties against phytopathogenic fungi. The biomass filtrate of *Streptomyces capillispiralis* isolated from the leaves of *Convolvulus arvensis* was used for the synthesis of CuNPs using 20 mM of $CuSO_4 \times 5H_2O$ under optimal conditions at a pH value of 9.0 and a temperature of 35°C for 4 days. The visible colour change of the filtrate from light blue to greenish brown was indicative of the surface plasmon resonance of the CuNPs that was obtained and further confirmed using ultraviolet-visible (UV-vis) absorption spectroscopy wherein the maximum absorption peak was observed at 600 nm. Transmission electron microscope (TEM) images showed the spherical-mono-dispersed morphology of the particles with a size range of 3.6–59.0 nm. X-ray diffraction (XRD) patterns then demonstrated the centered cubic form of metallic copper present in the CuNPs. The antimicrobial activity of the biogenic CuNPs was then evaluated which showed considerable growth inhibition against several pathogenic microbes such as *Bacillus dimenuta*, *Bacillus subtilis*, *Staphylococcus aureus*, *Escherichia coli*, *Pseudomonas aeruginosa*, *Candida albicans*, and *Aspergillus brasiliensis*. Moreover, the agar plug assay was conducted for investigating the biocontrol activity of phytopathogenic fungi wherein

Table 1: Nanoparticles for biocontrol of phytopathogens.

Source of nanoparticle synthesis	Type of nanoparticle	Size of nanoparticle (nm)	Shape of nanoparticle	Activity	Phytopathogen(s) inhibited	Efficiency of biocontrol	Reference
Biomass of *Streptomyces capillispiralis*	CuNPs	3.6–59.0	spherical	antimicrobial, larvicidal	*Alternaria* sp., *Aspergillus niger*, and *Pythium* sp.	57.14, 63.81, and 58.05%	Hassan et al. 2018
Biomass of *Streptomyces griseus*	CuNPs	30–50	spherical	antifungal	*Poria hypolateritia*	52.7%	Ponmurugan et al. 2016
Commercial	CuNPs	–	–	antifungal in conjugation with *Calothrix elenkinii*	*Fusarium solani*	76%	Mahawar et al. 2019
Biomass of *Fusarium oxysporum*	AgNPs	16–27	spherical and oval	antibacterial	*Pectobacterium carotovorum*	91.72%	Ghazy et al. 2021
Extracellular supernatant of *Bacillus cereus* SZT1	AgNPs	18–39	spherical	antibacterial	*Xanthomonas oryzae* pv. *oryzae*	72.51%	Ahmed et al. 2020
Biomass of *Pseudomonas* sp., *Achromobacter* sp., *Trichoderma* sp., and *Cephalosporium* sp.	AgNPs	20–50	–	antifungal	*Fusarium oxysporum* f. sp. *cicero*	73.33%	Kaur et al. 2018
Biomass of *Trichoderma* sp., *Rhizoctonia solani* Kuhn, and *Pseudomonas fluorescens*	AgNPs	–	–	antifungal	*Rhizoctonia solani* Kuhn	–	Chiranjeevi et al. 2018
Cell-free supernatant of *Bacillus pumilus*, *Bacillus persicus*, and *Bacillus licheniformis*	AgNPs	–	triangular, hexagonal, and spherical	Antiviral	Bean Yellow Mosaic Virus	66%	Elbeshehy et al. 2015
Flower extracts of *Allamanda cathartica* L.	SeNPs	51–68	spherical	antibacterial	*Pseudomonas marginalis* and *Pseudomonas aeruginosa*	–	Sarkar and Kalita 2022

Table 1 contd. ...

...Table 1 contd.

Source of nanoparticle synthesis	Type of nanoparticle	Size of nanoparticle (nm)	Shape of nanoparticle	Activity	Phytopathogen(s) inhibited	Efficiency of biocontrol	Reference
Bacillus megaterium	SeNPs	41.2	spherical	antifungal	*Rhizoctonia solani*	72.27%	Hashem et al. 2021
Trichoderma atroviride (Tri_AtJSB2)	SeNPs	–	–	antifungal	*Phytophthora infestans*	72.9%	Joshi et al. 2021
Chemical synthesis using sol-gel method	Ag-doped hollow TiO_2NPs	–	–	fungicidal	*Fusarium solani* and *Venturia inaequalis*	–	Boxi et al. 2016
Chemical synthesis using titanium(IV) isopropoxide	TiO_2NPs	10–80	spherical, rod-like, polyhedral, and tetragonal bipyramidal	antibacterial and antifungal	*Erwinia amylovora*, *Xanthomonas arboricola* pv. *juglandis*, *Pseudomonas syringae* pv. *tomato* and *Allorhizobium vitis*	–	Kőrösi et al. 2020
Leaf extract of *Moringa oleifera*	TiO_2NPs	40–65	irregular	antifungal	*Bipolaris sorokiniana*	80%	Satti et al. 2021
Chemical synthesis using citrate ions	AuNPs	3.151–31.67	spherical	antiviral	gold barley yellow dwarf virus-PAV	75.3%	Alkubaisi and Aref 2017
Chemical synthesis using quarternary microemulsions	MAS nanoparticles	100–250	spherical	antifungal	*Fusarium oxysporum*	–	Shenashen et al. 2016
Chemical synthesis by one-pot direct template approach	MSN	20–150 μm	–	antifungal	*Alternaria solani*	–	Derbalah et al. 2018
Leaf extracts of *Melia azedarach*	ZnONPs	30–40	hexagonal	fungicidal	*Cladosporium cladosporioides* and *Fusarium oxysporum*	–	Lakshmeesha et al. 2020

57.14, 63.81, and 58.05% inhibition was observed against *Alternaria* sp., *Aspergillus niger*, and *Pythium* sp., respectively, in the presence of 20 mM of CuNPs. In addition, the larvicidal bioassay of biosynthesized CuNPs was also performed against *Musca domistica* and *Culex pipens*, which showed complete mortality after 96 h and 24 h in the presence of 10 mM and 2 mM of CuNPs, respectively.

In another similar study, Ponmurugan et al. (2016) highlighted the antifungal activity of biogenic CuNPs against phytopathogenic fungi *Poria hypolateritia*, which causes red root-rot disease in tea plants. The CuNPs were prepared using the biomass of *Streptomyces griseus*, which reduced 1 mM of $CuSO_4$ under shaking conditions (150 rpm) at room temperature. UV-vis absorption spectra revealed maximum absorption at 592 nm whereas TEM images displayed a spherical shape of the particles with a size range of 30–50 nm. After its synthesis, the impact of CuNPs on the incidence of red root-rot disease in tea plants was studied which showed a 52.7% reduction in the disease incidence in the presence of 2.5 ppm of CuNPs. Moreover, the tea leaf yield was increased up to 2565 kg per hectare after the treatment which was the maximum as compared to plants treated with carbendazim and/or bulk copper. Further, the microbial population in the soil was reduced by 10% after 12 months of CuNPs treatment.

Similarly, Mahawar et al. (2019) examined the disease alleviating potential of CuNPs in *Solanum lycopersicum* L. (tomato) plants challenged with *Fusarium solani*. Commercially available CuNPs were used in this study in conjugation with *Calothrix elenkinii* cyanobacterium as biocontrol agents. The levels of the chitosanase enzyme were increased in tomato plants treated with CuNPs, cyanobacterium as well as the combination of both as compared to tomato plants only infected with the phytopathogen, which was suggested to increase the pathogenic resistance in the tomato plants. An evaluation of the physiology and biochemical activity of the pathogen-infected plants showed a significant reduction in the root length and fresh weight of the plant as compared to the healthy plants. The roots of the plants were disintegrated and decayed in the presence of *Fusarium solani*, which resulted in 2.0–2.9 and 1.4–2.2-fold reduction in the infected plants. Thereafter, pathogen-infected plants were treated with CuNPs and the cyanobacterium alleviated the plant stress, resulting in similar root lengths and fresh weight as that of healthy plants. The combination of both *Calothrix elenkinii* cyanobacterium and CuNPs displayed a disease control efficacy of 76%, which was higher than treatment with only CuNPs. In addition, the chitosanase activity of plants was also increased by 10% in the presence of CuNPs. Moreover, the analysis of rhizosphere soil after augmentation of *C. elenkinii* along with CuNPs demonstrated a 1.4–3.3-fold increase in the phospholipid fatty acid (PLFA) content, which was indicative of a positive effect in the soil microbial communities after CuNPs treatment.

Silver

Silver nanoparticles (AgNPs) are extensively highlighted to demonstrate efficacious antimicrobial activity in several studies, which include phytopathogens (Shinde et al. 2018). Biogenic AgNPs are small, stable and highly catalytic (Shende et al.

2017, 2018, Ghosh et al. 2016a, b). Ghazy et al. (2021) discussed the impact of AgNPs on controlling soft rot disease in *Beta vulgaris* L. (sugar beet) plants. AgNPs were synthesized using the biomass of *Fusarium oxysporum* isolated from tomato plants wherein the reaction mixture changed its colour from pale yellow to yellowish brown, which was indicative of AgNPs formation. The biosynthesis of AgNPs was further confirmed by UV-vis spectroscopy, which revealed maximum absorption at around 430 nm because of surface plasmon resonance (SPR) while TEM images showed spherical and oval morphologies of the AgNPs with a particle size range of 16–27 nm. Later, *Pectobacterium carotovorum* was grown in a nutrient agar medium, which is the causal bacterium for soft rot disease. The inhibitory effects of biogenic AgNPs were assessed on *P. carotovorum* and compared with that of the *Spirulina platensis* extracts and *Bacillus subtilis*. The maximum effect on the growth of the phytopathogen was observed with 50 ppm of AgNPs treatment, which exhibited the lowest pathogen growth rate along with an inhibition zone of 4.33 cm. Moreover, such efficient inhibitory activity was AgNPs was proposed to be a result of interaction between AgNPs and the sulphur-containing proteins present on the cell wall of the pathogen along with phosphorus-containing intracellular components such as deoxyribonucleic acid that facilitated cell lysis and death. Pots and field experiments then revealed 91.72% disease biocontrol efficiency using 100 ppm of AgNPs. Moreover, 37 mg/min of endogenous polyphenol oxidase enzyme activity was induced by 100 ppm of AgNPs after 81 days of sowing. Additionally, the sugar quality of the plants was increased by 20.5% after 100 ppm of AgNPs treatment.

In another study, Ahmed et al. (2020) biosynthesized AgNPs using *Bacillus cereus* SZT1 strain which was then used for the alleviation of damage caused by bacterial leaf blight pathogen in *Oryza sativa* L. (rice) plants. The extracellular supernatant of *Bacillus cereus* SZT1 strain was used for the AgNPs synthesis wherein the supernatant was reacted with 5 mM of $AgNO_3$ at 28 ± 2°C under shaking conditions for 24 h which resulted in the formation of a reddish-brown colour indicating formation of the AgNPs. The confirmation of biogenic AgNPs synthesis was obtained by UV-vis spectroscopy with a maximum absorption peak observed at 418.99 nm. X-ray diffraction (XRD) patterns further displayed the cubic silver structure of the AgNPs while scanning electron microscopy (SEM) and TEM images provided the structural characteristics of the biogenic AgNPs. The size of the particles ranged from 18 to 39 nm whereas the shape of the AgNPs was spherical. The prepared AgNPs were then investigated for their antibacterial activity against *Xanthomonas oryzae* pv. *oryzae* which is the causative agent for bacterial leaf blight disease as seen in Figure 1. The extent of the growth inhibition of the pathogen was directly proportional to the concentration of AgNPs with a maximum inhibition of 91.42% observed at the highest AgNPs concentration of 20 mg/mL. The impact of AgNPs on the lesions and inhibition of *X. oryzae* on rice plants was also investigated which highlighted a significant reduction in the length of the lesions after treatment with 100 mg/L of AgNPs suspension as compared to untreated plants. Likewise, the treatment with 100 mg/L of AgNPs resulted in a 72.51% inhibition of the phytopathogen in rice plants. The AgNPs treated plants showed a 16.18% and 11.23% increase in the root length and root dry weight of healthy rice plants. The extent of oxidative damage

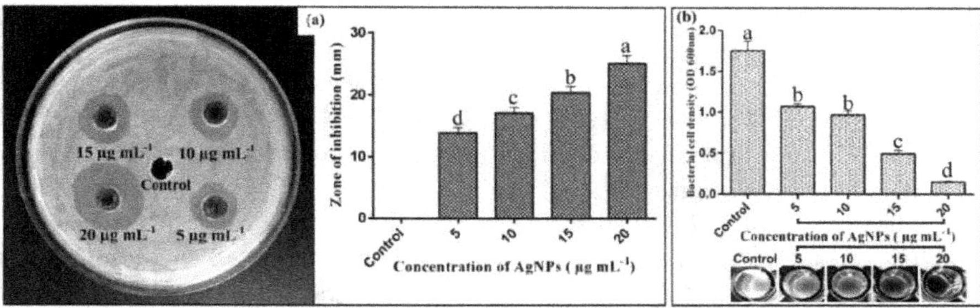

Figure 1: Antibacterial activity of silver nanoparticles (AgNPs) synthesized from *B. cereus* SZT1 in various concentrations against *Xanthomonas oryzae* pv. *oryzae* (*Xoo*). (a) Measurement of antibacterial activity through well-diffusion assay; (b) measurement of antibacterial activity through inhibition of bacterial cell density. Reprinted from Ahmed, T., Shahid, M., Noman, M., Niazi, M.B.K., Mahmood, F., Manzoor, I., Zhang, Y., Li, B., Yang, Y., Yan, C. and Chen, J. 2020. Silver nanoparticles synthesized by using *Bacillus cereus* SZT1 ameliorated the damage of bacterial leaf blight pathogen in rice. Pathogens 9: 160.

caused by the phytopathogen in rice plants was also alleviated after treatment with 100 mg/L of AgNPs along with reduction in the malondialdehyde (MDA) and H_2O_2 content by 48.63% and 60.12%, respectively, in the presence of the pathogen.

In another study, Kaur et al. (2018) reported the biocontrol properties of AgNPs against wilt disease caused in *Cicer arietinum* (chickpea) plants. The rhizospheric soil sample of chickpea plants was used for the isolation of bacteria such as *Pseudomonas* sp. and *Achromobacter* sp. as well as fungi such as *Trichoderma* sp. and *Cephalosporium* sp., which were then used for the synthesis of AgNPs. The bacterial strains were inoculated with 10 mmol/L of $AgNO_3$, which resulted in the formation of a brownish reaction mixture that provided the preliminary confirmation of AgNPs synthesis. However, no such colour change was observed using fungal strains but a precipitate was formed after $AgNO_3$ addition. UV-vis absorption spectral analysis then recorded SPR peaks of AgNPs ranging from 406–420 nm for all the bacterial and fungal strains. TEM images further demonstrated that the particle size of AgNPs was 44 nm and 22 nm when produced by *Pseudomonas* sp. and *Trichoderma* sp., respectively, whereas the size of AgNPs synthesized by *Achromobacter* sp. and *Cephalosporium* sp. was in the range of 20–50 nm. Thereafter, antifungal activity of the biosynthesized AgNPs was assessed against *Fusarium oxysporum* f. sp. *ciceri* (FOC) which is the causative agent of wilt disease of chickpea. AgNPs at a concentration of 100 mg/mL from all four microbial strains showed effective antifungal activity against FOC; however, maximum inhibition activity of 95% was demonstrated by AgNPs produced by *Trichoderma* sp. Moreover, *in vivo* biocontrol efficacy of AgNPs was also studied which showed almost 98% germinability of chickpea seeds after treatment with these biogenic AgNPs. In addition, pot experiments revealed 73.33% control of wilt incidence in chickpea as compared to control after treatment with 100 mg/mL of AgNPs.

The anti-phytopathogenic activity of biogenic AgNPs was also demonstrated by Chiranjeevi et al. (2018) wherein AgNPs were prepared using *Trichoderma* sp., *Rhizoctonia solani* Kuhn, and *Pseudomonas fluorescens.* The prepared AgNPs were then used against *R. solani* Kuhn, which causes rice sheath blight. The detached leaf technique was carried out to investigate the bioefficacy of the AgNPs at a concentration of 170 ppm wherein considerable disease reduction was attained after AgNPs treatment. The sensitivity of rice leaves towards AgNPs was also assessed wherein the leaves turned yellow in the presence of 170 ppm of AgNPs after three days of *in vitro* immersion.

Elbeshehy et al. (2015) also synthesized AgNPs using various isolates of *Bacillus* sp. and evaluated their activity against the Bean Yellow Mosaic Virus (BYMV), which infected *Vicia faba* (Faba bean) leaves. Three different strains of *Bacillus* namely, *Bacillus pumilus*, *Bacillus persicus*, and *Bacillus licheniformis* were used in this study for extracellular AgNPs synthesis which was evident with the change in the colour of the reaction mixture containing $AgNO_3$ from yellow to dark brown within a few hours. The zeta potential of the AgNPs obtained using the supernatant of *B. pumilus*, *B. persicus*, and *B. licheniformis* was –18.5, 16.6, and –21.3 mV, respectively. Moreover, TEM images revealed the monodispersed nature of all three AgNPs with triangular, hexagonal, and spherical morphologies without any agglomerations as evident from Figure 2. Thereafter, the antimicrobial activity of AgNPs was assessed against various pathogenic bacteria and fungi. AgNPs synthesized by *B. licheniformis* exhibited the most effective antimicrobial activity against Gram-negative bacteria such as *Escherichia coli*, *Klebsiella pneumoniae*, *Shigella sonnei*, *Pseudomonas aeruginosa* as well as Gram-positive bacteria such as *Staphylococcus epidermidis,* methicillin-resistant *Staphylococcus aureus* as well as *Streptococcus bovis.* Likewise, considerable antifungal activity against *Aspergillus flavus* and *Candida albicans* was also demonstrated by the biogenic AgNPs. The minimum inhibitory concentration (MIC) and minimum bactericidal concentrations (MBC) of AgNPs prepared from *B. licheniformis* supernatant were also lowest as compared to others. Faba bean leaves infected with BYMV showed significant alleviation from diseased symptoms after 24 h of treatment with all three AgNPs as evident from Figure 3. In addition, no toxic effects of AgNPs were observed after administration of the same in healthy Faba bean plants. Hence, a maximum of 66% infectivity of the plant virus was controlled by these biosynthesized AgNPs.

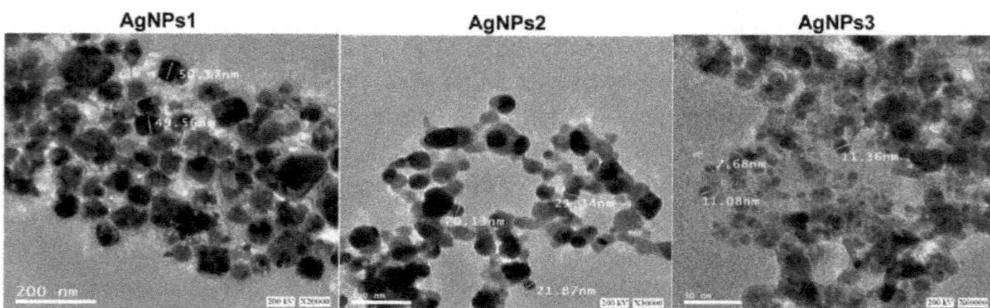

Figure 2: TEM images of the biosynthesized silver nanoparticles. A representative TEM image recorded from the drop coated film of the silver nanoparticles synthesized by extracellular agents of *Bacillus pumilus* (AgNPs1), *B. persicus* (AgNPs1), and *B. licheniformis* (AgNPs1) and estimation of nanoparticles diameter. Reprinted from Elbeshehy, E.K., Elazzazy, A.M. and Aggelis, G. 2015. Silver nanoparticles synthesis mediated by new isolates of *Bacillus* spp., nanoparticle characterization and their activity against Bean Yellow Mosaic virus and human pathogens. Front. Microbiol. 6: 453.

Figure 3: Effect of silver nanoparticles synthesized by extracellular agents of *Bacillus pumilus*, *B. persicus*, and *B. licheniformis* on BYMV infected *Vicia faba L.* cv. *Giza3* and non-infected plants. The nanoparticles were sprayed simultaneously with virus inoculation (experimental Group I); before inoculation (Group II) or after inoculation (Group III). Systemic severe mosaic, crinkling, and size reduction were observed in the case of infected leaves. These symptoms were reduced or completely disappeared when plants were treated with silver nanoparticles synthesized by the isolates, especially those of *B. licheniformis* of Group III. For details, authors are recommended to see text of the source reference. Reprinted from Elbeshehy, E.K., Elazzazy, A.M. and Aggelis, G. 2015. Silver nanoparticles synthesis mediated by new isolates of *Bacillus* spp., nanoparticle characterization and their activity against Bean Yellow Mosaic virus and human pathogens. Front. Microbiol. 6: 453.

Selenium

Sarkar and Kalita (2022) recently synthesized selenium nanoparticles (SeNPs) using *Allamanda cathartica* L. flower extracts, which promoted the growth of *Brassica campestris* (mustard) plants under salt stress and also inhibited the growth of phytopathogens. The formation of SeNPs was completed within 5 h and was confirmed by UV-vis absorption analysis wherein a distinct peak at 287 nm was attained because of the SPR of the SeNPs. The optimum flower extract concentration of 20% facilitated the maximum production of SeNPs in the presence of SeO_2 at a concentration of 25 mM. XRD results further confirmed the crystalline nature of the SeNPs along with the presence of phytochemicals, which were speculated to act as capping agents for the particles. SEM images then highlighted the spherical morphology along with an average size range of 51–68 nm. The green synthesized SeNPs were then analysed for their antimicrobial activity against phytopathogenic bacteria, namely, *Pseudomonas marginalis* and *P. aeruginosa* wherein clear zones of inhibition ranging from 10.67–23.67 mm were observed using varying concentrations of the particles. The germination percentage, shoot length, root length, and total chlorophyll content of salt-stressed mustard plants were also enhanced by 31%, 92%, 78%, and 49% in the presence of 25 mg/L of SeNPs.

Hashem et al. (2021) also reported the potential of SeNPs against the phytopathogen *R. solani* in Faba bean plants. The SeNPs used in this study were biogenic in nature and were prepared using extracellular supernatant of *Bacillus megaterium*, which acted as both reducing and stabilizing agent. The maximum UV-vis absorption of the prepared SeNPs was observed at 435 nm while the particles were spherical in shape and had an average size of 41.2 nm as observed in TEM images. Moreover, the crystalline nature and face-centered cubic (FCC) structure of the SeNPs was evident from the XRD patterns. The antifungal activity of the SeNPs was investigated wherein 1 mM of the metal nanoparticles exhibited maximum antifungal activity with a maximum zone of inhibition of 45 mm. The MIC of SeNPs for *R. solani* was thus calculated as 0.0625 mM. The *in vivo* biocontrol of *R. solani* by SeNPs was also evaluated which showed the best survival of 83.33% when the seeds were pre-soaked in 0.0625 mM of SeNPs and then foliar sprayed with the same after emergence. The biocontrol efficiency of the treatment was 72.27% which was the maximum compared to other treatments. Additionally, the total phenolic content of the roots and shoots of the pathogen-infected plants was considerably reduced after SeNPs treatment while the antioxidant defence system of the plants was enhanced after nanoparticle administration, which subsequently resulted in enhanced stress tolerance.

In another study, Joshi et al. (2021) prepared mycogenic SeNPs, which provided resistance against tomato late blight disease. *Trichoderma atroviride* AtJSB2 strain was used for the extracellular synthesis of SeNPs, which were then employed for priming of tomato cultivar Sagar seeds. The primed seeds demonstrated higher germination percentage as well as seedling vigour compared to the untreated control. Moreover, a positive effect on the growth of the tomato plants with respect to plant height, fruit weight, and flowering duration was seen after SeNPs priming. The systemic resistance of the tomato plants was also enhanced in the presence of SeNPs against *Phytopthora infestans*, which is the causative agent for the late blight of tomatoes. The highest induced disease protection of up to 72.9% was achieved after 28 days of sowing the tomato seeds primed with 100 ppm of SeNPs. The presence of high concentrations of lignin in the stems of tomato plants was attributed to the histological defence marker, which confirmed the SeNPs mediated induction of systemic resistance. Likewise, callose deposition in SeNPs-primed tomato leaves was higher indicating resistance against the phytopathogen. The enzymes responsible for the biochemical defence of the plants such as lipoxygenase, phenylalanine lyase, β-1,3-glucanase, and superoxide dismutase were also upregulated after SeNPs treatment. Therefore, SeNPs can be used as effective nano-biostimulant for inducing the defence system of the plants against pathogens.

Titanium oxide

Titanium dioxide nanoparticles (TiO_2NPs) were prepared by Boxi et al. (2016) and their applicability as an effective green fungicide was evaluated. Hollow TiO_2NPs were doped with Ag, which was evident in XRD spectral results wherein the major anatase peaks were shifted after doping. Energy dispersive X-ray (EDX) spectroscopy then revealed 9.02% doping of Ag in the TiO_2NPs. Thereafter, the antifungal activity of the TiO_2NPs was analysed in which Ag-doped hollow TiO_2NPs exhibited maximum toxicity toward *Fusarium solani* while minimum antifungal activity was achieved using pure titanium dioxide. Similar results were observed against *Venturia inaequalis*; however, *F. solani* was more sensitive to Ag-doped hollow TiO_2NPs. The pathogen and nanoparticle were proposed to electrostatically interact with each other as the Ag-doped hollow TiO_2NPs were positively charged, which could easily be attracted toward the negatively charged surface of the fungal cell wall. Moreover, the enhanced toxicity of Ag-doped hollow TiO_2NPs was attributed to the synergistic antimicrobial activity of Ag. Thereafter, the MIC of Ag-doped hollow TiO_2NPs was 0.43 and 0.75 mg/plate for *F. solani* and *V. inaequalis*, respectively. Additionally, visible light exposure at an intensity of 13,500 lux facilitated a 15% and 50% increase in the TiO_2NPs inhibition capacity of *F. solani* and *V. inaequalis*, respectively, which was attributed to the generation of hydroxyl radical which damaged the chitin and b-glucan structure of the fungal cell wall resulting in intracellular leakage. The formation of stable Ag-S bonds further accelerated the cell damage by the Ag-doped hollow TiO_2NPs. Therefore, the utilization of green nanofungicides for the biocontrol of phytopathogens was highlighted in this study. Hence, silver doping or surface functionalization strategy of nanoparticles can be promising for synergistic enhancement of their activity (Robkhob et al. 2020, Ghosh et al. 2022c).

Kőrösi et al. (2020) also reported the exploitation of TiO_2NPs for phytopathogen inactivation. The nanoparticles were synthesized using titanium(IV) isopropoxide and then treated hydrothermally at 250 or 310°C for varying time durations. The hydrothermal treatment increased the crystallinity of the TiO_2NPs and also induced the formation of the rutile phase. The specific surface area of the TiO_2NPs decreased with increasing synthesis time. TEM images of the TiO_2NPs demonstrated their polydispersity and varying morphology such as spherical, rod-like, polyhedral, and tetragonal bipyramidal. Moreover, the particle size ranged from 10–80 nm. The potential antimicrobial activity of the TiO_2NPs was then investigated against several phytopathogens which include *Erwinia amylovora*, *Xanthomonas arboricola* pv. *juglandis*, *Pseudomonas syringae* pv. *tomato* and *Allorhizobium vitis*. The maximum susceptibility of TiO_2NPs prepared at 310°C was observed against *A. vitis* after 20 min of UV irradiation. Later, the formation of hydroxyl or superoxide radicals upon UV photoexcitation was confirmed using electron paramagnetic resonance (EPR) spectroscopy.

In another recent study, Satti et al. (2021) reported the biocontrol property of TiO_2NPs against *Bipolaris sorokiniana*. The formation of TiO_2NPs was mediated by *Moringa oleifera* leaf extract at room temperature. The reaction mixture changed its colour from milky white to pink-brown because of the formation of TiO_2NPs, which was further confirmed through UV-vis absorption spectra wherein characteristic SPR peak of TiO_2NPs was observed in the range of 240–280 nm. SEM images of the TiO_2NPs demonstrated irregular morphology and anisotropic nature of the same with a size range of 40–65 nm. The disease incidence in *Triticum aestivum* L. (wheat) plants was lowest after treatment with 40 mg/L of the biosynthesized TiO_2NPs. The percent disease index value was decreased by 80% in infected wheat plants after TiO_2NPs treatment. Such significant antimicrobial activity was attributed to the small size of the biogenic TiO_2NPs, which facilitated easy transport of the same through the plasma cell membrane resulting in the destabilization of the outer covering of the fungal cell wall which, in turn, promotes cellular leakage. In addition, disease symptoms caused by the phytopathogen were also alleviated wherein the agro-morphological characteristics such as leaf and root surface area, plant fresh and dry weight along with physiological characteristics such as relative water content and membrane stability index were modified to tolerate the biotic stress.

Other nanoparticles

Gold nanoparticles (AuNPs) are also effective as biocontrol agents and thus, can mediate stress alleviation in diseased plants. Alkubaisi and Aref (2017), for example, demonstrated the biocontrol ability of AuNPs against gold barley yellow dwarf virus-PAV as seen in Figure 4. Citrate ions were used as the reducing agent for the preparation of AuNPs, which had a size range of 3.151–31.67 nm along with spherical morphology. TEM images also showed the spherical shape of the viral capsids. The virus-like particles (VLPs) were deteriorated and ruined after 24 h incubation with AuNPs while 48 h incubation completely lysed VLPs *in vivo*. Moreover, four unique types of VLPs, namely, puffed, deteriorated and decorated, ruined, and vanished were observed which was attributed to the capsid damage inflicted by the AuNPs. The smaller AuNPs with a particle size of 3.151 nm displayed 75.3% intensity of VLPs deterioration. Additionally, the number of ruined VLPs was higher in the local barley cultivar as compared to the Giza 121/Justo cultivar.

Shenashen et al. (2016) also reported a similar study highlighting the fungicidal efficacy of mesoporous alumina sphere (MAS) nanoparticles against phytopathogenic

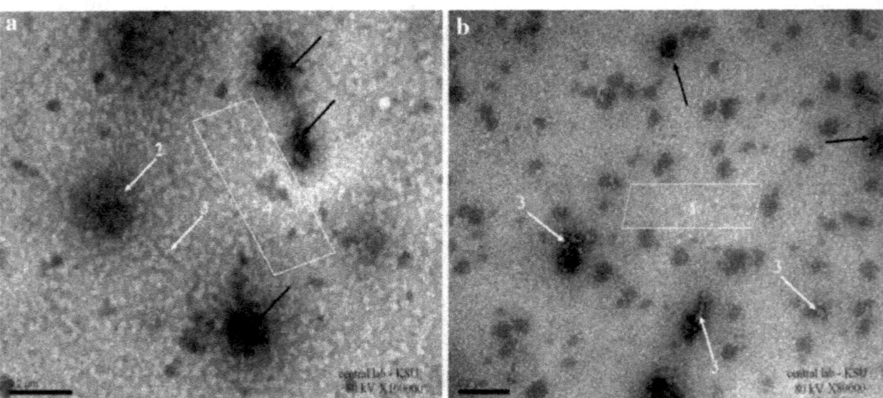

Figure 4: Electron micrographs illustrated the dual effect of AuNPs for two times processing on virus particles *in vivo/vitro* according to local—AuNPs/BYDV-PAV (NPs *in vivo*)—AuNPs—24 vs 48 h (NPs *in vitro*): (a) AuNPs incubation for 24 h indicated numerous aggregated VLPs, more deteriorated VLPs No. 2 and ruined VLPs No. 3. VLPs sink in a dark black area of AuNPs (black arrows). The large square illustrated the area of partially vanished VLPs No. 4; (b) AuNPs incubation for 48 h indicated major background square completely lysed VLPs No. 4, as well as numerous ruined corrosive VLPs No. 3 distributed in most micrographs with dark spots of AuNPs (black arrows). Scale bars 0.2 μm. Reprinted with permission from Alkubaisi, N.A. and Aref, N. 2017. Dispersed gold nanoparticles potentially ruin gold barley yellow dwarf virus and eliminate virus infectivity hazards. Appl. Nanosci. 7: 31–40. Copyright © 2016 The Author(s).

fungi *F. oxysporum* which is known to cause root rot disease in tomato plants. A quaternary microemulsion in liquid crystalline phase along with a surfactant/oil microemulsion phase was used for the synthesis of MAS nanoparticles. The crystalline nature of the MAS nanoparticles was evident from the wide-angle XRD analysis while N_2 adsorption isotherm results provided a specific surface area, pore volume, and average pore size of 256 m^2/g, 0.49 cm^3/g, and 5.9 nm, respectively. Field emission-SEM (FE-SEM) images further revealed the spherical morphology of the particles with a uniform particle size range of 100–250 nm. Thereafter, the efficacy of MAS nanoparticles to inhibit *F. oxysporum* growth was investigated wherein 78.75% growth inhibition was achieved after treatment with 400 mg/L of MAS nanoparticles. Additionally, tolclofos-methyl was used as the standard compound which showed almost similar antifungal activity as that of MAS nanoparticles. The *in vivo* applicability of MAS nanoparticles was then investigated which showed a considerable reduction in the disease symptoms with more than a 20-fold increase in the survival percentage of tomato plants. Furthermore, the growth characteristics of tomato plants such as the plant height, and fresh and dry weights were significantly increased after MAS nanoparticles treatment.

Similarly, Derbalah et al. (2018) fabricated mesoporous silica nanoparticles (MSN) which demonstrated antifungal activity against the early blight of tomatoes. A one-pot direct template method was carried out for the synthesis of MSN wherein small angle-XRD (SA-XRD) patterns showed the well-ordered mesoporous structure of the MSN with a cubic la3d nanophase domain. Then, N_2 adsorption isotherm studies demonstrated a typical type-IV sorption property of the MSN along with a specific surface area of 489 m^2/g. TEM and SEM images also confirmed the mesopore channels of the nanoparticles. The inhibitory action of MSN was then assessed against *Alternaria solani* in which treatment with 400 mg/L of MSN resulted in the highest growth inhibition. Moreover, the degree of severity of early blight of tomatoes was considerably reduced after MSN treatment while the growth parameters such as plant height and fresh and dry weight were also increased by almost 2-folds after MSN treatment. Therefore, such mesoporous nanomaterials also possess biocontrol properties, which are potentially applicable for inhibiting the growth of several phytopathogens.

Lakshmeesha et al. (2020) biofabricated zinc oxide nanoparticles (ZnONPs) using the leaf extract of *Melia azedarach* which showed a UV-vis absorption peak at 374 nm. In addition, XRD patterns revealed the hexagonal wurtzite phase of the particles while TEM and SEM micrographs displayed the hexagonal shape of the ZnONPs with a size range of 30–40 nm. The fungicidal activity of the biogenic ZnONPs was then investigated on common fungal pathogens found in the *Glycine max* (soybean) plants such as *Cladosporium cladosporioides* and *F. oxysporum* wherein the MIC and minimum fungicidal concentration (MFC) were 81.67 and 178.3 mg/mL, and 93.33 and 208.3 mg/mL, respectively, which were lower compared to Amphotericin B. The fungal biomass was also inhibited after ZnONPs treatment along with a considerable reduction in the ergosterol content. Moreover, the growth of *C. cladosporioides* and *F. oxysporum* was also controlled in the presence of 250 mg/mL of ZnONPs. Figure 5 demonstrates the possible underlying mechanism behind the antifungal activity of the ZnONPs. Contact of the ZnONPs with the fungal

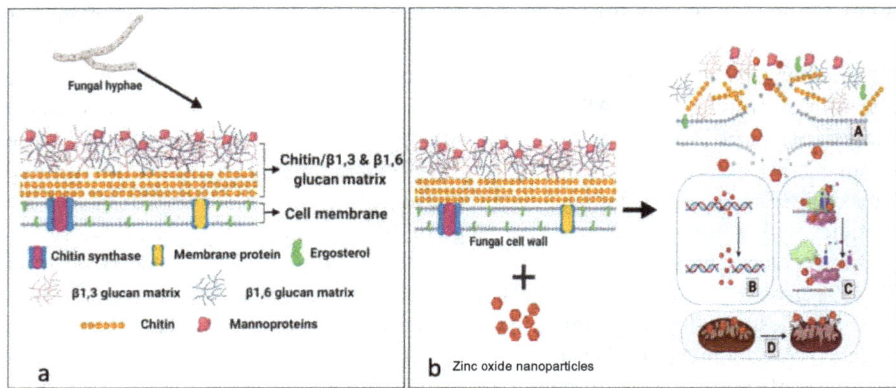

Figure 5: Mechanism of antifungal activity of ZnO NPs. (a) Fungal cell wall; (b) Mechanism of action; (A) Disruption of fungal cell wall; (B) DNA damage; (C) Inhibition of protein synthesis; (D) Mitochondria damage. Reprinted with permission from Lakshmeesha, T.R., Murali, M., Ansari, M.A., Udayashankar, A.C., Alzohairy, M.A., Almatroudi, A., Alomary, M.N., Asiri, S.M.M., Ashwini, B.S., Kalagatur, N.K. and Nayak, C.S. 2020. Biofabrication of zinc oxide nanoparticles from *Melia azedarach* and its potential in controlling soybean seed-borne phytopathogenic fungi. Saudi J. Biol. Sci. 27: 1923–1930. Copyright © 2020 The Author(s). Published by Elsevier B.V. on behalf of King Saud University.

cell wall results in generation of reactive oxygen species (ROS) that further attribute to oxidative damage to the fungal cell wall and cell components that include DNA, protein and mitochondria.

Conclusions and future perspectives

Nanotechnology-driven solution for controlling phytopathogen seems to be an attractive alternative to conventional methods that use hazardous chemicals. The chemical bactericidal and fungicidal agents are often refractory and can have severe detrimental effects on pollinating insects, cattle, and humans when they enter the food chain. Hence, nanomaterials can be used for controlling the microbial pathogens responsible for crop losses. Biogenic nanoparticles are considered more biocompatible and their synthesis is free from chemical reducing and stabilizing agents (Kharissova et al. 2013). Thus, bacteria, fungi, algae, and plants can be used for the synthesis of nanoparticles with enhanced antimicrobial and plant protecting activity (Ghosh et al. 2022d, Nitnavare et al. 2022b). The major limitation of biogenic synthesis is the control over the size and shape. The morphological features of the biogenic nanoparticles can be effectively controlled by careful optimization of the reaction parameters like duration, metal salt concentration, temperature, pH, and extract concentration.

The development of antimicrobial resistance can be controlled by the fabrication of a nanodelivery system of antibacterial and antifungal agents (Suresh et al. 2015). Strategies like liposomal entrapment, multifunctionalization on the metal and metal oxide surfaces, and polymeric encapsulation can ensure sustained release and stability of the bioactive agents (Tutaj et al. 2016). Resistant gene delivery into plants is another attractive strategy where nanocarriers can be used for inducing tolerance against heat, metal, salinity, and infections. In view of the background, it is evident that nanotechnology has immense scope in agriculture that can revolutionize the global crop protection regime.

Acknowledgement

Dr. Sougata Ghosh acknowledges Kasetsart University, Bangkok, Thailand for Post Doctoral Fellowship and funding under Reinventing University Program (Ref. No. 6501.0207/9219 dated 14th September, 2022).

References

Ahmed, T., Shahid, M., Noman, M., Niazi, M.B.K., Mahmood, F., Manzoor, I., Zhang, Y., Li, B., Yang, Y., Yan, C. and Chen, J. 2020. Silver nanoparticles synthesized by using *Bacillus cereus* SZT1 ameliorated the damage of bacterial leaf blight pathogen in rice. Pathogens 9: 160.

Alkubaisi, N.A. and Aref, N. 2017. Dispersed gold nanoparticles potentially ruin gold barley yellow dwarf virus and eliminate virus infectivity hazards. Appl. Nanosci. 7: 31–40.

Bhagwat, T.R., Joshi, K.A., Parihar, V.S., Asok, A., Bellare, J. and Ghosh, S. 2018. Biogenic copper nanoparticles from medicinal plants as novel antidiabetic nanomedicine. World J. Pharm. Res. 7(4): 183–196.

Boxi, S.S., Mukherjee, K. and Paria, S. 2016. Ag doped hollow TiO_2 nanoparticles as an effective green fungicide against *Fusarium solani* and *Venturia inaequalis* phytopathogens. Nanotechnology 27: 085103.

Chiranjeevi, N., Kumar, P.A., Jayalakshmi, R.S., Prasad, K.H. and Prasad, T.N. 2018. Bioefficacy of biogenic silver nanoparticles against rice sheath blight causing pathogen *Rhizoctonia solani* Kuhn. Int. J. Curr. Microbiol. App. Sci. 7(7): 4148–4160.

Derbalah, A., Shenashen, M., Hamza, A., Mohamed, A. and El Safty, S. 2018. Antifungal activity of fabricated mesoporous silica nanoparticles against early blight of tomato. Egypt. J. Basic Appl. Sci. 5: 145–150.

Elbeshehy, E.K., Elazzazy, A.M. and Aggelis, G. 2015. Silver nanoparticles synthesis mediated by new isolates of *Bacillus* spp., nanoparticle characterization and their activity against Bean Yellow Mosaic virus and human pathogens. Front. Microbiol. 6: 453.

Elmer, W., Ma, C. and White, J. 2018. Nanoparticles for plant disease management. Curr. Opin. Environ. Sci. 6: 66–70.

Ghazy, N.A., El-Hafez, A., Omnia, A., El-Bakery, A.M. and El-Geddawy, D.I. 2021. Impact of silver nanoparticles and two biological treatments to control soft rot disease in sugar beet (*Beta vulgaris* L). Egypt. J. Biol. Pest Control 31: 1–12.

Ghosh, S. 2018. Copper and palladium nanostructures: A bacteriogenic approach. Appl. Microbiol. Biotechnol. 101(18): 7693–7701.

Ghosh, S. and Kikani, B. 2022. Role of engineered nanomaterials in sustainable agriculture and crop production. pp. 371–387. *In*: Ghosh, S., Thongmee, S. and Kumar, A. (eds.). Agricultural Nanobiotechnology: Biogenic Nanoparticles, Nanofertilizers and Nanoscale Biocontrol Agents. Woodhead Publishing, Elsevier. Paperback ISBN: 9780323919081; eBook ISBN: 9780323999366.

Ghosh, S. and Sarkar, B. 2022a. Biogenic nanoparticles as novel biocontrol agents. pp. 301–322. *In*: Kumar, A (ed.). Microbial Biocontrol: Food security and post harvest management. Volume 2, Springer Nature, Switzerland. Print ISBN: 978-3-030-87511-4; eBook ISBN: 978-3-030-87512-1.

Ghosh, S. and Sarkar, B. 2022b. Crop-mediated synthesis of nanoparticles and their applications. pp. 23–54. *In*: Ghosh, S., Thongmee, S. and Kumar, A. (eds.). Agricultural Nanobiotechnology: Biogenic Nanoparticles, Nanofertilizers and Nanoscale Biocontrol Agents. Woodhead Publishing, Elsevier. Paperback ISBN: 9780323919081; eBook ISBN: 9780323999366.

Ghosh, S. and Sarkar, B. 2022c. Metal stress removal and nanotechnology driven solutions. pp. 129–153. *In*: Ghosh, S., Thongmee, S. and Kumar, A. (eds.). Agricultural Nanobiotechnology: Biogenic Nanoparticles, Nanofertilizers and Nanoscale Biocontrol Agents. Woodhead Publishing, Elsevier. Paperback ISBN: 9780323919081; eBook ISBN: 9780323999366.

Ghosh, S., Bhagwat, T., Kitture, R., Thongmee, S. and Webster, T.J. 2022c. Synthesis of graphene-hydroxyapatite nanocomposites for potential use in bone tissue engineering. J. Vis. Exp. e63985 (In Press).

Ghosh, S., Chacko, M.J., Harke, A.N., Gurav, S.P., Joshi, K.A., Dhepe, A., Kulkarni, A.S., Shinde, V.S., Parihar, V.S., Asok, A., Banerjee, K., Kamble, N., Bellare, J. and Chopade, B.A. 2016a. *Barleria prionitis* leaf mediated synthesis of silver and gold nanocatalysts. J. Nanomed. Nanotechnol. 7: 4.

Ghosh, S., Gurav, S.P., Harke, A.N., Chacko, M.J., Joshi, K.A., Dhepe, A., Charolkar, C., Shinde, V.S., Kitture, R., Parihar, V.S., Banerjee, K., Kamble, N., Bellare, J. and Chopade, B.A. 2016b. *Dioscorea oppositifolia* mediated synthesis of gold and silver nanoparticles with catalytic activity. J. Nanomed. Nanotechnol. 7: 5.

Ghosh, S., Sarkar, B. and Thongmee, S. 2022a. Nanoherbicides for field applications. pp. 439–463. *In*: Ghosh, S., Thongmee, S. and Kumar, A. (eds.). Agricultural Nanobiotechnology: Biogenic Nanoparticles, Nanofertilizers and Nanoscale Biocontrol Agents. Woodhead Publishing, Elsevier. Paperback ISBN: 9780323919081; eBook ISBN: 9780323999366.

Ghosh, S., Sarkar, B., Kaushik, A. and Mostafavi, E. 2022d. Nanobiotechnological prospects of probiotic microflora: Synthesis, mechanism, and applications. Sci. Total Environ. 838: 156212.

Ghosh, S., Sarkar, B., Kumar, A. and Thongmee, S. 2022b. Regulatory affairs, commercialization and economic aspects of nanomaterials used for agriculture. pp. 479–502. *In*: Ghosh, S., Thongmee, S. and Kumar, A. (eds.). Agricultural Nanobiotechnology: Biogenic Nanoparticles, Nanofertilizers and Nanoscale Biocontrol Agents. Woodhead Publishing, Elsevier. Paperback ISBN: 9780323919081; eBook ISBN: 9780323999366.

Hashem, A.H., Abdelaziz, A.M., Askar, A.A., Fouda, H.M., Khalil, A., Abd-Elsalam, K.A. and Khaleil, M.M. 2021. *Bacillus megaterium*-mediated synthesis of selenium nanoparticles and their antifungal activity against *Rhizoctonia solani* in Faba Bean Plants. J. Fungi 7: 195.

Hassan, S.E.D., Salem, S.S., Fouda, A., Awad, M.A., El-Gamal, M.S. and Abdo, A.M. 2018. New approach for antimicrobial activity and bio-control of various pathogens by biosynthesized copper nanoparticles using endophytic actinomycetes. J. Radiat. Res. Appl. Sci. 11(3): 262–270.

Hernández-Díaz, J.A., Garza-García, J.J., Zamudio-Ojeda, A., León-Morales, J.M., López-Velázquez, J.C. and García-Morales, S. 2021. Plant-mediated synthesis of nanoparticles and their antimicrobial activity against phytopathogens. J. Sci. Food Agric. 101: 1270–1287.

Joshi, S.M., De Britto, S. and Jogaiah, S. 2021. Myco-engineered selenium nanoparticles elicit resistance against tomato late blight disease by regulating differential expression of cellular, biochemical and defense responsive genes. J. Biotechnol. 325: 196–206.

Kaur, P., Thakur, R., Duhan, J.S. and Chaudhury, A. 2018. Management of wilt disease of chickpea *in vivo* by silver nanoparticles biosynthesized by rhizospheric microflora of chickpea (*Cicer arietinum*). J. Chem. Technol. Biotechnol. 93: 3233–3243.

Kharissova, O.V., Dias, H.V.R., Kharisov, B.I., Pérez, B.O. and Pérez, V.M.J. 2013. The greener synthesis of nanoparticles. Trends Biotechnol. 31: 240–248

Kőrösi, L., Pertics, B., Schneider, G., Bognár, B., Kovács, J., Meynen, V., Scarpellini, A., Pasquale, L. and Prato, M. 2020. Photocatalytic inactivation of plant pathogenic bacteria using TiO_2 nanoparticles prepared hydrothermally. Nanomaterials 10: 1730.

Kumar, A., Choudhary, A., Kaur, H., Guha, S., Mehta, S. and Husen, A. 2022. Potential applications of engineered nanoparticles in plant disease management: A critical update. Chemosphere 295: 133798.

Lakshmeesha, T.R., Murali, M., Ansari, M.A., Udayashankar, A.C., Alzohairy, M.A., Almatroudi, A., Alomary, M.N., Asiri, S.M.M., Ashwini, B.S., Kalagatur, N.K. and Nayak, C.S. 2020. Biofabrication of zinc oxide nanoparticles from *Melia azedarach* and its potential in controlling soybean seed-borne phytopathogenic fungi. Saudi J. Biol. Sci. 27: 1923–1930.

Mahawar, H., Prasanna, R. and Gogoi, R. 2019. Elucidating the disease alleviating potential of cyanobacteria, copper nanoparticles and their interactions in *Fusarium solani* challenged tomato plants. Plant Physiol. Rep. 24(4): 533–540.

Nitnavare, R., Bhattacharya, J. and Ghosh, S. 2022a. Nanoparticles for effective management of salinity stress in plants. pp. 189–216. *In*: Ghosh, S., Thongmee, S. and Kumar, A. (eds.). Agricultural Nanobiotechnology: Biogenic Nanoparticles, Nanofertilizers and Nanoscale Biocontrol Agents. Woodhead Publishing, Elsevier. Paperback ISBN: 9780323919081; eBook ISBN: 9780323999366.

Nitnavare, R., Bhattacharya, J., Thongmee, S. and Ghosh, S. 2022b. Utilizing photosynthetic microbes in nanobiotechnology: Applications and perspectives. Sci. Total Environ. 841: 156457.

Ponmurugan, P., Manjukarunambika, K., Elango, V. and Gnanamangai, B.M. 2016. Antifungal activity of biosynthesised copper nanoparticles evaluated against red root-rot disease in tea plants. J. Exp. Nanosci. 11(13): 1019–1031.

Rai, M., Yadav, A. and Gade, A. 2009b. Silver nanoparticles as a new generation of antimicrobials. Biotech. Adv. 27: 76–83.

Rai, M., Yadav, P., Bridge, P. and Gade, A. 2009a. Myconanotechnology: A new and emerging science. pp. 258–267. *In*: Rai, Bridge (ed.). Applied Mycology. CABI publication, UK.

Ranpariya, B., Salunke, G., Karmakar, S., Babiya, K., Sutar, S., Kadoo, N., Kumbhakar, P. and Ghosh, S. 2021. Antimicrobial synergy of silver-platinum nanohybrids with antibiotics. Front. Microbiol. 11: 610968.

Robkhob, P., Ghosh, S., Bellare, J., Jamdade, D., Tang, I.M. and Thongmee, S. 2020. Effect of silver doping on antidiabetic and antioxidant potential of ZnO nanorods. J. Trace Elem. Med. Biol. 58: 126448.

Sarkar, R.D. and Kalita, M.C. 2022. Se nanoparticles stabilized with *Allamanda cathartica* L. flower extract inhibited phytopathogens and promoted mustard growth under salt stress. Heliyon 8: e09076.

Satti, S.H., Raja, N.I., Javed, B., Akram, A., Mashwani, Z.U.R., Ahmad, M.S. and Ikram, M. 2021. Titanium dioxide nanoparticles elicited agro-morphological and physicochemical modifications in wheat plants to control *Bipolaris sorokiniana*. Plos One 16(2): e0246880.

Shenashen, M., Derbalah, A., Hamza, A., Mohamed, A. and El Safty, S. 2017. Antifungal activity of fabricated mesoporous alumina nanoparticles against root rot disease of tomato caused by *Fusarium oxysporium*. Pest Manag. Sci. 73: 1121–1126.

Shende, S., Joshi, K.A., Kulkarni, A.S., Charolkar, C., Shinde, V.S., Parihar, V.S., Kitture, R., Banerjee, K., Kamble, N., Bellare, J. and Ghosh, S. 2018. *Platanus orientalis* leaf mediated rapid synthesis of catalytic gold and silver nanoparticles. J. Nanomed. Nanotechnol. 9: 2.

Shende, S., Joshi, K.A., Kulkarni, A.S., Shinde, V.S., Parihar, V.S., Kitture, R., Banerjee, K., Kamble, N., Bellare, J. and Ghosh, S. 2017. *Litchi chinensis* peel : A novel source for synthesis of gold and silver nanocatalysts. Glob. J. Nanomedicine 3(1): 555603.

Shinde, S.S., Joshi, K.A., Patil, S., Singh, S., Kitture, R., Bellare, J. and Ghosh, S. 2018. Green synthesis of silver nanoparticles using *Gnidia glauca* and computational evaluation of synergistic potential with antimicrobial drugs. World J. Pharm. Res. 7(4): 156–171.

Singh, B.P., Ghosh, S. and Chauhan, A. 2021. Development, dynamics and control of antimicrobial-resistant bacterial biofilms: A review. Environmental Chemistry Letters. https://doi.org/10.1007/s10311-020-01169-5.

Suresh, D., Nethravathi, P.C., Udayabhanu Rajanaika, H., Nagabhushana, H. and Sharma, S.C. 2015. Green synthesis of multifunctional zinc oxide (ZnO) nanoparticles using Cassia fistula plant extract and their photodegradative, antioxidant and antibacterial activities. Mater. Sci. Semicond. Process. 31: 446–454.

Tutaj, K., Szlazak, R., Szalapata, K., Starzyk, J., Luchowski, R., Grudzinski, W., Osinska-Jaroszuk, M., Jarosz-Wilkolazka, A., Szuster-Ciesielska, A. and Gruszecki, W.I. 2016. Amphotericin B-silver hybrid nanoparticles: Synthesis, properties and antifungal activity. Nanomed. Nanotechnol. Biol. Med. 12: 1095–1103.

Chapter 5

Nanomaterials:

A Strategy for the Control of Phytopathogenic Fungi

Claudia A. Ramírez-Valdespino,[1] *Brenda Arriaga-García*,[1] *Socorro Héctor Tarango-Rivero*[2] and *Erasmo Orrantia-Borunda*[1,*]

Introduction

It is reported that plant diseases result in an annual estimated loss of 10–15% of the major crops around the world, and of these diseases, 70–80% are caused by pathogenic fungi (Chatterjee et al. 2016). These fungi developed strategies to attack plants, overcoming plant immune defenses, affecting the health and physiology of plants, and may result in systemic damage, including death (El-Baky et al. 2021). Among the fungi with the most scientific/economic importance are *Magnaporthe oryzae*, *Botrytis cinerea*, *Fusarium* spp., *Colletotrichum* spp., and *Ustilago maydis* (Dean et al. 2012). To overcome that problem, numerous agrochemicals are used, some of them may have negative environmental effects on soil organisms, plants, and may even be toxic to humans.

Recently, the use of agrochemicals containing nanostructured materials has emerged as an alternative in agriculture. Among them, nanoparticles (NPs) have been used, standing out are silver NPs (AgNPs), zinc oxide NPs (ZnONPs), and copper oxide NPs (CuONPs), which can be obtained by physical, chemical, and biological methods (Singh et al. 2021). On the other hand, carbon nanotubes (CNTs) may also be an alternative for the control of pathogens due to which they have strong antimicrobial activity (Maksimova 2019). This chapter focuses on nanoparticles and carbon nanostructures, which have already demonstrated antifungal activity against phytopathogens.

[1] Laboratorio de Nanotoxicología/Biohidrometalurgia, Departamento de Medio Ambiente y Energía, Centro de Investigación en Materiales Avanzados, S. C. Chihuahua, México.

[2] Campo Experimental Delicias, INIFAP, Chihuahua, México.

* Corresponding author: erasmo.orrantia@cimav.edu.mx

Among eukaryotes, fungi have the highest diversity with approximately 1.5 million species (Capote et al. 2012). Many of these fungi are beneficial to humans, animals and plants; however, some of them can cause diseases, including phytopathogenic fungi. Fungi represent the majority of pathogens posing a severe threat to plant health; however, most fungi are nonpathogenic and most plants are resistant to all but a few species developed strategies that allow them to cause disease. Dean and coworkers (2012) reported the top ten fungal plant pathogens; the list includes *M. oryzae*, *B. cinerea*, *Puccinia* spp., *Fusarium gaminearum*, and *F. oxysporum*, among others. All of them are organisms that attack crops of agricultural interest and their treatment, whether preventive or corrective, is of vital importance.

Since 1940, chemical and synthetic fungicides have been used to control phytopathogenic fungi and are commercially available. Fungicides can be classified according to their biochemical modes of action as fungicides that affect nucleic acid and/or amino acid and protein synthesis, mitosis, and cell division, fungal respiration, signal transduction, lipids and membrane synthesis, sterol biosynthesis in membranes, cell wall biosynthesis, melanin synthesis in the cell wall, host plant defense induction and multi-site contact activity (http://frac.info). Recently, the most used fungicides belong to the anilinopyrimidines, which include cyprodinil, mepanipyrim, and pyrimethanil. These substances act by inhibiting the biosynthesis of methionine and the secretion of hydrolytic enzymes. Fungicides such as imidazoles and triazoles are sterol biosynthesis inhibitors, preventing fungal germ tube and mycelium elongation. Other examples of fungicides used are the phenylpyrroles fludioxonil and fenpiclonil, which interfere with the transport processes of sugars and amino acids in the plasma membrane of fungi. These are some examples of fungicides used; however, fungi have developed tools that have allowed them to become resistant to these fungicides (frac.info 2012, El-Baki and Amara 2021).

Additionally, these substances can cause negative effects on human beings such as conditions related to the nervous system, congenital anomalies, and cancer (Asghar et al. 2016). Moreover, fungicides negatively affect the health of animals, human beings, and the environment, leading to the search for new and environmentally friendly approaches to fungal control. Here is where nanotechnology emerges as a strategy to control phytopathogens.

Characteristics of the nanostructured compounds

Nanostructured compounds are solid materials with dimensions between 1 and 100 nm and they are composed of several phases where at least one of the phases has one, two, or three dimensions in nanometer size (www.nano.gov). According to Younus and collaborators (2022), there are six physicochemical parameters that determine the functions of nanoparticles:

1. Size and surface area of the particles: These parameters are inversely proportional to each other as particle size decreases as the surface area increases, increasing its reactivity.

2. Particle shape: Depending on the shape of the NP, the NP will have specific physicochemical properties.
3. Surface charge: This parameter plays a significant role in determining the toxicity of NPs. Among the characteristics of NPs plasma protein binding, colloidal behaviour, and transmembrane permeability are mainly controlled by surface charges.
4. Composition and crystalline structure: They also have an effect on the toxicity of the material and they are very specific to NPs.
5. Aggregation and concentration: Aggregation of NPs influences the toxicity profile. On the other side, at higher concentrations the toxicity of NPs decreases.
6. Surface coating and surface roughness: These parameters have an impact on the cytotoxic properties and can alter the pharmacokinetics, distribution, accumulation, and toxicity of NPs.

Depending on their application, their mechanical and chemical stability, as well as their optical, catalytic, and electrical behaviour, can be controlled, which makes them suitable for the agricultural industry (Rajwade et al. 2022, Younus et al. 2022).

Different types of nanostructures such as carbon nanotubes, NPs of silver, zinc, copper, magnesium, and gold have been reported to exhibit antimicrobial activity (Dizaj et al. 2012, Shoala 2020).

It is well known that the toxicity of the NPs depends upon both dosage and particle size and, interestingly, in the case of some NPs their antimicrobial properties can be due to their ability to interact with cellular membranes (Ahmad et al. 2013). In the area of nanoagrotechnology, the nanoparticles that have been mostly studied are those based on Cu, Zn, and Ag, and some studies have focused on the study of carbon-based nanocomposites. Thus, the following are the most recent reports for some of the more studied nanoparticles for their antifungal activity, with emphasis on phytopathogenic fungi.

Nanomaterials used to counteract phytopathogenic fungi

The control of plant diseases using nanotechnology can be approached from the generation of tools for early-stage detection of pathogens, generating compounds that elicit the plant's immune response, or reduction of the pathogen to manageable levels (Rajwade et al. 2020).

On the other hand, the application of pesticides, fertilizers, and nutrients in the agricultural sector is generally by spraying the foliage of the plants or the soil through the irrigation system, so that they are then transported upwards or downwards through the vascular system of the plants. In modern agriculture, metallic or carbon-derived NPs can also be applied to the soil through fertigation systems, which provoke a variety of physiological and biochemical responses in plants (Lira-Saldivar et al. 2018). The following are some of the NPs used to reduce the number of phytopathogenic fungi of interest in agriculture, through various mechanisms that include oxidative stress, membrane damage, and interaction with enzymes, among others (Figure 1).

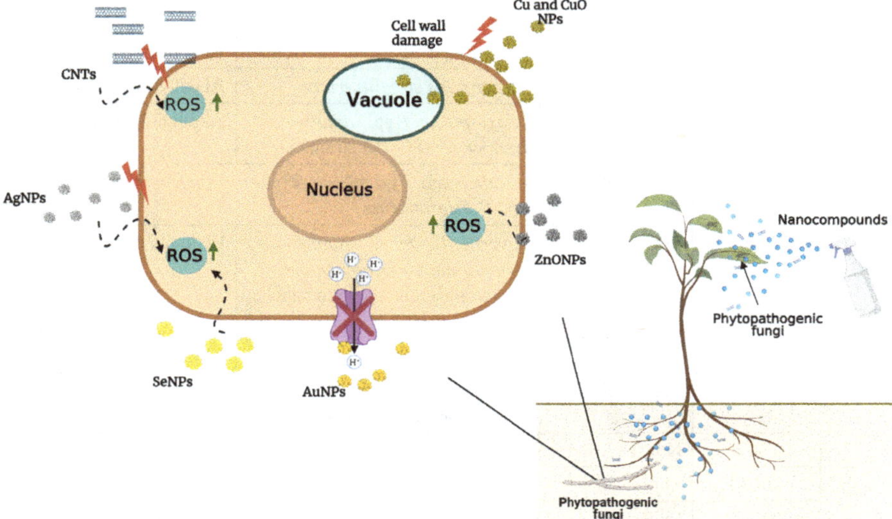

Figure 1: Mechanism showing interaction of nanoparticles with pathogenic fungus including oxidative stress, membrane damage, and interaction with enzymes.

Nanoparticles

Cu and CuONPs

Due to their chemical nature, copper NPs have emerged as unique broad-spectrum antibacterial and antifungal agents against phytopathogens, because they have high selectivity against microorganisms at low concentrations, low toxicity to humans, and their production costs are lower compared with the synthesis of other NPs (Pham et al. 2019, Maqsood et al. 2020).

CuNPs have been studied on the phytopathogens *Fusarium* and *Aspergillus*, and on pathogens of maize, tomato, and other crops, and it has been shown that the efficiency of the inhibition of fungal growth depends on factors such as concentration and NP size. Thus, at concentrations of 450 and 500 pp, *Fusarium* and *Aspergillus* showed growth inhibition percentages of up to 93.98 and 97%, respectively (Viet et al. 2016, Pariona et al. 2019). Interestingly, when the size of the NPs is considered, it is possible to obtain a high inhibition percentage with much lower concentrations, as is shown in the study of Pham et al. (2019) who obtained an inhibition percentage of 72% at a concentration of 20 ppm when the NPs is 53 nm in average size.

On the other side, CuONPs are used as fertilizers, plant growth regulators, and pesticides (Anderson et al. 2018, Gogos et al. 2012). There are studies reporting the cytotoxicity of the CuONPs including pathogenic fungi such as *Alternaria alternata*, *Pyricularia oryzae*, and *Sclerotinia sclerotiorum* (Consolo et al. 2020). Vera-Reyes and collaborators (2019) reported that CuONPs were effective against *A. solani* and *F. oxysporum,* with an inhibition of 95 and 96%, respectively. Another study showed that CuONPs significantly inhibited the growth of *Alternaria mali*, *Diplodia seriata*, and *Botryosphaeria dothidea* (Ahmad et al. 2020a). Table 1 shows other examples of the antifungal activity of Cu and CuONPs.

Table 1: Recent studies on the use of nanoparticles to counteract phytopathogenic fungi.

NPs	Size (nm)	Shape	Tested against	Reference
Ag	10–15	Spherical	*Aspergillus* sp., *Fusarium*	Abdel-Azeem et al. 2020
	3, 5, 8 and 10 nm	Not specified	*F. oxysporum f.* sp. *radicis-lycopersici* (FORL) strains	Akpinar et al. 2021
	5–18	Spherical	*Al. alternata*, *Pyricularia oryzae* and *S. sclerotiorum*	Consolo et al. 2020
	10	Spherical	*F. verticillioides*, *F. moniliforme*, *Penicillium brevicompactum*, *Helminthosporium oryzae* and *Pyricylaria grisea*	Elamawi et al. 2018
	36	Round	*A. alternata*	El-Gazzar et al. 2020
	40–60	Spherical	*F. solani*, *F. semitectum*, *F. oxysporum* and *F. roseum*	El-Wakil 2020
	< 100 nm	Not specified	*F. culmorum* and *A. alternata*	Ganash et al. 2018
	20–30	Spherical	*S. sclerotiorum*	Guilger et al. 2017
	88–182	Spherical - Rod and Prism	*S. sclerotiorum*	Guilger-Casagrande et al. 2019
	59.66	Spherical	*S. rolfsii*	Hirpara and Gajera 2019
	12.7	Not specified	*A. solani*	Ismail et al. 2016
	7–25	Not specified	18 fungi, including: *Fusarium* spp., *Alternaria* spp., *Glomerella cingulata* and *Pythium* spp.	Kim et al. 2012
	2	Not specified	*F. graminearum*	Jian et al. 2021
	< 100 nm	Not specified	*B. cinerea*, *A. alternata*, *M. fructicola*, *C. gloeosporioides*, *F. solani*, *F. oxysporum f.* sp *Radicis Lycopersici*, and *V. dahliae*	Malandrakis et al. 2019
	100–250	Spherical	*F. moliniforme* and *R. solani*	Manikandaselvi et al. 2020
	32–47	Spherical	*Macrophomina phaseolina*	Mohamed and Elshahawy 2022
	50.56	Spherical	*Alternaria solani*, *Corynespora cassiicola*, and *Fusarium* spp.	Tyagi et al. 2020
	5–50	Spherical and oval	*S. slerotiorum*	Tomah et al. 2020
ZnO	52–70	Spherical	*Alternaria mali*, *Botryosphaeria dothidea* and *Diplodia seriata*	Ahmad et al. 2020b
	6–21	No specified	*A. alternata*, *R. solani* and *B. cinerea*	Al-Dhabaan et al. 2017
	< 100 nm	Spherical	*Mycena citricolor* and *Colletotrichum* sp.	Arciniegas-Grijalba et al. 2019

Table 1 contd. ...

...Table 1 contd.

NPs	Size (nm)	Shape	Tested against	Reference
	134–200	Fan/bouquet structure	*A. alternata, P. oryzae* and *S. sclerotiorum*	Consolo et al. 2020
	20–50	Short and long rods or cylinders	*R. solani, S. rolfsii* and *F. oxysporum f.* sp. *lycopersici* (Fol)	Guerrero et al. 2020
	17.41	Spherical to hexagonal	*B. cinerea*	Issam et al. 2021
	< 50	Not specified	*B. cinerea, A. alternata, M. fructicola, C. gloeosporioides, F. solani, F. oxysporum f.* sp. *Radicis Lycopersici* and *V. dahliae*	Malandrakis et al. 2019
	20–70	Spheroidal	*Colletotrichum* sp.	Mosqueda-Sánchez et al. 2020
	35–50	Quasi-spherica	*F. oxysporum, F. solani* and *C. gloeosporioides*	Pariona et al. 2021
	40–100	Not specified	*F. oxysporum f.* sp. *lycopersici* and *A. solani*	Parveen and Siddiqui 2021
Se	78	Spherical and irregular	*A. alternata*	El-Gazzar et al. 2020
	41.2	Spherical	*R. solani*	Hashem et al. 2021
	60	Irregular	*Fusarium, Alternaria*	Hu et al. 2019
	Not specified	Not specified	*A. solani*	Ismail et al. 2016
	60–123	Spherical	*Pyricularia grisea, Colletotrichum capsici* and *A. solani*	Joshi et al. 2019
	49.5–312.5	Hexagonal, near spherical and irregular	*S. graminicola*	Nandini et al. 2017
Cu	6–26	Spherical	*F. oxysporum, F. solani* and *C. curvulata*	Abboud 2021
	50–100	Not specified	*A. alternata, R. solani* and *B. cinerea*	Al-Dhabaan et al. 2017
	50–300	Spherical	*Fusarium* spp.	Bramhanwade et al. 2016
	25	Not specified	*B. cinerea, A. alternata, M. fructicola, C. gloeosporioides, F. solani, F. oxysporum f.* sp. *Radicis Lycopersici* and *V. dahliae*	Malandrakis et al. 2019
	150–200 nm	Not specified	*F. oxysporum, F. solani* and *C. gloeosporioides*	Pariona et al. 2021

Table 1 contd. ...

...*Table 1 contd.*

NPs	Size (nm)	Shape	Tested against	Reference
CuO	dic-38	Spherical	*F. oxysporum* f. sp. *lycopersici*	Ashraf et al. 2021
	38–77	Dispersed and elongated fibres	*A. alternata, P. oryzae* and *S. sclerotiorum*	Consolo et al. 2020
	50–80	Short and elongated particles	*R. solani, S. rolfsii* and *F. oxysporum f.* sp. *lycopersici* (Fol)	Guerrero et al. 2020
	78	Spherical	*F. oxysporum, P. ultimum, A. niger* and *A. alternata*	Hassan et al. 2019
	< 50	Not specified	*B. cinerea, A. alternata, M. fructicola, C. gloeosporioides, F. solani, F. oxysporum f.* sp. *Radicis Lycopersici* and *V. dahliae*	Malandrakis et al. 2019
	20–100	Spherical	*Phytophthora parasitica* (oomycete)	Sawake et al. 2022
	Not specified	Not specified	*Alternaria solani*	Zakharova et al. 2019
Au	5–40	Triangular, hexagonal, spherical, decahedral and pyramidal	*B. cinerea*	Castro et al. 2014
	2–30	Spherical and triangular	*A. alternata*	Sarkar et al. 2012
	10–100	Spherical	*R. solani*	Soltani et al. 2022
CNTs	Diameter of 30–50 nm, length of 10 to 20 μm	Multilayer	*F. oxysporum*	González-García et al. 2022

It has been reported that part of the fungal growth inhibition of CuNPs is because they cause changes in morphology, ranging from deformations, rugosities, and weakening in the mycelium, or even non-germinated conidia, indicating a deleterious effect on the cell wall that leads to the expulsion of intracellular components and ultimately, cell death. These effects might possibly be due to disruption of the synthesis of cell wall compounds, such as chitin, causing a loss of cell wall stability and generating deformations and osmotic imbalances, thus facilitating the permeation of CuNPs (Pariona et al. 2019). Moreover, copper can be absorbed on the cell surface by components that contain negatively charged groups in order to have an affinity for metal ions, or another theory is that copper transporting proteins could be present and internalize into the cell so that it can be used in metabolic processes or accumulated in the vacuole. The latter possibility would imply a highly strict transport regulation since high concentrations of copper inside the cell could cause cell death (Ramírez-Valdespino and Orrantia-Borunda 2021).

AuNPs

Gold NPs possess different physical and chemical properties that make them excellent structures for the fabrication of novel chemicals with antimicrobial activity (Das et al. 2011). Moreover, AuNPs can be used as catalysts, antimicrobial agents, or in the biomedical field as diagnostics and therapy, biomolecule detection, and nano drug carriers (do Nascimento et al. 2021). For example, as it is shown in Table 1, several studies have reported that AuNPs have antimicrobial activity against phytopathogens such as *Aspergillus* and *Fusarium* species (Jayaseelan et al. 2013, Lipsa et al. 2020).

Studies have shown that AuNPs have an antifungal effect against *A. niger* and *A. flavus*, even better than commercial antifungal drugs (Jayaseelan et al. 2013). Interestingly, it is known that when these NPs are applied in conjugation with other materials such as chitosan, their antifungal effect is enhanced. For example, the combination of these two nanomaterials inhibited up to 100% of the growth of *F. oxysporum* when treated with doses up to 5ml with 25, 50, or 75 µg/ml. The antifungal activity of these NPs may be due to the fact that they have the ability to interact with enzymes involved in the regulation of the proton gradient to the extracellular medium, causing their inactivation, which would result in cell death (Lipsa et al. 2020).

ZnONPs

Among the NPs that have been most widely used in agriculture as a nutrient and antimicrobial, ZnONPs stand out as beneficial to plants and the soil microbiota (Shobha et al. 2020). Among the advantages of ZnONPs are the following: low cost, low toxicity to human cells, stability and having an antimicrobial activity that includes phytopathogens (Table 1) such as *Fusarium* spp., *B. cinerea*, *Penicillium expansum*, *A. niger*, and *Rhizopus stolonifera* (Sirelkhatim et al. 2015).

In a study by Dimkpa and co-workers (2013) where the antifungal effect of ZnONPs was compared with the ZnO micro-particles on *F. graminearum*, it was shown that ZnONPs exhibit a greater inhibition in fungal growth as opposed to zinc oxide micro-particles, which was due to the fact that NPs present a greater surface area compared with microparticles, which causes a greater interaction with the fungus. There is no consensus on the exact mechanism that generates toxicity induced by the ZnONPs; however, the most accepted mechanism is that their toxicity is usually mediated by ROS that would be generated on the surface of the particle.

AgNPs

Silver NPs are of great interest due to their physicochemical and high bioactivity and broad antimicrobial spectrum, including plant pathogens, which can improve the efficiency of agrochemicals as well as reduce the use of pesticides and biocides (Ahmed and Dutta 2019, Tomah et al. 2020).

Studies have shown that the use of AgNPs as antifungals reduces, even by 90%, the growth of some plant pathogens (Table 1), such as *F. verticillioides*,

F. moniliforme, *Penicillium brevicompactum*, *Helminthosporium oryzae*, and *Pyricularia grisea* (Elamawi et al. 2018). In *Rhizopus* sp. and *Aspergillus* sp., the AgNPs inhibited the hyphal growth, in addition to the detrimental effects on conidia germination and other deformations in the cell membrane structure, and the inhibition of the normal development of both strains was observed (Medda et al. 2015). Another study reported that AgNPs almost completely inhibited the growth of *Pythium aphanidermatum* and *S. sclerotiurum* (Mahdizadeh et al. 2015).

The effect of these NPs is possibly due to the high affinity of AgNPs to interact and bind to external membrane proteins, affecting their integrity, forming ROS, and causing cell death (Ouda 2014, Mahdizadeh et al. 2015, Medda et al. 2015, Elamawi et al. 2018).

SeNPs

Selenium is a micronutrient that has beneficial effects at low concentrations on plant and animals metabolism, and is incorporated into the structure of enzymes such as glutathione peroxidase, iodothyronine disiodase, and thioredoxin reductase, involved in the antioxidant response, detoxification processes, and cell growth, respectively (Forootanfar et al. 2014, Shakibaie et al. 2015). SeNPs have biomedical applications, such as treatments for cancer, diabetes, or inflammatory disorders, and have antioxidant, antimicrobial, and antiviral properties (Chen et al. 1992, Tseng 2004, Messarah et al. 2012, Khurana et al. 2019).

Its antifungal effect was evaluated in *A. niger*, where at concentrations between 250 and 1,000 μg/ml, diminished fungal growth; conversely, at subinhibitory concentrations (10 μg/ml) *A. niger* growth was stimulated, possibly since being a micronutrient, small concentrations of NPs can be used as cofactors by the enzymes mentioned above (Kazempour et al. 2013). The generation of ROS, penetration of the NPs into the cell and disruption of cell survival pathways are suggested as probable mechanisms of antimicrobial activity of SeNPs (Abdel-Moneim et al. 2022).

Carbon nanocompounds

Carbon nanotubes (CNTs) are laminar structures of hybridized carbon atoms (sp^2), arranged hexagonally, and formed cylindrical tubes, with a diameter approximately of a few nanometers. They can be structured as single-walled CNTs (SWCNTs) or multiwalled CNTs (MWCNTs) (Ijima 1991, Anzar et al. 2020) and are of particular interest to researchers due to their thermal, mechanical, and electrical properties as well as a wide range of technological applications (Lu et al. 2009). Moreover, the antifungal effect of CNTs has been reported, including phytopathogens such as *A. alternata*, and *Fusarium* species (Sarlak et al. 2014, Wang et al. 2014).

González-García and co-workers (2021) reported that MWCNTs applied at 100 mg/L decreased the incidence and severity of *A. solani*, and increased the fruit yield of tomato crop and dry shoot biomass. Another study reported that MWCNTs with different surface groups (OH–, COOH– and NH_2–) showed an inhibition of the spore germination of *F. graminearum* (Wang et al. 2017).

Regarding the mechanism of toxicity of CNTs, it has been suggested that it is through physical damage to the cell membrane and cytoplasm separation or chemical effects such as oxidative stress and reactive oxygen species (ROS) (Wang et al. 2017).

Conclusion

In conclusion, resistance to pesticides by phytopathogenic fungi has been increasing at an alarming rate, so the search for new strategies for their control is of worldwide importance. Thus, the use of nanostructured compounds has emerged as an alternative to counteract diverse microorganisms, including phytopathogenic fungi.

However, although several nanostructured compounds show promising results in the management of these fungi, it is important to elucidate which are the mechanisms of action for these compounds; although some have already been described (Figure 1), some remain as hypotheses. This knowledge will allow us to suggest the effects that these compounds may have on the soil microbiome and on crops of agricultural interest. Therefore, this is presented as a solid line of research, the results of which could be the basis for deciding which nanostructured compounds are the most suitable to be used, in what doses and towards which target organisms.

The use of nanomaterials for the control of phytopathogenic fungi is an alternative that can replace the agrochemicals that are currently used, and that generate collateral damage to the environment and human health; however, molecular studies on the microbiota associated with various crops must be carried out to evaluate the changes in their composition, to avoid damaging the microorganisms that provide some benefit to the plant, regardless of whether they are biocontrollers or provide nutrients.

References

Abboud, M.A.A. 2021. A novel biological approach to copper nanoparticles synthesis: Characterization and its application against phytopathogenic fungi. 30 November 2021, PREPRINT (Version 2) available at Research Square. https://doi.org/10.21203/rs.3.rs-125001/v2.

Abdel-Azeem, A., Nada, A.A., O'donovan, A., Thakur, V.K. and Elkelish, A. 2020. Mycogenic silver nanoparticles from endophytic *Trichoderma atroviride* with antimicrobial activity. J. Renew. Mater. 8(2): 171. Doi: 10.32604/jrm.2020.08960.

Abdel-Moneim, A.M.E., El-Saadony, M.T., Shehata, A.M., Saad, A.M., Aldhumri, S.A., Ouda, S.M. and Mesalam, N.M. 2022. Antioxidant and antimicrobial activities of *Spirulina platensis* extracts and biogenic selenium nanoparticles against selected pathogenic bacteria and fungi. Saudi J. Biol. Sci. 29(2): 1197–1209. https://doi.org/10.1016/j.sjbs.2021.09.046.

Ahmed, A.A. and Dutta, P. 2019. *Trichoderma asperellum* mediated synthesis of silver nanoparticles: Characterization and its physiological effects on tea [*Camellia sinensis* (L.) *Kuntze var. assamica (J. Masters) Kitam.*]. Int. J. Curr. Microbiol. App. Sci. 8: 1215–1229. Doi: 10.20546/ijcmas.2019.804.140.

Ahmad, H., Venugopal, K., Bhat, A.H., Kavitha, K., Ramanan, A., Rajagopal, K., Srinivasan, R. and Manikandan, E. 2020a. Enhanced biosynthesis synthesis of copper oxide nanoparticles (CuO-NPs) for their antifungal activity toxicity against major phyto-pathogens of apple orchards. Pharm. Res. 37(12): 1–12. https://doi.org/10.1007/s11095-020-02966-x.

Ahmad, H., Venugopal, K., Rajagopal, K., De Britto, S., Nandini, B., Pushpalatha, H.G., Konappa, N., Udayashankar, A.C., Geetha, N. and Jogaiah, S. 2020b. Green synthesis and characterization of zinc oxide nanoparticles using *Eucalyptus globules* and their fungicidal ability against pathogenic fungi of apple orchards. Biomolecules 10(3): 425. https://doi.org/10.3390/biom10030425.

Ahmad, T., Wani, I.A., Manzoor, N., Ahmed, J. and Asiri, A.M. 2013. Biosynthesis, structural characterization and antimicrobial activity of gold and silver nanoparticles. Colloids Surf. B Colloid Surface B 107: 227–234. https://doi.org/10.1016/j.colsurfb.2013.02.004.

Akpinar, I., Unal, M. and Sar, T. 2021. Potential antifungal effects of silver nanoparticles (AgNPs) of different sizes against phytopathogenic *Fusarium oxysporum* f. sp. radicis-lycopersici (FORL) strains. SN Appl. Sci. 3(4): 1–9. https://doi.org/10.1007/s42452-021-04524-5.

Al-Dhabaan, F.A., Shoala, T., Ali, A.A., Alaa, M., Abd-Elsalam, K. and Abd-Elsalam, K. 2017. Chemically-produced copper, zinc nanoparticles and chitosan-bimetallic nanocomposites and their antifungal activity against three phytopathogenic fungi. Int. J. Agric. Technol. 13(5): 753–769.

Anderson, A.J., Mclean, J.E., Jacobson, A.R. and Britt, D.W. 2018. CuO and ZnO nanoparticles modify interkingdom cell signaling processes relevant to crop production. J. Agric. Food Chem. 66(26): 6513–6524. https://doi.org/10.1021/acs.jafc.7b01302.

Anzar, N., Hasan, R., Tyagi, M., Yadav, N. and Narang, J. 2020. Carbon nanotube—A review on synthesis, properties and plethora of applications in the field of biomedical science. SI 1: 100003. https://doi.org/10.1016/j.sintl.2020.100003.

Arciniegas-Grijalba, P.A., Patiño-Portela, M.C., Mosquera-Sánchez, L.P., Sierra, B.G., Muñoz-Florez, J.E., Erazo-Castillo, L.A. and Rodríguez-Páez, J.E. 2019. ZnO-based nanofungicides: Synthesis, characterization and their effect on the coffee fungi *Mycena citricolor* and *Colletotrichum* sp. Mater. Sci. Eng. C 98: 808–825. https://doi.org/10.1016/j.msec.2019.01.031.

Asghar, U., Malik, M.F. and Javed, A. 2016. Pesticide exposure and human health: A review. J. Ecosys. Ecograph. S 5: 2. Doi: 10.4172/2157-7625.S5-005.

Ashraf, H., Anjum, T., Riaz, S., Ahmad, I.S., Irudayaraj, J., Javed, S., Qaiser, U. and Naseem, S. 2021. Inhibition mechanism of green-synthesized copper oxide nanoparticles from *Cassia fistula* towards *Fusarium oxysporum* by boosting growth and defense response in tomatoes. Environ. Sci. Nano 8(6): 1729–1748. https://doi.org/10.1039/D0EN01281E.

Bramhanwade, K., Shende, S., Bonde, S., Gade, A. and Rai, M. 2016. Fungicidal activity of Cu nanoparticles against *Fusarium* causing crop diseases. Environ. Chem. Lett., 14(2): 229–235. https://doi.org/10.1007/s10311-015-0543-1.

Capote, N., Pastrana, A.M., Aguado, A. and Sánchez-Torres, P. 2012. Molecular tools for detection of plant pathogenic fungi and fungicide resistance. Plant Pathol. 151–202.

Castro, M.E., Cottet, L. and Castillo, A. 2014. Biosynthesis of gold nanoparticles by extracellular molecules produced by the phytopathogenic fungus *Botrytis cinerea*. Mater. Lett. 115: 42–44. https://doi.org/10.1016/j.matlet.2013.10.020.

Chatterjee, S., Kuang, Y., Splivallo, R., Chatterjee, P. and Karlovsky, P. 2016. Interactions among filamentous fungi *Aspergillus niger*, *Fusarium verticillioides* and *Clonostachys rosea*: fungal biomass, diversity of secreted metabolites and fumonisin production. BMC Microbiol. 16(1): 1–13. https://doi.org/10.1186/s12866-016-0698-3.

Chen, C.J., Chen, C.W., Wu, M.M. and Kuo, T.L. 1992. Cancer potential in liver, lung, bladder and kidney due to ingested inorganic arsenic in drinking water. Br. J. Cancer 66: 888–892. Doi: 10.1038/bjc.1992.380.

Consolo, V.F., Torres-Nicolini, A. and Alvarez, V.A. 2020. Mycosinthetized Ag, CuO and ZnO nanoparticles from a promising *Trichoderma harzianum* strain and their antifungal potential against important phytopathogens. Sci. Rep. 10(1): 1–9. https://doi.org/10.1038/s41598-020-77294-6.

Das, M., Shim, K.H., An, S.S.A. and Yi, D.K. 2011. Review on gold nanoparticles and their applications. Toxicol. Environ. Health Sci. 3: 193–205. Doi: 10.1007/s13530-011-0109-y.

Dean, R., Van Kan, J.A., Pretorius, Z.A., Hammond-Kosack, K.E., Di Pietro, A., Spanu, P.D., Rudd, J.J., Dickman, M., Kahmann, R., Ellis, J. and Foster, G.D. 2012. The top 10 fungal pathogens in molecular plant pathology. Mol. Plant Pathol. 13(4): 414–430. https://doi.org/10.1111/j.1364-3703.2011.00783.x.

Dimkpa, C.O., McLean, J.E., Britt, D.W. and Anderson, A.J. 2013. Antifungal activity of ZnO nanoparticles and their interactive effect with a biocontrol bacterium on growth antagonism of the plant pathogen *Fusarium graminearum*. Biometals 26: 913–924. Doi: 10.1007/s10534-013-9667-6.

Dizaj, S.M., Lotfipour, F., Barzegar-Jalali, M., Zarrintan, M.H. and Adibkia, K. 2014. Antimicrobial activity of the metals and metal oxide nanoparticles. Mater. Sci. Eng. 44: 278–284. https://doi.org/10.1016/j.msec.2014.08.031.

do Nascimento, J.M., Cruz, N.D., de Oliveira, G.R., Sa, W.S., de Oliveira, J.D., Ribeiro, P.R.S. et al. 2021. Evaluation of the kinetics of gold biosorption processes and consequent biogenic synthesis of AuNPs mediated by the fungus *Trichoderma harzianum*. Environm. Technol. Innov. 21: 101238. doi: 10.1016/j.eti.2020.101238.

Elamawi, R.M., Al-Harbi, R.E. and Hendi, A.A. 2018. Biosynthesis and characterization of silver nanoparticles using *Trichoderma longibrachiatum* and their effect on phytopathogenic fungi. Egyptian J. Biol. Pest Control 28: 1–11. Doi: 10.1186/s41938-018-0028-1.

El-Baky, N.A. and Amara, A.A.A.F. 2021. Recent approaches towards control of fungal diseases in plants: An updated review. J. Fungi 7(11): 900. https://doi.org/10.3390/jof7110900.

El-Baky, N.A., Abdel Rahman, R.A., Sharaf, M.M. and Amara, A.A.A.F. 2021. The development of a phytopathogenic fungi control Trial: *Aspergillus flavus* and *Aspergillus niger* infection in Jojoba tissue culture as a model. Sci. World J. 2021. https://doi.org/10.1155/2021/6639850.

El-Gazzar, N. and Ismail, A.M. 2020. The potential use of Titanium, Silver and Selenium nanoparticles in controlling leaf blight of tomato caused by *Alternaria alternata*. Biocatal. Agric. Biotechnol. 27: 101708. https://doi.org/10.1016/j.bcab.2020.101708.

El-Wakil, D.A. 2020. Antifungal activity of silver nanoparticles by *Trichoderma* species: Synthesis, characterization and biological evaluation. Egypt. J. Phytopathol. 48: 71–80. Doi: 10.21608/ejp.2020.49395.1015.

Forootanfar, H., Adeli-Sardou, M., Nikkhoo, M., Mehrabani, M., Amir-Heidari, B., Shahverdi, A.R. et al. 2014. Antioxidant and cytotoxic effect of biologically synthesized selenium nanoparticles in comparison to selenium dioxide. J. Trace Elements Med. Biol. 28: 75–79. Doi: 10.1016/j.jtemb.2013.07.005.

FRAC. 2012. Mode of action of fungicides. Available Online: http://frac.info (accessed on 20 February 2022).

Ganash, M., Ghany, A. and Omar, A.M. 2018. Morphological and biomolecules dynamics of phytopathogenic fungi under stress of silver nanoparticles. BioNanoScience 8(2): 566–573. https://doi.org/10.1007/s12668-018-0510-y.

Gogos, A., Knauer, K. and Bucheli, T.D. 2012. Nanomaterials in plant protection and fertilization: Current state, foreseen applications, and research priorities. J. Agric. Food Chem. 60(39): 9781–9792. https://doi.org/10.1021/jf302154y.

González-García, Y., Cadenas-Pliego, G., Alpuche-Solís, Á.G., Cabrera, R.I. and Juárez-Maldonado, A. 2021. Carbon nanotubes decrease the negative impact of *Alternaria solani* in tomato crop. Nanomaterials 11(5): 1080. https://doi.org/10.3390/nano11051080.

González-García, Y., Cadenas-Pliego, G., Alpuche-Solís, Á.G., Cabrera, R.I. and Juárez-Maldonado, A. 2022. Effect of carbon-based nanomaterials on *Fusarium* wilt in tomato. Sci. Hortic 291: 110586. https://doi.org/10.1016/j.scienta.2021.110586.

Guerrero, J.J.G., Songkumarn, P., Dalisay, T.U., Pangga, I.B. and Organo, N.D. 2020. Toxicity of CuO and ZnO nanoparticles and their bulk counterparts on selected soil-borne fungi. Agric. Nat. Resour. 54(3): 325–332.

Guilger, M., Pasquoto-Stigliani, T., Bilesky-Jose, N., Grillo, R., Abhilash, P.C., Fraceto, L.F. and De Lima, R. 2017. Biogenic silver nanoparticles based on *Trichoderma harzianum*: Synthesis, characterization, toxicity evaluation and biological activity. Sci. Rep. 7: 1–13. Doi: 10.1038/srep44421.

Guilger-Casagrande, M. and Lima, R.D. 2019. Synthesis of silver nanoparticles mediated by fungi: A review. Front. Bioeng. Biotechnol. 7: 287. Doi: 10.3389/fbioe.2019.00287.

Hashem, A.H., Abdelaziz, A.M., Askar, A.A., Fouda, H.M., Khalil, A., Abd-Elsalam, K.A. and Khaleil, M.M. 2021. *Bacillus megaterium*-mediated synthesis of selenium nanoparticles and their antifungal activity against *Rhizoctonia solani* in Faba Bean Plants. J. Fungi 7(3): 195. https://doi.org/10.3390/jof7030195.

Hassan, S.E.D., Fouda, A., Radwan, A.A., Salem, S.S., Barghoth, M.G., Awad, M.A., Abdo, A.M. and El-Gamal, M.S. 2019. Endophytic actinomycetes *Streptomyces* spp mediated biosynthesis of copper oxide nanoparticles as a promising tool for biotechnological applications. J. Biol. Inorg. Chem., 24(3): 377–393. https://doi.org/10.1007/s00775-019-01654-5.

Hirpara, D.G. and Gajera, H.P. 2020. Green synthesis and antifungal mechanism of silver nanoparticles derived from chitin induced exometabolites of *Trichoderma interfusant*. Appl. Organ. Chem. 34: e5407. Doi: 10.1002/aoc.5407.

Hu, D., Yu, S., Yu, D., Liu, N., Tang, Y., Fan, Y. et al. 2019. Biogenic *Trichoderma harzianum*-derived selenium nanoparticles with control functionalities originating from diverse recognition metabolites against phytopathogens and mycotoxins. Food Control 106: 106748. Doi: 10.1016/j.foodcont.2019.106748.

Iijima, S. 1991. Helical microtubules of graphitic carbon. Nature 354(6348): 56–58. https://doi.org/10.1038/354056a0.

Ismail, A.W.A., Sidkey, N.M., Arafa, R.A., Fathy, R.M. and El-Batal, A.I. 2016. Evaluation of *in vitro* antifungal activity of silver and selenium nanoparticles against *Alternaria solani* caused early blight disease on potato. Br. Biotechnol. J. 12(3): 1. Doi: 10.9734/BBJ/2016/24155.

Issam, N., Naceur, D., Nechi, G., Maatalah, S., Zribi, K. and Mhadhbi, H. 2021. Green synthesised ZnO nanoparticles mediated by *Olea europaea* leaf extract and their antifungal activity against *Botrytis cinerea* infecting faba bean plants. Arch. Phytopathol. 54(15-16): 1083–1105. https://doi.org/10.1080/03235408.2021.1889859.

Jayaseelan, C., Ramkumar, R., Rahuman, A.A. and Perumal, P. 2013. Green synthesis of gold nanoparticles using seed aqueous extract of *Abelmoschus esculentus* and its antifungal activity. Ind. Crops Prod. 45: 423–429. Doi: 10.1016/j.indcrop.2012.12.019.

Jian, Y., Chen, X., Ahmed, T., Shang, Q., Zhang, S., Ma, Z. and Yin, Y. 2021. Toxicity and action mechanisms of silver nanoparticles against the mycotoxin-producing fungus *Fusarium graminearum*. J. Adv. Res. https://doi.org/10.1016/j.jare.2021.09.006.

Joshi, S.M., De Britto, S., Jogaiah, S. and Ito, S.I. 2019. Mycogenic selenium nanoparticles as potential new generation broad spectrum antifungal molecules. Biomolecules 9(9): 419. https://doi.org/10.3390/biom9090419.

Kazempour, Z.B., Yazdi, M.H., Rafii, F. and Shahverdi, A.R. 2013. Sub-inhibitory concentration of biogenic selenium nanoparticles lacks post antifungal effect for *Aspergillus niger* and *Candida albicans* and stimulates the growth of *Aspergillus niger*. Iran. J. Microbiol. 5: 81.

Khurana, A., Tekula, S., Saifi, M.A., Venkatesh, P. and Godugu, C. 2019. Therapeutic applications of selenium nanoparticles. Biomed. Pharmacother. 111: 802–812. Doi: 10.1016/j.biopha.2018.12.146.

Kim, S.W., Jung, J.H., Lamsal, K., Kim, Y.S., Min, J.S. and Lee, Y.S. 2012. Antifungal effects of silver nanoparticles (AgNPs) against various plant pathogenic fungi. Mycobiology 40(1): 53–58. https://doi.org/10.5941/MYCO.2012.40.1.053.

Lipsa, F.D., Ursu, E.L., Ursu, C., Ulea, E. and Cazacu, A. 2020. Evaluation of the antifungal activity of gold–chitosan and carbon nanoparticles on *Fusarium oxysporum*. Agronomy 10: 1143. Doi: 10.3390/agronomy10081143.

Lira Saldivar, R.H., Méndez Argüello, B., Santos Villarreal, G.D.L. and Vera Reyes, I. 2018. Potential de la nanotecnología en la agricultura. Acta Universitaria 28(2): 9–24. https://doi.org/10.15174/au.2018.1575.

Lu, F., Gu, L., Meziani, M.J., Wang, X., Luo, P.G., Veca, L.M., Cao, L. and Sun, Y.P. 2009. Advances in bioapplications of carbon nanotubes. Adv. Mater. 21(2): 139–152. https://doi.org/10.1002/adma.200801491.

Mahdizadeh, V., Safaie, N. and Khelghatibana, F. 2015. Evaluation of antifungal activity of silver nanoparticles against some phytopathogenic fungi and *Trichoderma harzianum*. J. Crop Prot. 4: 291–300.

Maksimova, Y.G. 2019. Microorganisms and carbon nanotubes: Interaction and applications. Appl. Biochem. Microbiol. 55(1): 1–12. https://doi.org/10.1134/S0003683819010101.

Malandrakis, A.A., Kavroulakis, N. and Chrysikopoulos, C.V. 2019. Use of copper, silver and zinc nanoparticles against foliar and soil-borne plant pathogens. Sci. Total Environ. 670: 292–299. https://doi.org/10.1016/j.scitotenv.2019.03.210.

Manikandaselvi, S., Sathya, V., Vadivel, V., Sampath, N. and Brindha, P. 2020. Evaluation of bio control potential of AgNPs synthesized from *Trichoderma viride*. Adv. Nat. Sci.: Nanosci. Nanotechnol., 11(3): 035004. https://doi.org/10.1088/2043-6254/ab9d16.

Maqsood, S., Qadir, S., Hussain, A., Asghar, A., Saleem, R., Zaheer, S. et al. 2020. Antifungal properties of copper nanoparticles against *Aspergillus niger*. Scholars Int. J. Biochem. 3: 87–91. Doi: 10.36348/sijb.2020.v03i04.002.

Medda, S., Hajra, A., Dey, U., Bose, P. and Mondal, N.K. 2015. Biosynthesis of silver nanoparticles from Aloe vera leaf extract and antifungal activity against *Rhizopus* sp. and *Aspergillus* sp. Appl. Nanosci. 5: 875–880. Doi: 10.1007/s13204-014-0387-1.

Messarah, M., Klibet, F., Boumendjel, A., Abdennour, C., Bouzerna, N., Boulakoud, M.S. et al. 2012. Hepatoprotective role and antioxidant capacity of selenium on arsenic-induced liver injury in rats. Exp. Toxicol. Pathol. 64: 167–174. Doi: 10.1016/j.etp.2010.08.002.

Mohamed, Y.M.A. and Elshahawy, I.E. 2022. Antifungal activity of photo-biosynthesized silver nanoparticles (AgNPs) from organic constituents in orange peel extract against phytopathogenic *Macrophomina phaseolina*. Eur. J. Plant Pathol. 1–14. https://doi.org/10.1007/s10658-021-02434-1.

Mosquera-Sánchez, L.P., Arciniegas-Grijalba, P.A., Patiño-Portela, M.C., Guerra–Sierra, B.E., Muñoz-Florez, J.E. and Rodríguez-Páez, J.E. 2020. Antifungal effect of zinc oxide nanoparticles (ZnO-NPs) on *Colletotrichum* sp., causal agent of anthracnose in coffee crops. Biocatal. Agric. Biotechnol., 25: 101579. https://doi.org/10.1016/j.bcab.2020.101579.

Nandini, B., Hariprasad, P., Prakash, H.S., Shetty, H.S. and Geetha, N. 2017. Trichogenic-selenium nanoparticles enhance disease suppressive ability of *Trichoderma* against downy mildew disease caused by *Sclerospora graminicola* in pearl millet. Sci. Rep. 7: 1–11. Doi: 10.1038/s41598-017-02737-6.

NNI. Nanotechnology. Available online: www.nano.gov (accessed on 24 February 2022).

Ouda, S.M. 2014. Antifungal activity of silver and copper nanoparticles on two plant pathogens, *Alternaria alternata* and *Botrytis cinerea*. Res. J. Microbiol. 9: 34. Doi: 10.3923/jm.2014.34.42.

Pariona, N., Cereceda, S.B., Mondaca, F., Carrión, G. and Mtz-Enriquez, A.I. 2021. Antifungal activity and degradation of methylene blue by ZnO, Cu, and Cu2O/Cu nanoparticles, a comparative study. Mater. Lett. 301: 130182. https://doi.org/10.1016/j.matlet.2021.130182.

Pariona, N., Mtz-Enriquez, A.I., Sánchez-Rangel, D., Carrión, G., Paraguay-Delgado, F. and Rosas-Saito, G. 2019. Green-synthesized copper nanoparticles as a potential antifungal against plant pathogens. RSC Adv. 9: 18835–18843. Doi: 10.1039/C9RA03110C

Parveen, A. and Siddiqui, Z.A. 2021. Zinc oxide nanoparticles affect growth, photosynthetic pigments, proline content and bacterial and fungal diseases of tomato. Arch. Phytopathol. 54(17-18): 1519–1538. https://doi.org/10.1080/03235408.2021.1917952.

Pham, N.D., Duong, M.M., Le, M.V. and Hoang, H.A. 2019. Preparation and characterization of antifungal colloidal copper nanoparticles and their antifungal activity against *Fusarium oxysporum* and *Phytophthora capsici*. Comptes Rendus Chimie 22: 786–793. Doi: 10.1016/j.crci.2019.10.007.

Rajwade, J.M., Chikte, R.G. and Paknikar, K.M. 2020. Nanomaterials: New weapons in a crusade against phytopathogens. Appl. Microbiol. Biotechnol. 104(4): 1437–1461. https://doi.org/10.1007/s00253-019-10334-y.

Rajwade, J.M., Chikte, R.C., Singh, N. and Paknikar, K.M. 2022. Copper-based nanostructures: Antimicrobial properties against agri-food pathogens. In Copper Nanostructures: Next-Generation of Agrochemicals for Sustainable Agroecosystems, pp. 477–503. Elsevier. https://doi.org/10.1016/B978-0-12-823833-2.00002-7.

Ramírez-Valdespino, C.A. and Orrantia-Borunda, E. 2021. *Trichoderma* and nanotechnology in sustainable agriculture. A review. Front. Fungal Biol. 61. Doi: 10.3389/ffunb.2021.764675.

Sarkar, J., Ray, S., Chattopadhyay, D., Laskar, A. and Acharya, K. 2012. Mycogenesis of gold nanoparticles using a phytopathogen *Alternaria alternata*. Bioprocess Biosyst. Eng. 35(4): 637–643. https://doi.org/10.1007/s00449-011-0646-4.

Sarlak, N., Taherifar, A. and Salehi, F. 2014. Synthesis of nanopesticides by encapsulating pesticide nanoparticles using functionalized carbon nanotubes and application of new nanocomposite for plant disease treatment. J. Agric. Food Chem. 62(21): 4833–4838. https://doi.org/10.1021/jf404720d.

Sawake, M.M., Moharil, M.P., Ingle, Y.V., Jadhav, P.V., Ingle, A.P., Khelurkar, V.C., Paithankar, D.H., Bathe, G.A. and Gade, A.K. 2022. Management of *Phytophthora parasitica* causing gummosis in citrus using biogenic copper oxide nanoparticles. J. Appl. Microbiol. https://doi.org/10.1111/jam.15472.

Shakibaie, M., Mohazab, N.S. and Mousavi, S.A.A. 2015. Antifungal activity of selenium nanoparticles synthesized by Bacillus species Msh-1 against *Aspergillus fumigatus* and *Candida albicans*. Jundishapur. J. Microbiol. 8: 1–4. Doi: 10.5812/jjm.26381.

Shoala, T. 2020. Carbon nanostructures: Detection, controlling plant diseases and mycotoxins. In Carbon Nanomaterials for Agri-Food and Environmental Applications, pp. 261–277. Elsevier. https://doi.org/10.1016/B978-0-12-819786-8.00013-X.

Shobha, B., Lakshmeesha, T.R., Ansari, M.A., Almatroudi, A., Alzohairy, M.A., Basavaraju, S. et al. 2020. Mycosynthesis of ZnO nanoparticles using *Trichoderma* spp. isolated from rhizosphere soils and its synergistic antibacterial effect against *Xanthomonas oryzae pv. oryzae*. J. Fungi 6: 181. Doi: 10.3390/jof6030181.

Singh, R.P., Handa, R. and Manchanda, G. 2021. Nanoparticles in sustainable agriculture: An emerging opportunity. Journal of Controlled Release 329: 1234–1248.

Sirelkhatim, A., Mahmud, S., Seeni, A., Kaus, N.H.M., Ann, L.C., Bakhori, S.K.M. et al. 2015. Review on zinc oxide nanoparticles: Antibacterial activity and toxicity mechanism. Nano-micro Lett. 7: 219–242. Doi: 10.1007/s40820-015-0040-x.

Soltani Nejad, M., Samandari Najafabadi, N., Aghighi, S., Pakina, E. and Zargar, M. 2022. Evaluation of *Phoma* sp. biomass as an endophytic fungus for synthesis of extracellular gold nanoparticles with antibacterial and antifungal properties. Molecules 27(4): 1181. https://doi.org/10.3390/molecules27041181.

Tomah, A.A., Alamer, I.S.A., Li, B. and Zhang, J.Z. 2020. Mycosynthesis of silver nanoparticles using screened *Trichoderma* isolates and their antifungal activity against *Sclerotinia sclerotiorum*. Nanomaterials 10: 1955. Doi: 10.3390/nano10101955.

Tseng, C.H. 2004. The potential biological mechanisms of arsenicinduced diabetes mellitus. Toxicol. Appl. Pharmacol. 197: 67–83. Doi: 10.1016/j.taap.2004.02.009.

Tyagi, P.K., Mishra, R., Khan, F., Gupta, D. and Gola, D. 2020. Antifungal effects of silver nanoparticles against various plant pathogenic fungi and its safety evaluation on *Drosophila melanogaster*. Biointerface Res. Appl. Chem. 10: 6587–6596. https://doi.org/10.33263/BRIAC106.65876596.

Vera-Reyes, I., Esparza-Arredondo, I.J.E., Lira-Saldivar, R.H., Granados-Echegoyen, C.A., Alvarez-Roman, R., Vásquez-López, A., De los Santos-Villarreal, G. and Díaz-Barriga Castro, E. 2019. *In vitro* antimicrobial effect of metallic nanoparticles on phytopathogenic strains of crop plants. J. Phytopathol. 167(7-8): 461–469. https://doi.org/10.1111/jph.12818.

Viet, P.V., Nguyen, H.T., Cao, T.M. and Hieu, L.V. 2016. *Fusarium* antifungal activities of copper nanoparticles synthesized by a chemical reduction method. J. Nanomater., 1–7. Doi: 10.1155/2016/1957612.

Wang, X., Liu, X., Chen, J., Han, H. and Yuan, Z. 2014. Evaluation and mechanism of antifungal effects of carbon nanomaterials in controlling plant fungal pathogen. Carbon 68: 798–806. https://doi.org/10.1016/j.carbon.2013.11.072.

Wang, X., Zhou, Z. and Chen, F. 2017. Surface modification of carbon nanotubes with an enhanced antifungal activity for the control of plant fungal pathogen. Materials 10(12): 1375. https://doi.org/10.3390/ma10121375.

Younus, I., Khan, S.J., Maqbool, S. and Begum, Z. 2021. Anti-fungal therapy via incorporation of nanostructures: A systematic review for new dimensions. Phys. Scr. 97: 012001.

Zakharova, O., Kolesnikov, E., Shatrova, N. and Gusev, A. 2019. The effects of CuO nanoparticles on wheat seeds and seedlings and *Alternaria solani* fungi: *In vitro* study. IOP Conf. Ser. Earth Environ. Sci. 226(1): 012036. IOP Publishing.

Chapter 6

Nanoparticles Used for Management of Oomycetes Diseases

Edith Garay-Serrano,[1,2] *Alfredo Reyes-Tena,*[3]
Gerardo Rodríguez-Alvarado,[4] *Alejandro Soto-Plancarte,*[4]
Nuria Gómez-Dorantes[4] and *Sylvia P. Fernández-Pavía*[4,*]

Introduction

Nanotechnology provides new agrochemical agents and promises to reduce the use of pesticides which, when used indiscriminately, affect biodiversity by reducing populations of beneficial microorganisms as well as favoring the development of resistance in plant pathogens. The use of nanoparticles (NPs) as fungicides and bactericides for suppression and management of plant pathogens has gained great interest due to their small size (1–100 nm) and high reactivity (Elmer and White 2018). NPs' molecules can serve as a vehicle for the application of active substances and increase their efficacy to inhibit the growth of phytopathogens (Sekhon 2014, Kutawa et al. 2021). A considerable number of molecules have been used in conventional agrochemical formulations for disease management; however, its encapsulation in biodegradable polymer NPs increases the ability to adhere to pathogens, which could reduce its indiscriminate use (Fukamachi et al. 2019). There are several reports of the application of NPs in growth suppression and reduction of diseases caused by bacteria and fungi (Alghuthaymi et al. 2015, Cruz-Luna et al. 2021). Elements such as Ag, Cu and Zn metal oxides have been evaluated against species of the fungi *Aspergillus*,

[1] Instituto de Ecología, A. C. Pátzcuaro, Michoacán, México. Red de Diversidad Biológica del Occidente Mexicano. Avenida Lázaro Cárdenas 253, Pátzcuaro, Michoacán, México, 61600.
[2] CONACYT. Avenida Insurgentes Sur 1582, Ciudad de México, México, 03940.
[3] Instituto de Investigaciones en Ecosistemas y Sustentabilidad, UNAM, Antigua Carretera a Pátzcuaro 8701, Morelia, Michoacán, Mexico, 58190.
[4] Instituto de Investigaciones Agropecuarias y Forestales, Universidad Michoacana de San Nicolás de Hidalgo, Km. 9.5 Carr. Morelia-Zinapécuaro, Tarímbaro, Michoacán, México, 58880.
* Corresponding author: patricia.pavia@umich.mx

Fusarium, *Penicillium* and *Verticillium*, *in vitro* and under field conditions with promising results (Moussa et al. 2013, Elmer et al. 2021). In addition, the efficacy of NPs produced from cell filtrates of beneficial microorganisms such as *Trichoderma longibrachiatum* has been evaluated, registering inhibitions of up to 90% of colony-forming units (CFU) of the pathogenic fungi *Fusarium verticillioides*, *Penicillium brevicompactum*, *Helminthosporium oryzae* and *Pyricularia grisea* (Elamawi et al. 2018). This could favor the development of environmentally friendly tools for the control of plant diseases by reducing the application of conventional agrochemicals.

There are other nanomaterials with high potential for its use in the control of plant diseases. Thus, the metal oxides CeO, MgO, TiO_2 have shown to reduce the disease severity caused by *F. oxysporum* and the phytopathogenic bacteria *Ralstonia solanacearum* on tomatoes. Additional effects on experimental plants included an increase in the levels of chlorophyll, lycopene, catalases, peroxidases, polyphenol-oxidases, fruit production and plant biomass (Imada et al. 2016, Adisa et al. 2018). Elmer and White (2018) showed that metalloids with boron can increase plant growth of *Glycine max* by up to 34% in the presence of *F. virguliforme* (syn. *F. solani* f. sp. *glycines*). Moreover, carbon nanotubes, fullerenes, and graphene oxides have been reported to exhibit antimicrobial activity against phytopathogenic bacteria and fungi, while promoting plant growth (Wang et al. 2013, 2014, Chen et al. 2017, Hao et al. 2017). Nanoliposomes molecules composed of a vesicle and a phospholipid bilayer, when combined with flavonoids with antiviral activity such as quercetin, have been shown to inhibit the disease caused by tobacco mosaic virus in the field (Wang et al. 2022). The combination of liposomes with amphotericin B reduced the severity of the disease caused by *F. oxysporum* f. sp. *ciceris* in pea plants at a concentration of 10 $\mu g/mL^{-1}$ (Pérez-de-Luque et al. 2012). Other nanomaterials such as dendrimers, which are synthetic polymers composed of a central core and branches with different terminal groups, have been successfully used as a vehicle for the administration of drugs and genes in human medicine and their application in phytopathology has been proposed (Khairnar et al. 2010, Nair and Kumar 2013). The aim of the present chapter is to focus on the advances of the use of nanoparticles with phytopathogenic oomycetes diseases, caused by *Phytophthora*, *Pythium* and some downy mildew genera. In this chapter, we emphasize the type and effects of NPs used in the detection and management studies of these diseases.

Importance of phytopathogenic Oomycetes

Downy mildews

Oomycetes in the Peronosporaceae family (Peronosporales, Oomycota) are plant pathogens that cause downy mildew diseases in a diversity of plants, including ornamentals, vegetables, and grasses, causing great economic losses worldwide (Thines and Choi 2016). Usually, they affect the leaves causing chlorotic angular spots which become necrotic and can cause defoliation in a few days. Some other symptoms can be confused with nutritional disorders or mosaic effects (Salcedo et al. 2020, Standish et al. 2020). This group of pathogenic mildews is one of the largest of the Oomycota, comprising more than 700 registered species distributed in 19 genera (Thines and Choi 2016). Among the most commonly found mildews in herbarium records from North America, including Mexico, are species within the genera *Peronospora*, *Plasmopara*, *Hyaloperonospora*, *Plasmoverna*, and *Pseudoperonospora* (Davis and Crouch 2022). These organisms are obligate biotrophs, which means that they are completely dependent on their host for development, nutrition, and reproduction (Thines and Choi 2016).

Pythium

This genus comprises nearly 150 species (http://www.indexfungorum.org). *Pythium* species are found as saprophytic, fungi-parasitic or plant pathogenic. Some of them are of significant economic importance as plant pathogens cause substantial losses in a wide variety of crops by reducing their yield and quality (Figure 1). Symptomatology includes rotting of plant tissues (fruit, stem, root) and significant damage to seedlings and seeds (pre and post-emergence). *Pythium* is considered one of the most important soil-borne pathogens (Uzuhashi et al. 2010, Schroeder et al. 2013) on cereal crops and turf grasses. In addition, some species are able to infect different organisms, including mammals, such as *P. insidiosum* which causes the disease known as Pythiosis, on canines, bovines, equines, felines, and humans (De Souza et al. 2018). *Pythium* includes many saprophytic species that have been discovered in ecosystems different from the forest or cultivated areas (Uzuhashi et al. 2010).

Figure 1: Examples of diseases caused by *Pythium aphanidermatum*, collar rot on *Carica papaya* (left) and crown and stem rot on *Opuntia ficus-indica* (right) [Rodríguez-Alvarado et al. 2001a, b, photos taken by G. Rodríguez-Alvarado].

Phytophthora

This genus is the most important among the phytopathogenic oomycetes due to its destructive capacity, hence the meaning of the genus in Greek (Phyton, plant; phthora, destroyer) (Ho 2018), and because of the large number of pathogenic species, as well as the wide range of hosts (Erwin and Ribeiro 1996, https://idtools.org/id/phytophthora/

about.php). In this genus (as of 2018), 182 species had been described (https://idtools.org/id/phytophthora/about.php). Few species are saprophytic (Hansen et al. 2012, Chepsergon et al. 2020), and others have been isolated from marine plants (Govers et al. 2016), however, most of the species are pathogenic organisms that induce diseases in both herbaceous plants and trees, including monocotyledonous and dicotyledonous. *Phytophthora* is present in plant nurseries worldwide (Bienapfl and Balci 2014). Due to the increase in commercial trade of plants produced in nurseries, there is a potential threat to the dispersion of *Phytophthora* species to places where it was not originally present (Figure 2), including natural ecosystems (Simamora et al. 2018). This oomycete affects seeds, roots, stems, leaves, flowers and fruits. The symptoms, depending on the species, are associated with the aerial tissues or the root, and can manifest in plants as blight, rot, spots or cankers (Hansen et al. 2012, Ho 2018). Some of the diseases they cause include potato and tomato late blight, chili pepper wilt, soybean root and stem rot, avocado dieback, and strawberry rot, caused by *P. infestans*, *P. capsici*, *P. sojae*, *P. cinnamomi*, and *P. cactorum*, respectively (Tyler 2007, Lamour et al. 2012, Fry 2016, Armitage et al. 2018).

Figure 2: Symptoms observed in Mexican nurseries on ornamental plants, caused by *Phytophthora*, on the following a) *Catharanthus roseus*, b) *Cyclamen persicum*, c) *Rhododendron*, d) *Pentas lanceolata*, e) *Euphorbia pulcherrima*, f) *Capsicum pubescens*. (Photos taken by A. Soto-Plancarte and A. Reyes-Tena).

Nanoparticles used in phytopathogenic Oomycetes studies

Among the organic and inorganic nanoparticles studied for the detection and management of oomycetes with promising results are copper (Pham et al. 2019), curcumin (Khatua et al. 2020), gold (Zhan et al. 2018), silver (Nath et al. 2008, Kim et al. 2012, Ali et al. 2015, Mahdizadeh et al. 2015, Schwenkbier et al. 2015a, b, Luan and Xo 2018, Bibi et al. 2021, Sharma et al. 2022), selenium (Siddaiah et al. 2018, Perfileva et al. 2021), zinc (Wang et al. 2017), and different oxides such as copper oxide (Giannousi et al. 2013, Sawake et al. 2022), tin oxide (Fang et al. 2014), graphene oxide (Wang et al. 2017), magnesium oxide (Chen et al. 2020), titanium oxide (Fang et al. 2014), and zinc oxide (Sharma et al. 2011, Zabrieski et al. 2015, Wang et al. 2017, Nandhini et al. 2019). Also, some composite nanoparticles have been investigated - AgNPs and polyphenols (Ali et al. 2015, Matei et al. 2018), chitosan-copper (Vanti et al. 2020), copper NPs and copper oxychloride (Banik and Pérez-de-Luque 2017), silica-silver (Park et al. 2006), silver in combination with chitosan oligomers and propolis (Silva-Castro et al. 2018). In addition, nanoparticles have been used as transporters of compounds such as cyazofamid encapsulated in NPs poly lactic-co-glycolic acid (Fukamachi et al. 2019), or salicylic acid encapsulated in mesoporous silica nanoparticles (Lu et al. 2019). On the other hand, detection devices have been developed, for example, DNA-based electrochemical nanobiosensor (Franco et al. 2019). Studies carried out with nanoparticles with phytopathogenic oomycetes are listed in Table 1; this research has been conducted *in vitro* and *in planta* (Figure 3).

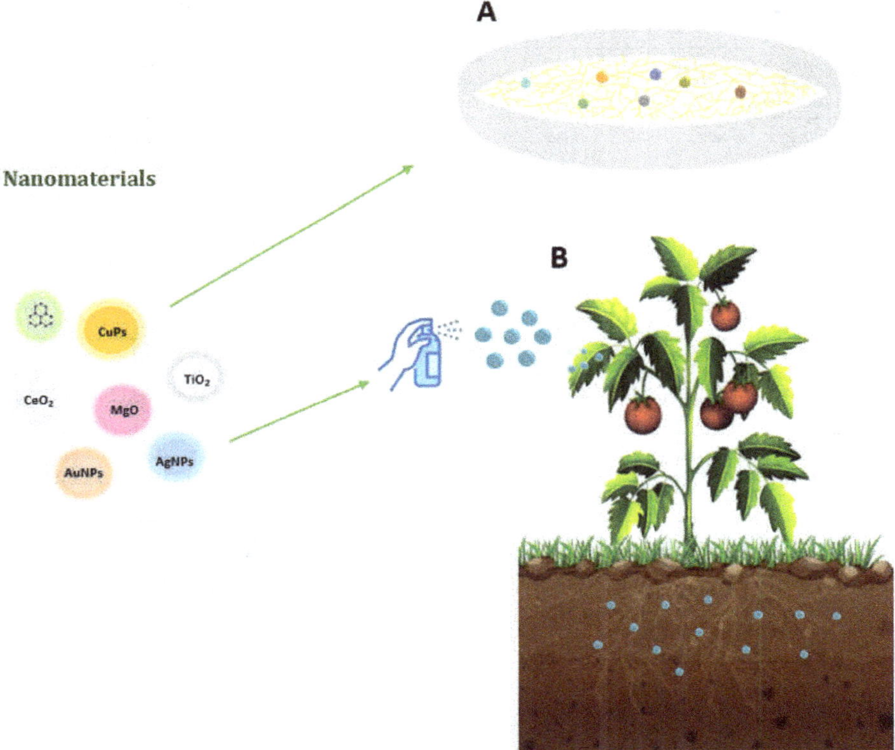

Figure 3: Studies with oomycetes using different nanomaterials have been conducted *in vitro* (A) and *in planta* sprayed to the aerial part and/or applied to the soil to reach the lower part of the plant (B).

Table 1: Nanoparticles for phytopathogenic oomycetes' detection and management of the diseases they cause.

Oomycete	**Nanomaterial used**	**Reference**
Phytophthora alni	Silver Nanoparticles (AgNPs) in combination with chitosan oligomers and propolis	Silva-Castro et al. 2018
Phytophthora cactorum	Titanium oxide (TiO_2), Tin oxide (SnO_2), Selenium Nanoparticles (SeNPs)	Fang et al. 2014, Perfileva et al. 2021
Phytophthora cambivora	AgNPs in combination with chitosan oligomers and propolis	Silva-Castro et al. 2018
Phytophthora capsici	Colloidal Copper Nanoparticles (CuNPs) AgNPs	Pham et al. 2019, Luan and Xo 2018
Phytophthora cinnamomi	AgNPs and polyphenols CuNPs Mesoporous silica nanoparticles with salicylic acid	Ali et al. 2015, Banik and Pérez-de-Luque 2017, Matei et al. 2018, Lu et al. 2019
Phytophthora infestans	Gold Nanoparticles (AuNPs) Cyazofamid encapsulated in NPs poly lactic-co-glycolic acid CuNPs, Copper oxides (CuO, Cu_2O)	Giannousi et al. 2013, Zhan et al. 2018, Fukamachi et al. 2019
Phytophthora kernoviae	AgNPs	Schwenkbier et al. 2015b
Phytophthora nicotianae	Magnesium oxide Nanoparticles	Chen et al. 2020
Phytophthora palmivora	DNA-based electrochemical nanobiosensor	Franco et al. 2019
Phytophthora parasitica	AgNPs CuONPs	Ali et al. 2015, Bibi et al. 2021, Sawake et al. 2022
Phytophthora plurivora	AgNPs in combination with chitosan oligomers and propolis	Silva-Castro et al. 2018
Phytophthora spp.	AgNPs	Schwenkbier et al. 2015a
Peronospora tabacina	Zinc NPs and zinc oxide NPs (ZnO NPs)	Wang et al. 2017
Plasmopara viticola	Nanoparticles of graphene oxides (GO) Fe_3O_4	Wang et al. 2017
Pythium aphanidermatum	AgNPs Chitosan-copper nanoparticles (Ch-CuNPs)	Kim et al. 2012, Mahdizadeh et al. 2015, Vanti et al. 2020, Sharma et al. 2022
Pythium debarynum	ZnONPs	Sharma et al. 2011
Pythium spinosum	AgNPs	Kim et al. 2012
Pythium spp.	ZnONPs	Zabrieski et al. 2015
Pythium ultimum	Silica-Siver	Park et al. 2006
Pythium ultimum var. *ultimum*	Curcumin Nanoparticles (CurNPs) AgNPs	Nath et al. 2008, Khatua et al. 2020,
Sclerospora graminicola	ZnONPs SeNPs	Siddaiah et al. 2018, Nandhini et al. 2019

Nanotechnology for plant pathogens' detection

One of the main challenges in plant pathology is the correct detection and identification of the pathogens that affect a crop. There are various methods for diagnosis and quantification of pathogens that use diverse methods, from the recognition of signs and symptoms in plants, the isolation of pathogens in culture media, as well as identification through observation of morphological structures using a microscope, biochemical tests and even the highly assisted molecular and immunological techniques (Ray et al. 2017). Lately, the ability to diagnose, detect and discriminate different phytopathogens has been explored through biosensors, which are devices that work through a physicochemical transducer that generates a signal that comes into contact with a biological element that works as a bioreceptor, and whose function is to be a specific pathogen recognition molecule; this can be an antibody, volatile compounds, enzymes, nucleic acids, or some type of cell or tissue (Ray et al. 2017, Dyussembayev et al. 2021). Biosensors are classified according to the type of transducer and can be optical, volatile (by detecting volatile compounds), electrochemical, mass sensitive, based on paper-based diagnostic technologies, or even based on nanomaterials (Ray et al. 2017). In the latter case, the materials used include metal and metal oxide nanoparticles, quantum dots, magnetic nanoparticles, carbon nanomaterials (for example graphene), as well as polymeric nanomaterials (Ray et al. 2017). One of the properties of nanoparticles for diagnosis is that they improve sensitivity since nanomaterials have good electronic or optical properties and are considered rapid diagnostic biomarkers for the detection of phytopathogens (Ray et al. 2017, Dyussembayev et al. 2021).

Several advantages are offered by the use of biosensors for the detection of phytopathogens, including the ability to be used *in situ*, their low cost, high sensitivity to the detection of the target pathogen, are fast and easy to perform, do not require an exhaustive staff training, and the results can be assessed visually (Cao et al. 2011, Dyussembayev et al. 2021).

Nanoparticles for *Phytophthora* detection

Diseases caused by *Phytophthora* species include some of the most important and devastating to vegetable production worldwide (Erwin and Ribeiro 1996, https://idtools.org/id/phytophthora/). Some of those diseases are placed in the category of quarantined species (https://www.eppo.int/QUARANTINE/listA2.htm), because they are considered very destructive. Therefore, developing a *Phytophthora* detection protocol that is fast, accurate, and inexpensive, is of the utmost importance. In this section, we will focus on describing the advances in the development of nanotechnology for the detection of *Phytophthora*.

Various nanomaterials have been used for biomolecular recognition because they are highly sensitive to transduce various types of signals such as electronic or optical ones (Cao et al. 2011). Among these materials are gold nanoparticles (AuNPs) that efficiently transmit electronic conductivity (Cao et al. 2011); however, other materials have also been explored as being less expensive than AuNPs (Fang et al. 2014).

Electrochemical technology with nanoparticles for the detection of pathogenic oomycetes is non-destructive and detection is in real-time. The process is conducted through electrochemical sensors that are highly sensitive and precise. Fang et al. (2014) developed an electrochemical device that detects the volatiles of p-ethylguaiacol, a compound that forms in tissues of fruits and plants infected by *P. cactorum*. The device includes titanium oxide (TiO_2) and tin oxide (SnO_2) nanoparticles, which are cheaper than noble metal nanoparticles, arranged on carbon electrodes. The TiO_2 and SnO_2 nanoparticles showed high sensitivity to p-ethylguaiacol (174–188 $\mu A\ cm^{-2}\ mM^{-1}$).

Zhan et al. (2018) developed a rapid and sensitive detection biosensor for *P. infestans* that is based on the combination of chromatography and immunology techniques in a lateral flow assay. The protocol consists of extracting *P. infestans* DNA from samples of infected tissues or from the cultured mycelium, and perform PCR with a universal oligonucleotide and a pair of oligonucleotides specific for this oomycete (this technique is known as asymmetric PCR). Subsequently, a drop of the amplicon is applied to the sample pad of the lateral flow biosensor, the DNA hybridizes with the gold nanoparticle probes in the conjugate pad forming a complex that migrates by capillarity through the nitrocellulose membrane, where the test and control line are marked with streptavidin-biotinylated capture probe. Once the complex is captured, one red band is showed for control and a second line for positive *P. infestans* detection in the test sample. The biosensor they developed was able to detect amounts of *P. infestans* genomic DNA as small as 0.1 pg/µl.

Several DNA-based electrochemical biosensors have been implemented that use the hybridization of a probe for the specific detection of the oomycete. Among them is the nanobiosensor developed by Franco et al. (2019), which detects the DNA of *P. palmivora*, the causal agent of black pod rot of cocoa. The biosensor works by hybridizing a single-stranded DNA (ssDNA) analyte with a *P. palmivora*-specific probe labeled with electrically active magnetic nanoparticles (EAM), then the hybrids are detected electrochemically on the surface of the streptavidin-modified electrode by cyclical voltammetry. The nanosensor was able to detect up to 0.30 ng of *P. palmivora* DNA.

Schwenkbier et al. (2015a) also used DNA to detect various species of *Phytophthora*. They used an isothermal helicase-dependent amplification (HDA) PCR whose DNA products hybridized on a miniaturized chip for the detection of selected *Phytophthora* species. Subsequently, with the deposition of silver nanoparticles, the results could be obtained visually and electrically. Schwenkbier et al. (2015b) used a similar technique to detect *P. kernoviae*, a species of quarantine importance that attacks *Rhododendron* and *Fagus* species in England. In this work, the detection of the amplicons on the chip was performed by successive hybridization and enzymatic binding. They utilized enzymatically generated deposits of silver nanoparticles for signal expression that enabled visual readout. The hybridization assay allowed the reliable detection of 10 pg/µL of target DNA.

The development of technologies for the detection of phytopathogenic species with nanomaterials is in vogue and still being explored with the aim of achieving specific sensors for phytopathogenic species, of lower cost, greater sensitivity, reliability and ease of use.

Nanoparticles and management of Oomycetes diseases

The leading strategies to control and manage the diseases caused by oomycetes include the search for resistant and/or tolerant varieties (Barchenger et al. 2018), biological control through antagonistic microorganisms (Sánchez et al. 2019, Martins et al. 2021), and the use of chemical pesticides (Miao et al. 2016, Siegenthaler and Hansen 2021). However, in most cases, an effective control method has not yet been found. It is necessary to continue the search for biological materials and active substances with better performance and friendly to the environment. In this sense, the application of specific molecules against oomycetes using NPs as a vehicle could increase the efficacy of the fungicides used and reduce the dose and frequency of application.

Downy mildews

Oomycetes in the Peronosporaceae family that cause downy mildew diseases are currently detected and identified using molecular techniques, primarily by PCR amplification of mitochondrial and nuclear DNA regions (Crandall et al. 2018). Thus, once the pathogen is identified, appropriate disease management strategies can be applied (Salcedo et al. 2021). Due to the complexity and economic importance of the downy mildew diseases, it is important to establish management plans that include different techniques to control this pathogenic problem. Several types of fungicides with different forms of action are currently used (Lebeda and Cohen 2011, Salcedo et al. 2021); however, the integration of new technologies that can provide better control of these diseases is essential.

There are few reports on the use of NPs in the management and control of oomycetes that cause downy mildew. However, in recent years different nano-composites have been evaluated with promising results. Wang et al. (2017) reported that the application of NPs of graphene oxides (GO) Fe_3O_4 in leaves of *Vitis vinifera* induced a protective effect by reducing downy mildew caused by *Plasmopara viticola* in the field. In this study, the inhibition of sporangia germination was recorded, preventing the development of the disease using doses of 50 to 250 μg/mL. On the other hand, it has been reported that NPs composed of Zn and ZnO inhibit the germination of *Peronospora tabacina* spores *in vitro* and decrease their infective capacity in tobacco leaves using doses of 8 and 10 mg/L. In addition, ZnO NPs can be biosynthesized from plant extracts; in this sense, Nandhini et al. (2019) synthesized ZnO NPs from the aqueous extract of *Eclipta alba* and reported the inhibition of zoospore germination of *Sclerospora graminicola*, the causal agent of downy mildew on *Pennisetum glaucum*, using doses of 50 ppm. Additionally, the use of ZnONPs promoted plant growth and induced a systemic defense response by expressing transcripts of genes related to defense enzymes. Siddaiah et al. (2018) reported similar results in the same pathosystem using chitosan NPs, which increased acetylation levels and promoted changes in gene expression of different plant defense enzymes. It has been reported that the use of selenium nanoparticles (SeNPS) can increase the efficacy of *Trichoderma* to promote plant growth and decrease the level of damage caused by *S. graminicola* in *P. glaucum*.

Pythium spp.

Pythium is considered a complex and diverse group. The delimitation of species within this genus is complicated due to similar morphological structures among the species (Lévesque and De Cock 2004). An additional complication for the farmers is that this oomycete can remain dormant during adverse conditions such as cold, heat or drought, and once favorable conditions are present, including susceptible plants, the onset of the disease starts (Schroeder et al. 2013). A lack of effective control of the diseases caused by *Pythium* species using traditional cultural practices requires the development of new technologies.

Recently, studies using NPs and its effects on *Pythium* species *in vitro* have been reported. Results of tests probing biogenic AgNPs (synthesized by the fungi *Fusarium*) against *P. insidiosum*, a pathogenic oomycete on mammals, showed activity by damaging hyphae, deforming their external surface, and forming multiple areas of retraction in the cell walls, causing loss of hyphal continuity and rupture of the cell wall in some areas (De Souza et al. 2018). Sharma et al. (2022) demonstrated the antimicrobial activity *in vitro* of AgNPs (synthesized by four species of *Lactobacillus*) against *P. aphanidermatum* (which causes damping-off in a variety of plants). Mahdizadeh et al. (2015) also showed the antagonistic effect of AgNPs against *P. aphanidermatum* applied at different concentrations, resulting in the total inhibition of the mycelial growth of the pathogen after 10 days of incubation with the nanoparticles. Kim et al. (2012) reported inhibitory effects of AgNPs against various plant pathogens, including *P. aphanidermatum* and *P. spinosum*, on *in vitro* tests carried out in different culture media. These authors found that the concentration of the NPs and the incubation time are important factors to consider, because the inhibition of the mycelium was greater with increasing incubation time and showed similar patterns in different culture media.

Zinc oxide nanoparticles (ZnONPs) have also been shown to have an inhibitory effect on the growth of *Pythium* strains. Zabrieski et al. (2015) evaluated ZnONPs at a concentration between 50 and 500 mg/L. According to the authors, the antagonistic effect caused by ZnONPs could be caused by competition for sequestration and transport of iron ions between the oomycetes and the competing ZnONPs. Sharma et al. (2011) reported that *P. debarynum*, exposed to different concentrations of ZnONPs, showed a significant decrease in mycelial growth as the concentration of ZnONPs was increased.

Pythium species have also been evaluated on their response to exposition of NPs composed of different nanomaterials to those mentioned above. Park et al. (2006) developed a formulation of nanometric-sized silica-silver particles that were exposed to *P. ultimum* isolates. The results indicated 100% inhibition of mycelial growth at a concentration of 10 ppm when evaluated in PDA medium. Khatua et al. (2020) reported that curcumin nanoparticles (CurNPs), when tested against *P. ultimum* var. *ultimum*, which causes blight disease of barley, wheat, and vegetables such as tomato and cucumber, presented biocidal activity by inhibiting mycelial growth. Accordingly, the inhibitory effects were caused by an increased generation of intracellular reactive oxygen species (ROS), and changes in the potential of the mitochondrial membrane (MMP). Using the same oomycete, Nath et al. (2008) tested AuNPs biosynthesized from *Pongamia pinnata* leaf extracts, reporting toxicity and antifungal activity at a concentration of 0.8 mg/mL.

Nanoparticles have also been combined with other compounds in order to find activity against pathogens such as *P. aphanidermatum*. Vanti et al. (2020) evaluated copper nanoparticles (CuNPs) coupled to chitosan (Ch), detecting between 88.1 ± 3.8% and 98.3 ± 2.0% of mycelial growth inhibition at a concentration of 0.05% and 0.1% nanoparticles, respectively.

Phytophthora spp.

Phytophthora is the most devastating oomycete plant pathogen, both in agroecosystems and in natural ecosystems. Thus, most of the research with NPs for the management of oomycetes has been carried out with this phytopathogen. Fukamachi and collaborators (2019) showed that the encapsulation of cyazofamide, a fungicide of the class of imidazoles, in polylactic-co-glycolic acid (PLGA)NPs improved the adhesion of this compound and the suppression of the growth of *P. infestans* in tomato plants under rainy conditions. Cyazofamide has a systemic action mechanism in plant tissue and is particularly effective in preventing infections caused by this pathogen in potato and tomato crops (Mitani et al. 2001). NPs have also been used as vehicles for the encapsulation and dosage of compounds for the induction of resistance. Lu et al. (2019) synthesized 20-nm mesoporous silica nanoparticles (MSN) loaded with salicylic acid (SA), a compound capable of inducing induced resistance in many plants. The MSNNPs applied to pineapple plants reduced the size of lesions induced by *P. cinnamomi*, which causes by heart and root rot disease of pineapple plants, and also promoted plant root growth.

Nanometals have been synthesized and evaluated for the control of *P. infestans*. Giannousi et al. (2013) reported the synthesis, characterization and evaluation of nanoparticles (11–55 nm) based on copper (CuO, Cu_2O) to reduce field infections in tomato crops. Higher efficacy (54.29–69.61%) was found with a lower amount of active ingredient compared to conventional agrochemicals containing copper. Colloidal nanoparticles of this metal have also been reported for the control of *P. capsici*. Pham et al. (2019) reported that different synthesis methods and different sizes of NPs influenced the level of efficacy in inhibiting mycelial growth. The authors indicated that NPs of 53 nm at a concentration of 10 ppm inhibited 100% of pathogen growth one day after incubation in potato-dextrose-agar (PDA) medium. Also, the efficacy of copper NPs decreased with increasing NP diameter; however, this effect was compensated by adding a higher concentration (20 ppm). Copper NPs were also evaluated against *P. cinnamomi* by Banik and Pérez-de-Luque (2017), who reported around 60% inhibition of mycelial growth *in vitro* at a concentration of 100 mg/L. When NPs were combined with copper oxychloride at 50 mg/L, a synergistic effect was observed and inhibition of 76% was obtained.

Silver nanoparticles (AgNPs) obtained through γ (gamma) irradiation have also been evaluated against *P. capsici* under *in vitro* and *in vivo* conditions. A decrease in pathogen infection of up to 95% in pre-infected plants has been reported (Luan and Xo 2018). In this same study, 91.7% of the plants pre-infected with the pathogen did not develop symptoms after being treated with a concentration of 50 ppm. Moreover, Matei et al. (2018) reported inhibition of up to 90% of the growth of *P. cinnamomi* using AgNPs and polyphenols under *in vitro* conditions. The possible mechanisms of action of AgNPs against oomycetes have been poorly studied; however, *in vitro* exposure of *P. parasitica* to AgNPs directly affects the expression of genes related to oxidative stress, particularly the heat shock protein gene, ABC transporter, and glutathione-S-transferase, which negatively affects the growth of the pathogen (Bibi et al. 2021). The decrease in mycelial growth, sporangia formation, number and germination of zoospores in *P. capsici* due to the effect of the photocatalytic inactivation through the non-metallic-based ultrathin graphitic carbon nitride nanosheets (g-C_3N_4 nanosheets) has also been related to the modification of various mechanisms related to the regulation of oxidative stress (Cai et al. 2021).

AgNPs have also been reported as inhibitors of the growth of *P. capsici* and *P. parasitica* under *in vitro* conditions, registering 100% inhibition of pathogen growth with only 8.3 μg/mL of a solution made with aqueous extract from *Artemisia absinthium* (Ali et al. 2015). The synthesis of NPs from plant extracts could be an economical, effective, and environmentally friendly alternative, by avoiding the use of physical production methods, which can produce toxic substances during the process, and their high cost could limit its application (Borase et al. 2014). The production of NPs from fungal extracts that produce secondary metabolites against oomycetes could be an important source of new biocontrol products. Tongon et al. (2018) reported that the use of NPs from the crude extract of *Chaetomium brasilense* decreased root rot of *Durio zibithenus* (a tropical fruit tree from Southeast Asia) by up to 40% and reduced mycelial growth and spore production under *in vitro* conditions with doses of 8.68 and 1.08 μg/mL, respectively. Other microorganisms that have been used against *Phytophthora* as biocontrol agents include *Pseudomonas fluorescens* and *Trichoderma viride*, through the biogenic synthesis of copper oxide NPs (CuONPs) in cell-free extracts (Sawake et al. 2022). In the *in vitro* tests with *P. parasitica*, the highest inhibition of mycelial growth was obtained with NPs from *T. viride* (78.7%) at a concentration of 100 mg/L of CuONPs. *In vivo* assays applying CuONPs, on citrus gummosis lesions, registered a greater reduction of the lesion with *T. viride* (100 mg/L) than with Bordeaux mixture (1%), 21.1%, and 17.3% respectively.

The effect of selenium nanocomposites obtained from natural polysaccharide solutions has been successfully evaluated in reducing the damage caused by *P. cactorum* in potato (*Solanum tuberosum*) crops (Perfileva et al. 2021). In addition, chitosan and propolis oligomers in combination with AgNPs have been shown to be effective in inhibiting the growth of *P. cambivora*, *P. alni* and *P. plurivora*, which are oomycetes that cause important diseases in forest species (Hansen et al. 2012, Silva-Castro et al. 2018).

The fungicidal activity of magnesium oxide (MgONPs), both in nanoparticles and in a large scale (mMGO), has also been evaluated against *Phytophthora nicotianae* that induces black shank in tobacco plants (Chen et al. 2020). The MgONPs, compared to the MgO particles at macroscale, had a significantly higher inhibitory effect since they decreased mycelial growth and affected the development of sporangia more effectively. In addition, there was a toxicity effect of NPs and MgO on hyphae when the compounds came into direct contact with the mycelia. During greenhouse testing, the effects of MgONPs on the tobacco black leg disease were reduced by 36% compared to the control (Chen et al. 2020).

Conclusions

The use of nanoparticles and other nanocomposites for the management of diseases caused by oomycetes is promising since they are effective at low doses against different species *in vitro* and *in planta* conditions. Its mechanisms of action are novel and due to the action on various genes and signaling pathways, it is possible that the risk of resistance is very low compared to conventionally used agrochemicals. The combined application of nanoparticles with plant extracts and antagonistic microorganisms is a promising tool in the search for new active substances that are friendly to the environment. However, the high costs involved in the current use of NPs prevent them from being used by producers of intermediate to low socioeconomic levels. It is also necessary to expand the number of investigations on the impact of NPs on the soil microflora and the environment, as well as its possible toxicity in plants. Additionally, a greater number of phytopathogenic oomycete species need to be tested, *in vitro* and in the field to determine their efficacy in the substitution of toxic fungicides.

References

Adisa, I.O., Pullagurala, V.L., Rawat, S., Hernández-Viezcas, J.A., Dimpka, C., Elmer, W.H., White, J.C., Peralta-Videa, J.R. and Gardea-Torresdey, J.L. 2018. Role of cerium compounds in Fusarium wilt suppression and growth enhancement in tomato (*Solanum lycopersicum*). J. Agric. Food Chem. 66: 5959–5970.

Alghuthaymi, M.A., Almoammar, H., Rai, M., Said-Galiev, E. and Abd-Elsalam, K.A. 2015. Myconanoparticles: Synthesis and their role in phytopathogens management. Biotechnol. Biotechnol. Equip. 29: 221–236.

Ali, M., Kim, B., Belfield, K.D., Norman, D., Brennan, M. and Ali, G.S. 2015. Inhibition of *Phytophthora parasitica* and *P. capsici* by silver nanoparticles synthesized using aqueous extract of *Artemisia absinthium*. Phytopathology 105: 1–32.

Armitage, A.D., Lysøe, E., Nellist, C.F., Lewis, L.A., Cano, L.M., Harrison, R.J. and Brurberg, M.B. 2018. Bioinformatic characterisation of the effector repertoire of the strawberry pathogen *Phytophthora cactorum*. PLoS ONE 13: 1–24.

Banik, S. and Pérez-de-Luque, A. 2017. *In vitro* effects of copper nanoparticles on plant pathogens, beneficial microbes and crop plants. Span. J. Agric. Res. 15(2): e1005–e1005.

Barchenger, D.W., Lamour, K.H. and Bosland, P.W. 2018. Challenges and strategies for breeding resistance in *Capsicum annuum* to the multifarious pathogen, *Phytophthora capsici*. Front. Plant Sci. 9: 1–16.

Bibi, S., Huguet-Tapia, J.C., Naveed, Z.A., El-Sayed, A.S.A., Timilsina, S., Jones, J.B. and Ali, G.S. 2021. Study of silver nanoparticle effects on some molecular responses and metabolic pathways of *Phytophthora parasitica*. Int. J. Nanomater. Nanotechnol. Nanomed. 7: 047–056.

Bienapfl, J.C. and Balci, Y. 2014. Movement of *Phytophthora* spp. in Maryland's nursery trade. Plant Dis. 98: 134–144.

Borase, H.P., Salunke, B.K., Salunkhe, R.B., Patil, C.D., Hallsworth, J.E., Kim, B.S. and Patil, S.V. 2014. Plant extract: A promising biomatrix for ecofriendly, controlled synthesis of silver nanoparticles. Appl. Biochem. Biotechnol. 173: 1–29.

Cai, L., Wei, X., Feng, H., Fan, G., Gao, C., Chen, H. and Sun, X. 2021. Antimicrobial mechanisms of g-C_3N_4 nanosheets against the oomycetes *Phytophthora capsici*: Disrupting metabolism and membrane structures and inhibiting vegetative and reproductive growth. J. Hazard Mater. 417: 126121. Doi: 10.1016/j.jhazmat.2021.126121.

Cao, X., Ye, Y. and Liu, S. 2011. Gold nanoparticle-based signal amplification for biosensing. Anal. Biochem. 417: 1–16.

Chen, J., Li, S., Luo, J., Zhang, Y. and Ding, W. 2017. Graphene oxide induces toxicity and alters energy metabolism and gene expression in *Ralstonia solanacearum*. J. Nanosci. Nanotechnol. 17: 186–195.

Chen, J., Wu, L., Lu, M., Lu, S., Li, Z. and Ding, W. 2020. Comparative study on the fungicidal activity of metallic MgO nanoparticles and macroscale MgO against soilborne fungal phytopathogens. Front. Microbiol. 11: 365. Doi: 10.3389/fmicb.2020.00365.

Chepsergon, J., Motaung, T.E., Bellieny-Rabelo, D. and Moleleki, L.N. 2020. Organize, don't agonize: Strategic success of *Phytophthora* species. Microorganisms 8(6): 917.

Crandall, S.G., Rahman, A., Quesada-Ocampo, L.M., Martin, F.N., Bilodeau, G.J. and Miles, T.D. 2018. Advances in diagnostics of downy mildews: Lessons learned from other oomycetes and future challenges. Plant Dis. 102: 265–275.

Cruz-Luna, A.R., Cruz-Martínez, H., Vásquez-López, A. and Medina, D.I. 2021. Metal nanoparticles as novel antifungal agents for sustainable agriculture: Current advances and future directions. J. Fungi 7: 1033.

Davis, W.J. and Crouch, J.A. 2022. Analysis of digitized herbarium records and community science observations provides a glimpse of downy mildew species diversity of North America, reveals potentially undescribed species, and documents the need for continued digitization and collecting. Fungal Ecol. 55: 101126. Doi: 10.1016/j.funeco.2021.101126.

De Souza, S.V.J., Braga, C.Q., Brasil, C.L., Baptista, C.T., Reis, G.F., Panagio, L.A., Nakazato, G., De Oliveira Hübner, S., Soares, M.P., De Avila Botton, S. and Pereira, D.I. 2018. *In vitro* anti-*Pythium insidiosum* activity of biogenic silver nanoparticles. Med. Mycol. 57(7): 858–863.

Dyussembayev, K., Sambasivam, P., Bar, I., Brownlie, J.C., Shiddiky, M.J. and Ford, R. 2021. Biosensor technologies for early detection and quantification of plant pathogens. Front. Chem. 9: 144.

Elamawi, R.M., Al-Harbi, R.E. and Hendi, A.A. 2018. Biosynthesis and characterization of silver nanoparticles using *Trichoderma longibrachiatum* and their effect on phytopathogenic fungi. Egypt. J. Biol. Pest Control 28: 1–11.

Elmer, W. and White, J.C. 2018. The future of nanotechnology in plant pathology. Annu. Rev. Phytopathol. 56: 111–133.

Elmer, W.H., Zuverza-Mena, N., Triplett, L.R., Roberts, E.L., Silady, R.A. and White, J.C. 2021. Foliar application of copper oxide nanoparticles suppresses Fusarium wilt development on chrysanthemum. Environ. Sci. Technol. 55: 10805–10810.

Erwin, D.C. and Ribeiro, O.K. 1996. Phytophthora Diseases Worldwide. American Phytopathological Society Press. St. Paul, Minnesota, 562 p.

Fang, Y., Umasankar, Y. and Ramasamy, R.P. 2014. Electrochemical detection of p-ethylguaiacol, a fungi infected fruit volatile using metal oxide nanoparticles. Analyst 139: 3804–3810.

Franco, A.J.D.M., Merca, F.E., Rodriguez, M.S., Balidion, J.F., Migo, V.P., Amalin, D.M., Alocilja, E.C. and Fernando, L.M. 2019. DNA-based electrochemical nanobiosensor for the detection of *Phytophthora palmivora* (Butler) Butler, causing black pod rot in cacao (*Theobroma cacao* L.) pods. Physiol. Mol. Plant Pathol. 107: 14–20.

Fry, W.E. 2016. *Phytophthora infestans*: New tools (and old ones) lead to new understanding and precision management. Annu. Rev. Phytopathol. 54: 529–547.

Fukamachi, K., Konishi, Y. and Nomura, T. 2019. Disease control of *Phytophthora infestans* using cyazofamid encapsulated in poly lactic-co-glycolic acid (PLGA) nanoparticles. Colloids Surf., A Physicochem. Eng. Asp. 577: 315–322.

Giannousi, K., Avramidis, I. and Dendrinou-Samara, C. 2013. Synthesis, characterization and evaluation of copper based nanoparticles as agrochemicals against *Phytophthora infestans*. RSC Advances 3: 21743–21752.

Govers, L.L., Man in 't Veld, W.A., Meffert, J.P., Bouma, T.J., van Rijswick, P.C., Heusinkveld, J.H., Orth, R.J., van Katwik, M.M. and van der Heide, T. 2016. Marine *Phytophthora* species can hamper conservation and restoration of vegetated coastal ecosystems. Proc. R Soc. B: Biol. Sci. 283(1837): 20160812.

Hansen, E.M., Reeser, P.W. and Sutton, W. 2012. *Phytophthora* beyond agriculture. Annu. Rev. Phytopathol. 250: 359–378.

Hao, Y., Cao, X., Ma, C., Zhang, Z., Zhao, N., Ali, A., Hou, T., Xiang, Z., Zhuang, J., Wu, S., Xing, B., Zhang, Z. and Rui, Y. 2017. Potential applications and antifungal activities of engineered nanomaterials against gray mold disease agent *Botrytis cinerea* on rose petals. Front. Plant Sci. 8: 1–9.

Ho, H.H. 2018. The taxonomy and biology of *Phytophthora* and *Pythium*. J. Bacteriol. Mycol. 6: 40–45.

Imada, K., Sakai, S., Kajihara, H., Tanaka, S. and Ito, S. 2016. Magnesium oxide nanoparticles induce systemic resistance in tomato against bacterial wilt disease. Plant Pathol. 65: 551–560.

Khairnar, G.A., Chavan-Patil, A.B., Palve, P.R., Bhise, S.B., Mourya, V.K. and Kulkarni, C.G. 2010. Dendrimers: Potential tool for enhancement of antifungal activity. Int. J. Pharmtech. Res. 2(1): 736–739.

Khatua, A., Prasad, A., Priyadarshini, E., Virmani, I., Ghosh, L., Paul, B., Meena, R., Barabadi, H., Patel, A.K. and Saravanan, M. 2020. CTAB-PLGA curcumin nanoparticles: Preparation, biophysical characterization and their enhanced antifungal activity against phytopathogenic fungus *Pythium ultimum*. Chemistry Select 5(34): 10574–10580.

Kim, S.W., Jung, J.H., Lamsal, K., Kim, Y.S., Min, J.S. and Lee, Y.S. 2012. Antifungal effects of silver nanoparticles (AgNPs) against various plant pathogenic fungi. Mycobiology 40(1): 53–58.

Kutawa, A.B., Ahmad, K., Ali, A., Hussein, M.Z., Abdul Wahab, M.A., Adamu, A., Ismaila, A.A., Tiran-Gunasena, M., Ziaur Rahman, M. and Imam-Hossain, Md. 2021. Trends in nanotechnology and its potentialities to control plant pathogenic fungi: A review. Biology 10: 881.

Lamour, K.H., Stam, R., Jupe, J. and Huitema, E. 2012. The oomycete broad-host-range pathogen *Phytophthora capsici*. Mol. Plant Pathol. 13: 329–337.

Lebeda, A. and Cohen, Y. 2011. Cucurbit downy mildew (*Pseudoperonospora cubensis*)—biology, ecology, epidemiology, host-pathogen interaction and control. Eur. J. Plant Pathol. 129(2): 157–192.

Lévesque, A.C. and De Cock, A.W. 2004. Molecular phylogeny and taxonomy of the genus *Pythium*. Mycol. Res. 108(12): 1363–1383.

Lu, X., Sun, D., Rookes, J.E., Kong, L., Zhang, X. and Cahill, D.M. 2019. Nanoapplication of a resistance inducer to reduce *Phytophthora* disease in pineapple (*Ananas comosus* L.). Front. Plant Sci. 1238. Doi: 10.3389/fpls.2019.01238.

Luan, L.Q. and Xo, D.H. 2018. *In vitro* and *in vivo* fungicidal effects of γ-irradiation synthesized silver nanoparticles against *Phytophthora capsici* causing the foot rot disease on pepper plant. J. Plant Pathol. 100: 241–248.

Mahdizadeh, V., Safaie, N. and Khelghatibana, F. 2015. Evaluation of antifungal activity of silver nanoparticles against some phytopathogenic fungi and *Trichoderma harzianum*. J. Crop Prot. 4(3): 291–300.

Martins, J., Veríssimo, P. and Canhoto, J. 2021. Isolation and identification of *Arbutus unedo* L. fungi endophytes and biological control for *Phytophthora cinnamomi in vitro*. Protoplasma. Doi: 10.1007/s00709-021-01686-2.

Matei, P.M., Martín-Gil, J., Michaela Iacomi, B., Pérez-Lebeña, E., Barrio-Arredondo, M.T. and Martín-Ramos, P. 2018. Silver nanoparticles and polyphenol inclusion compounds composites for *Phytophthora cinnamomi* mycelial growth inhibition. Antibiotics 7: 76. Doi: 10.3390/antibiotics7030076.

Miao, J., Dong, X., Lin, D., Wang, Q., Liu, P., Chen, F., Du, Y. and Liu, X. 2016. Activity of the novel fungicide oxathiapiprolin against plant-pathogenic oomycetes. Pest. Manag. Sci. 72: 1572–1577.

Mitani, S., Araki, S., Yamaguchi, T., Takii, Y., Ohshima, T. and Matsuo, N. 2001. Antifungal activity of the novel fungicide cyazofamid against *Phytophthora infestans* and other plant pathogenic fungi *in vitro*. Pestic. Biochem. Phys. 70: 92–99.

Moussa, S.H., Tayel, A.A., Alsohim, A.S. and Abdallah, R.R. 2013. Botryticidal activity of nanosized silver-chitosan composite and its application for the control of gray mold in strawberry. J. Food Sci. 78: 1589–1594.

Nair, R. and Kumar, D. 2013. Plant diseases, control and remedy through nanotechnology. *In*: Tuteja, N. and Gill, S. (eds.). Crop Improvement Under Adverse Conditions. Springer, 4633 p.

Nandhini, M., Rajini, S.B., Udayashankar, A.C., Niranjana, S.R., Lund, O.S., Shetty, H.S. and Prakash, H.S. 2019. Biofabricated zinc oxide nanoparticles as an eco-friendly alternative for growth promotion and management of downy mildew of pearl millet. Crop Prot. 121: 103–112.

Nath, S., Kaittanis, C., Tinkham, A. and Perez, J.M. 2008. Dextran-coated gold nanoparticles for the assessment of antimicrobial susceptibility. Anal. Chem. 80(4): 1033–1038.

Park, H.J., Kim, S.H., Kim, S.J. and Choi, S.H. 2006. A new composition of nanosized silica-silver for control of various plant diseases. Plant Pathol. J. 22: 295–302.

Pérez-de-Luque, A., Cifuentes, Z., Beckstead, J.A., Sillero, J.C., Ávila, C., Rubio, J. and Ryan, R.O. 2012. Effect of amphotericin B nanodisks on plant fungal diseases. Pest. Manag. Sci. 68: 67–74.

Perfileva, A.I., Tsivileva, O.M., Nozhkina, O.A., Karepova, M.S., Graskova, I.A., Ganenko, T.V., Sukhov, B.G. and Krutovsky, K.V. 2021. Effect of natural polysaccharide matrix-based selenium nanocomposites on *Phytophthora cactorum* and rhizospheric microorganisms. Nanomaterials 11: 2274. Doi: 10.3390/nano11092274.

Pham, N.D., Duong, M.M., Le, M.V. and Hoang, H.A. 2019. Preparation and characterization of antifungal colloidal copper nanoparticles and their antifungal activity against *Fusarium oxysporum* and *Phytophthora capsici*. Comptes Rendus Chimie. 22: 786–793.

Ray, M., Ray, A., Dash, S., Mishra, A., Achary, K.G., Nayak, S. and Singh, S. 2017. Fungal disease detection in plants: Traditional assays, novel diagnostic techniques and biosensors. Biosens. Bioelectron. 87: 708–723.

Rodríguez-Alvarado, G., Fernández-Pavía, S.P., Geraldo-Verdugo, J.A. and Landa-Hernández, L. 2001a. *Pythium aphanidermatum* causing collar rot on papaya in Baja California Sur, Mexico. Plant Dis. 85: 444.

Rodríguez-Alvarado, G., Fernández-Pavía, S.P. and Landa-Hernández, L. 2001b. First report of *Pythium aphanidermatum* causing crown and stem rot on *Opuntia ficus-indica*. Plant Dis. 85: 231.

Salcedo, A., Hausbeck, M., Pigg, S. and Quesada-Ocampo, L.M. 2020. Diagnostic guide for cucurbit downy mildew. Plant Health Prog. 21(3): 166–172.

Salcedo, A.F., Purayannur, S., Standish, J.R., Miles, T., Thiessen, L. and Quesada-Ocampo, L.M. 2021. Fantastic downy mildew pathogens and how to find them: Advances in detection and diagnostics. Plants 10(3): 435. Doi: 10.3390/plants10030435.

Sánchez, A.D., Ousset, M.J. and Sosa, M.C. 2019. Biological control of *Phytophthora* collar rot of pear using regional *Trichoderma* strains with multiple mechanisms. Biol. Control 135: 124–134.

Sawake, M.M., Moharil, M.P., Ingle, Y.V., Jadhav, P.V., Ingle, A.P., Khelurkar, V.C., Paithankar, D.H., Bathe, G.A. and Gade, A.K. 2022. Management of *Phytophthora parasitica* causing gummosis in citrus using biogenic copper oxide nanoparticles. J. Appl. Microbiol. Doi: 10.1111/jam.15472.

Schroeder, K.L., Martin, F.N., De Cock, A.W., Lévesque, C.A., Spies, C.F., Okubara, P.A. and Paulitz, T.C. 2013. Molecular detection and quantification of *Pythium* species: Evolving taxonomy, new tools, and challenges. Plant Dis. 97(1): 4–20.

Schwenkbier, L., Pollok, S., König, S., Urban, M., Werres, S., Cialla-May, D., Weber, K. and Popp, J. 2015a. Towards on-site testing of *Phytophthora* species. Anal. Methods 7: 211–217.

Schwenkbier, L., Pollok, S., Rudloff, A., Sailer, S., Cialla-May, D., Weber, K. and Popp, J. 2015b. Non-instrumented DNA isolation, amplification and microarray-based hybridization for a rapid on-site detection of devastating *Phytophthora kernoviae*. Analyst 140: 6610–6618.

Sekhon, B.S. 2014. Nanotechnology in agri-food production: An overview. Nanotechnol. Sci. Appl. 7: 31–53.

Sharma, D., Sharma, S., Kaith, B., Rajput, J. and Kaur, M. 2011. Synthesis of ZnO nanoparticles using surfactant free in-air and microwave method. Appl. Surf. Sci. 257(22): 9661–9672.

Sharma, S., Sharma, N. and Kaushal, N. 2022. A comparative account of biogenic synthesis of silver nanoparticles using in-house potential probiotics and their antimicrobial activity against challenging antibiotic resistant pathogens. BioNanoScience 12: 833–840.

Siddaiah, C.N., Prasanth, K.V.H., Satyanarayana, N.R., Mudili, V., Gupta, V.K., Kalagatur, N.K., Satyavani, T., Dai, X.F., Chen, J.Y., Mocan, A., Singh, B.P. and Srivastava, R.K. 2018. Chitosan nanoparticles having higher degree of acetylation induce resistance against pearl millet downy mildew through nitric oxide generation. Sci. Rep. 8(1): 1–14.

Siegenthaler, T.B. and Hansen, Z.R. 2021. Sensitivity of *Phytophthora capsici* from Tennessee to mefenoxam, fluopicolide, oxathiapiprolin, dimethomorph, mandipropamid and cyazofamid. Plant Dis. 105: 3000–3007. Doi: 10.1094/PDIS-08-20-1805-RE.

Silva-Castro, I., Martín-García, J., Diez, J.J., Flores-Pacheco, J.A., Martín-Gil, J. and Martín-Ramos, P. 2018. Potential control of forest diseases by solutions of chitosan oligomers, propolis and nanosilver. Eur. J. Plant Pathol. 150: 401–411.

Simamora, A.V., Paap, T., Howard, K., Stukely, M.J.C., Hardy, G.E.St.J. and Burgess, T.I. 2018. *Phytophthora* contamination in a nursery and its potential dispersion in the natural environment. Plant Dis. 102: 132–139.

Standish, J.R., Raid, R.N., Pigg, S. and Quesada-Ocampo, L.M. 2020. A Diagnostic guide for basil downy mildew. Plant Health Prog. 21(2): 77–81.

Thines, M. and Choi, Y.J. 2016. Evolution, diversity, and taxonomy of the Peronosporaceae, with focus on the genus *Peronospora*. Phytopathology 106(1): 6–18.

Tongon, R., Soytong, K., Kanokmedhakul, S. and Kanokmedhakul, K. 2018. Nano-particles from *Chaetomium brasiliense* to control *Phytophthora palmivora* caused root rot disease in durian var Montong. Int. J. Agric. Technol. 14: 2163–2170.

Tyler, B.M. 2007. *Phytophthora sojae*: root rot pathogen of soybean and model oomycete. Mol. Plant Pathol. 8: 1–8.

Uzuhashi, S., Kakishima, M. and Tojo, M. 2010. Phylogeny of the genus *Pythium* and description of new genera. Mycoscience 51(5): 337–365.

Vanti, G.L., Masaphy, S., Kurjogi, M., Chakrasali, S. and Nargund, V.B. 2020. Synthesis and application of chitosan-copper nanoparticles on damping off causing plant pathogenic fungi. Int. J. Biol. Macromol. 156: 1387–1395.

Wang, J., Hao, K., Yu, F., Shen, L., Wang, F., Yang, J. and Su, C. 2022. Field application of nanoliposomes delivered quercetin by inhibiting specific hsp70 gene expression against plant virus disease. J. Nanobiotechnol. 20: 1–16.

Wang, X., Cai, A., Wen, X., Jing, D., Qi, H. and Yuan, H. 2017. Graphene oxide-Fe3O4 nanocomposites as high-performance antifungal agents against *Plasmopara viticola*. Sci. China Mater. 60(3): 258–268.

Wang, X., Liu, X. and Han, H. 2013. Evaluation of bacterial effects of carbon nanomaterials against copper-resistant *Ralstonia solanacearum*. Colloids Surf. B: Biointerfaces 103: 136–142.

Wang, X., Liu, X., Chen, J., Han, H. and Yuan, Z. 2014. Evaluation and mechanism of antifungal effects of carbon nanomaterials in controlling plant fungal pathogen. Carbon 68: 798–806.

Zabrieski, Z., Morrell, E., Hortin, J., Dimkpa, C., McLean, J., Britt, D. and Anderson, A. 2015. Pesticidal activity of metal oxide nanoparticles on plant pathogenic isolates of *Pythium*. Ecotoxicol. 24(6): 1305–1314.

Zhan, F., Wang, T., Iradukunda, L. and Zhan, J. 2018. A gold nanoparticle-based lateral flow biosensor for sensitive visual detection of the potato late blight pathogen, *Phytophthora infestans*. Anal. Chim. Acta 1036: 153–161.

Chapter 7

Biogenic Metal Oxide Nanoparticles for the Control of Plant Pathogens

Elsherbiny A. Elsherbiny[1,*] and *Sabrien A. Omar*[2]

Introduction

Plant diseases cause substantial economic losses in food crop production both in terms of quality and quantity (Strange and Scott 2005). Global yield losses associated with pests and diseases on food crops are reaching on average 21.5% in wheat, 30.3% in rice, 22.6% in maize, 17.2% in potato, and 21.4% in soybean (Avila-Quezada et al. 2022). The management of plant diseases is very difficult due to their ability to spread on a continental scale, adapt to environmental changes, and develop strains resistant to chemical pesticides (Alghuthaymi et al. 2021). Furthermore, the tremendous and indiscriminate application of synthetic pesticides causes voluminous damage to the environment and high toxicity to living beings (Ristaino et al. 2021). In this context, the use of pesticides for crop protection reaches two million tons more worldwide each year: 24% in the United States, 45% in Europe, and 31% in the rest of the world. Only less than 0.1% of three million tons of active ingredients in these pesticides successfully control plant pathogens and the rest is lost in the soil, water, air, and by photodegradation (Sabzevari and Hofman 2022, Khan et al. 2022). Therefore, innovative and appropriate technologies for plant disease management need to be developed and harnessed to ensure a good supply of crop yields in the longer term (Savary et al. 2019).

Nanobiotechnology offers a promising alternative for the control of plant disease with numerous advantages over conventional products and approaches (Ali et al. 2020). Nanoparticles are materials with a range between 10 to 100 nanometers (nm) in at least one dimension with a high surface-to-volume ratio which increases their

[1] Plant Pathology Department, Faculty of Agriculture, Mansoura University, Mansoura 35516, Egypt.

[2] Agricultural Microbiology Department, Faculty of Agriculture, Mansoura University, Mansoura 35516, Egypt.

* Corresponding author: sherbiny@mans.edu.eg

reactivity and potential biochemical activity (Rajwade et al. 2020). The nanoparticles can be synthesized by chemical, physical, and biological methods using top-down and bottom-up approaches (Worrall et al. 2018). Great emphasis has been placed on the use of biological materials such as plant extracts, fungi, bacteria, yeast, and algae to design nanoparticles due to their unique advantages like less energy consumption and without the use of toxic chemicals as well as cost-effective and eco-friendly technique (Bukhari et al. 2021, Kutawa et al. 2021). Moreover, nanoparticles can be directly applied to seeds, foliage, or roots for the control of pests and plant pathogens (Fu et al. 2020). The effect of nanoparticles on plants is related to their size, composition, surface charge, and concentration. The analytical techniques such as microscopy, magnetic resonance imaging, and fluorescence spectroscopy help to understand better the action mechanisms of nanoparticles on plants (Raj et al. 2021, Kumar et al. 2022).

Metal oxide nanoparticles play a vital role in the fields of medicine, agriculture, information technology, biomedical, materials chemistry, optical, electronics, catalysis, environment, and energy. Also, the metal oxide nanoparticles have gained remarkable interest for their unique magnetic, electronic, and chemical properties, especially the high density and limited size of corners and edges on their surface (Chavali and Nikolova 2019). Besides, metal oxide nanoparticles have other advantages such as high stability, porosity, easy preparation methods, simple engineering for the required size, and shape, easy merging with both hydrophobic and hydrophilic systems, and easy functionalization (Nikolova and Chavali 2020). The present chapter is intended to focus on the biosynthesis, characterization, and biological activities of metal oxide nanoparticles with an emphasis on their role in the management of plant pathogens. Additionally, this chapter describes the potential benefits, and applications of metal oxide nanoparticles based on zinc, copper, magnesium, titanium, iron, and zirconium as promising and effective tools for the control of fungal and bacterial plant diseases.

Biosynthesis of metal oxide nanoparticles

The synthesis of metal oxide nanoparticles (MO NPs) using biological methods has been considered of great interest. The materials prepared in these ways have voluminous applications in different fields including biomedical, pharmaceutical, and agriculture (Achari and Kowshik 2018, Marouzi et al. 2021). The biosynthesis of MO NPs possesses the potential to improve the biological properties of metal oxide nanoparticles and reduce toxicity in many cases (Gour and Jain 2019, Jeevanandam et al. 2022). The MO NPs synthesis via this approach is mainly performed using plants, fungi, bacteria, yeast, and algae (Figure 1). The properties of MO NPs obtained by biological methods depended on the kind of biological structure used, its concentration, temperature, pH, the metal ion concentration, reaction time, and the type of reducing agent (Gour and Jain 2019). The biological synthesis of metal oxide nanoparticles provides environmentally friendly, clean, cost-effective, and less toxic techniques for sustainable development (Nair et al. 2022).

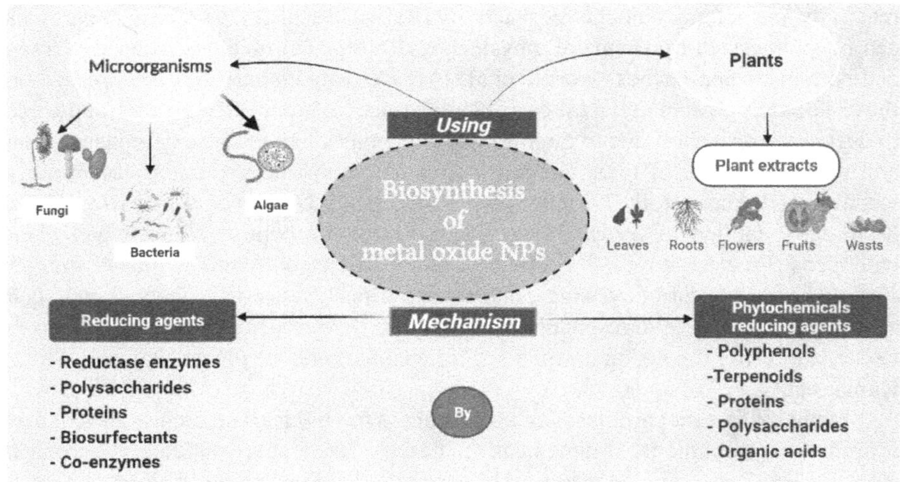

Figure 1: Biosynthesis of metal oxide nanoparticles.

Biosynthesis by plants

Plants are known to have numerous metabolites that are used for the biological synthesis of metal oxide nanoparticles. Plant extracts have been used as an effective reducing agent for the biosynthesis of these nanoparticles. The biosynthesis of metal oxide nanoparticles by plant extracts has great physicochemical properties because these extracts are a rich source of phenols, flavonoids, biomolecules, and other significant metabolites that act as bio-reductant capping agents (Jeevanandam et al. 2016). Plants such as *Thymus vulgaris* (Nasrollahzadeh et al. 2016), *Aerva javanica* (Amin et al. 2021), *Punica granatum* (Vidovix et al. 2019), *Annona squamosa* (Singh et al. 2021), *Psidium guajava* (Sathiyavimal et al. 2021), and *Canthium coromandelicum* (Selvam et al. 2022) have been used to produce CuO nanoparticles. Likewise, ZnO nanoparticles have been synthesized using *Mimosa pudica* (Balogun et al. 2020), *Coffea arabica* (Abel et al. 2021), *Eriobotrya japonica* (Nazir et al. 2021), *Bixa orellana* (Gharpure et al. 2022), and *Carica papaya* (Dulta et al. 2022). Different plant extracts have been utilized for the biosynthesis of TiO_2 nanoparticles including *Artemisia haussknechtii* (Alavi et al. 2018), *Carica papaya* (Kaur et al. 2019), *Azadirachta indica* (Thakur et al. 2019), *Mentha arvensis* (Ahmad et al. 2020), *Cola nitida* (Akinola et al. 2020), and *Cuminum cyminum* (Mathew et al. 2021).

Biosynthesis by fungi

Fungi are considered unique agents for the biosynthesis of metal oxide nanoparticles, both intracellular and extracellular. The fungi give a high yield of nanoparticles due to a large amount of fungal biomass and a high yield of extracellular compounds (Gahlawat and Choudhury 2019). In addition, fungi used for nanobiotechnological purposes have multiple advantages compared to other microorganisms. These include easy isolation and preservation in the laboratory, secretion of extracellular enzymes, high intracellular metal uptake, metal biological accumulation potential, economic feasibility, and ecological friendliness (Adebayo et al. 2021). The extracellular fungal metabolites such as enzymes, polysaccharides, polypeptides, proteins, carbohydrates, quinones, and other macromolecules contribute to the biosynthesis of metal oxide nanoparticles by the reduction and precipitation of metal ions (Mughal et al. 2021). The metal ions transform into nanoparticles by the enzymes in the cytoplasm and cell wall of fungi where the biosynthesis occurs by the attraction of fungi to the positive charge possessed by the metal ions (Rana et al. 2020).

Many fungal species have been employed to produce different kinds of metal oxide nanoparticles including *Aspergillus flavus* (Raliya et al. 2015), *A. fumigatus* (Raliya et al. 2016), *Alternaria alternata* (Sarkar et al. 2017), *Trichoderma viride* (Chinnaperumal et al. 2018), *A. nidulans* (Vijayanandan and Balakrishnan 2018), *Cochliobolus geniculatus* (Kadam et al. 2019), *Fusarium keratoplasticum,* and *A. niger* (Mohamed et al. 2019), *T. asperellum* (Saravanakumar et al. 2019), *T. harzianum* (Consolo et al. 2020), *F. solani* (El Sayed and El Sayed 2020), *Periconium* sp. (Ganesan et al. 2020), *F. oxysporum* (Gupta and Chundawat 2020), *Fomitopsis pinicola* (Rehman et al. 2020), *Penicillium chrysogenum* (Mohamed et al. 2021), and *Rhizopus oryzae* (Hassan et al. 2021).

Biosynthesis by bacteria

Bacteria are a large group of single-celled, prokaryotic microorganisms with diameters of about 15 μm. Most bacteria are unicellular or can form complex multicellular groups during their life cycles. Bacteria lack membrane-bound organelles, and the cell walls contain peptidoglycan (Fariq et al. 2017). Bacteria can synthesize metal oxide nanoparticles by either intracellular or extracellular mechanisms. Extracellular synthesis is more efficient and easier for the synthesis of MO NPs that occurs outside the bacterial cell using the bacterial suspension, culture filtrates, or cell-free supernatant (Mughal et al. 2021). Prokaryotic bacteria are capable of producing less toxic metal oxide nanoparticles such as ZnO, CuO, TiO_2, ZrO, and iron oxide (Marouzi et al. 2021). Bacteria have gained considerable attention as means of synthesizing MO NPs due to their excellent advantages, including abundant presence in the environment, rapid growth rates, easy adaptation to extreme conditions, low-cost production, inexpensive culture, and easily controllable growth conditions (Koul et al. 2021).

The major bacterial species used for the synthesis of metal oxide nanoparticles include *Enterococcus faecalis* (Ashajyothi et al. 2016), *Bacillus amyloliquefaciens* (Khan and Fulekar 2016), *Desulfovibrio* sp. (Das et al. 2018), *B. cereus* (Fatemi et al. 2018), *Lactobacillus paracasei* (Król et al. 2018), *Streptomyces zaomyceticus* and *S. pseudogriseolus* (Hassan et al. 2019), *Acinetobacter* sp. (Suriyaraj et al. 2019), *Streptomyces* sp. (Ağçeli et al. 2020), *L. plantarum* (Yusof et al. 2020), *Pseudomonas aeruginosa* (Abdo et al. 2021), *Enterobacter* sp. (Ahmed et al. 2021b), and *B. cereus* (Parthasarathy et al. 2022).

Biosynthesis by algae

Algae are aquatic filamentous eukaryotic organisms without the multicellular structures of the plant and include autotrophic and aquatic photosynthetic organisms. Algae are broadly classified into either microalgae or macroalgae (Mughal et al. 2021). Algae are highly significant, economically and ecologically, as a source of food, fodder, fertilizers, medicine, and industrially important products as well as the source of oxygen generated through photosynthesis for most aquatic forms of life (Chen et al. 2019). The algae can synthesize metal oxide nanoparticles by the inclusion of active compounds in the cell walls such as alginate and laminarin which contain reactive groups (Sharma et al. 2016). This reaction can be either intracellular or extracellular depending on the location of the nanoparticles. Intracellular production occurs within the algal cell wherein NADPH or NADPH act as reducing agents during algal metabolic processes. Extracellular synthesis of nanoparticles occurs by attaching the metal ions to the surface of the cells. In this case, several metabolites act as reducing agents such as proteins, nucleic acids, lipids, pigments, carbohydrates, and enzymes (Agarwal et al. 2019).

Most of the major algal groups have been used to efficiently synthesize metal oxide nanoparticles such as *Bifurcaria bifurcata* (Abboud et al. 2014), *Sargassum muticum* (Azizi et al. 2014), *Padina pavonica*, and *Sargassum acinarium* (El-Kassas et al. 2016), *Sargassum wightii* (Kumaresan et al. 2018), *Ulva flexuosa* (Mashjoor et al. 2018), *Colpomenia sinuosa* and *Pterocladia capillacea* (Salem et al. 2019), *Sargassum vulgare*, *Ulva fasciata* and *Jania rubens* (Salem et al. 2020), *Ulva fasciata* (Alsaggaf et al. 2021), *Arthrospira platensis* (El-Belely et al. 2021), and *Petalonia fascia*, *Colpomenia sinuosa*, and *Padina pavonica* (El-Sheekhet al. 2021).

Application of biogenic metal oxide nanoparticles for plant pathogens control

Plant diseases are one of the most serious challenges that the agriculture sector is facing, alongside climate change and water scarcity, which might influence crop yields and food security (Savary et al. 2017). Plant pathogens negatively impact the quality and quantity of agricultural produces with yield losses of 20–40% in the susceptible cultivars over the world, which leads to adverse effects on the economy (Peng et al. 2021). Unfortunately, the prolonged and excessive applications of synthetic pesticides have led to severe problems for human health and the environment. Therefore, the use of nanomaterials in crop protection is the most promising application of nanotechnology, where it provides effective mechanisms for plant protection against pests and diseases, monitoring pathogens, and detecting plant diseases, which in turn leads to the improvement of crops (Mishra et al. 2017).

Plant pathogenic fungi

Plant pathogenic fungi are the key causative agent among phytopathogens which cause persistent and significant losses in crop yield and quality worldwide. These fungi are responsible for 70–80% of plant diseases and represent a serious challenge for sustainable agriculture in the future (Peng et al. 2021). The lifestyles of phytopathogenic fungi vary greatly, some of them are necrotrophic, while others are biotrophic, or hemibiotrophic depending on modes of interaction with their host plants (Doehlemann et al. 2017). The top ten fungal pathogens in plants are listed as follows: *Magnaporthe oryzae*, *Botrytis cinerea*, *Puccinia* spp., *Fusarium graminearum*, *Fusarium oxysporum*, *Blumeria graminis*, *Mycosphaerella graminicola*, *Colletotrichum* spp., *Ustilago maydis*, and *Melampsora lini* (Dean et al. 2012). The application of chemical fungicides is the major tool to control fungal pathogens in different crops. However, the extensive use of these fungicides in long-term doses has led to toxic effects on human health and the environment as well as the development of resistant strains of fungal phytopathogens which complicate the control of plant disease (Elsherbiny et al. 2021). Therefore, there is a need to develop effective, non-toxic, eco-friendly, and viable strategies rather than synthetic fungicides. Consequently, the use of nanoparticles for the control of plant diseases has gained a lot of attention worldwide for their positive impacts on all ecological systems. Table 1 contains some biogenic metal oxide nanoparticles that are used for the control of phytopathogenic fungi.

Zinc oxide nanoparticles (ZnO NPs)

ZnO nanoparticles show promising applications in various fields (Osuntokun et al. 2019, Chennimalai et al. 2021, Kangralkar et al. 2021, Resmi et al. 2021, Islam et al. 2022) due to their intrinsic physicochemical properties such as highly crystalline structure, good electrochemical behavior, mechanical properties, chemical stability, a wide range of radiation absorption, and a large surface-to-volume ratio (Kołodziejczak-Radzimska et al. 2014, Czyżowska and Barbasz 2020, Gur et al. 2022, Suarez and León-Molina 2022). Moreover, zinc oxide has been classified as GRAS (generally recognized as safe) by Food and Drug Administration (FDA-USA) (Batool et al. 2021).

Several studies have provided evidence for the tremendous benefits of ZnO NPs in the plant protection area. Arciniegas-Grijalba et al. (2019) used two types of zinc oxide nanoparticles; the first was synthesized by a chemical route and the second was nanobiohybrid obtained by the green synthesis in garlic extract (*Allium sativum*). ZnO-based nanobiohybrid showed remarkable antifungal activity by 97% inhibition in the growth of *Mycena citricolor*, and 93% for *Colletotrichum* sp. Non-toxic ZnO nanoparticles were produced using the flower bud extract of *Syzygium aromaticum* to control the growth and mycotoxins of *Fusarium graminearum*, which causes *Fusarium* head blight, and stalk rot in cereals. The biosynthesized ZnO NPs reduced the growth and production of deoxynivalenol (DON) and zearalenone (ZEA) of *F. graminearum*. Moreover, ZnO NPs treatments enhanced ROS accumulation, lipid peroxidation, and ergosterol content in the mycelia, and strong damage to the fungal membrane integrity. The macroconidia treated with ZnO NPs was shown to be wrinkled, rough and shrank surface under SEM analysis (Lakshmeesha et al. 2019).

The ZnO nanoparticles, synthesized by the fungus *Trichoderma harzianum*, caused 100% inhibition of the growth of soil-borne pathogens, including *Fusarium* sp., *Rhizoctonia solani*, and *Macrophomina phaseolina* at the concentrations of

Table 1: Effect of biogenic metal oxide nanoparticles on plant pathogenic fungi.

NP	Production source	Size (nm) and shape	Effective concentration	Pathogen	Host	Effect	References
ZnO	*Allium sativum*	< 100 Spherical	12 mmol L^{-1}	*Mycena citricolor*, *Colletotrichum* sp.	Coffee	Antifungal activity	Arciniegas-Grijalba et al. (2019)
ZnO	*Syzygium aromaticum*	30–40 Polymorphic	140 μg mL^{-1}	*Fusarium graminearum*	Cereals	Antifungal activity and antimycotoxin activity	Lakshmeesha et al. (2019)
ZnO	*Trichoderma harzianum*	8–25 Polymorphic	20 μg mL^{-1} 200 μg mL^{-1}	*Fusarium* sp., *Rhizoctonia solani*, *Macrophomina phaseolina*	Cotton	Maximum efficiency in damping-off control	Zaki et al. (2021)
ZnO	*Cinnamomum camphora*	13.92–21.13 Spherical	20–160 mg L^{-1}	*Alternaria alternata*	Tomato	Antifungal activity	Zhu et al. (2021)
ZnO	*Penicillium expansum*	3.5–67.3 Hexagonal	1000 μg mL^{-1}	*Fusarium oxysporum*	Eggplant	Reducing the severity of *Fusarium* wilt disease by 75%	Abdelaziz et al. (2022)
ZnO	*Citrus limon*	33.92 Spherical	100 mg mL^{-1}	*Alternaria citri*	Citrus	Antifungal activity	Sardar et al. (2022)
CuO	*Streptomyces zaomyceticus* Oc-5 *S. pseudogriseolus* Acv-11	78–80 Spherical	20 mM	*Alternaria alternata*, *Fusarium oxysporum*, *Pythium ultimum*, *Aspergillus niger*	–	Antifungal activity	Hassan et al. (2019)
CuO	*Trichoderma harzianum*	38–77 Polymorphic	20 ppm	*Alternaria Alternata* *Pyricularia oryzae*	–	Antifungal activity	Consolo et al. (2020)
CuO	*Cassia fistula*	12–38 Spherical	350 μg mL^{-1}	*Fusarium oxysporum* f. sp. *lycopersici*	Tomato	Decrease the *Fusarium* wilt incidence and severity	Ashraf et al. (2021)
CuO	*Ginkgo biloba*	– Short rod	200 μg mL^{-1}	*Bipolaris maydis*	Maize	Antifungal activity	Huang et al. (2021)

CuO	*Citrus limon*	25 Spherical	100 mg mL^{-1}	*Alternaria citri*	Citrus	Antifungal activity	Sardar et al. (2022)
MgO	*Burkholderia rinojensis*	26.70 Spherical	15.36 µg mL^{-1} 1.92 µg mL^{-1}	*Fusarium oxysporum* f. sp. *lycopersici*	Tomato	Antifungal activity Antibiofilm activity	Abdel-Aziz et al. (2020)
MgO	*Carica papaya*	100 Spherical	500 µg mL	*Phytophthora nicotianae* *Thielaviopsis basicola*	Tobacco	Decrease in tobacco black shank and black root rot disease	Chen et al. (2020)
TiO_2	*Moringa oleifera*	20–100 Polymorphic	40 mg L^{-1}	*Bipolaris sorokiniana*	Wheat	Decrease of spot blotch disease severity	Satti et al. (2021)
TiO_2	*Luffa acutangula*	10–49 Hexagonal	40 mg mL^{-1}	*Aspergillus flavus* *Aspergillus niger* *Rhizopus oryzae* *Sclerotium rolfsii*	–	Antifungal activity	Anbumani et al. (2022)
Fe_2O_3	*Trichoderma harzianum*	17.78 Crystalline	1 mg mL^{-1}	*Fusarium oxysporum*	Apple	Significantly reduced the brown rot severity on apple fruits	Akbar et al. (2022)
Fe_3O_4	*Spinacia oleracea*	20 Spherical	15 µg mL^{-1}	*Fusarium oxysporum* f. sp. *lycopersici*	Tomato	Substantial reduction in *Fusarium* wilt severity	Ashraf et al. (2022)
ZrO	*Enterobacter* sp.	33–75 Spherical	20 µg mL^{-1}	*Pestalotiopsis versicolor*	Bayberry	Significantly inhibited twig blight in detached leaf assay	Ahmed et al. (2021b)
ZrO_2	*Tinospora cordifolia*	73 Crystalline	15 mg mL^{-1}	*Aspergillus fumigatus*, *Aspergillus niger*	–	Antifungal activity	Joshi et al. (2021)

20, 40, and 100 μg mL^{-1} under laboratory conditions. In the greenhouse experiments, seeds of Giza-90 treated with ZnO NPs at 200 μg mL^{-1} showed the maximum efficiency in controlling the damping-off of cotton seedlings by 91.1% survival, according to Zaki et al. (2021). Similarly, Zhu et al. (2021) synthesized ZnO NPs using leaf extract of *Cinnamomum camphora* with the average sizes of 13.92, 15.19, and 21.13 nm at pH 7, pH 8, and pH 9, respectively. The treatments of ZnO NPs at 20–160 mg L^{-1} significantly inhibited the mycelia growth, spore germination, and germ tube elongation of *Alternaria alternata* which caused early blight disease in tomato (*Solanum lycopersicum*) with MIC value recorded as 20 mg L^{-1}. Besides, ZnO nanoparticles induced excessive accumulation of malondialdehyde (MDA) in *A. alternata* mycelia as an indicator of cellular oxidative damage with great damage in the fungal cell membrane, leading to the leakage of protein and nucleic acid.

Recently, Abdelaziz et al. (2022) reported on the synthesis of ZnO NPs with an average size ranging between 3.5 and 67.3 nm using cell filtrate of the fungus *Penicillium expansum* ATCC 7861. These nanoparticles showed promising antifungal activity against *Fusarium oxysporum* at the concentration of 1000 μg mL^{-1} as well as reducing the severity of *Fusarium* wilt disease by 75% in cultivated eggplants (*Solanum melongena*). Currently, Sardar et al. (2022) assessed the antifungal activity of ZnO and CuO nanoparticles synthesized with lemon peel extract against *Alternaria citri*, the pathogen of citrus black rot. The inhibition zone was 50, 51.5, and 53 mm by CuO NPs, ZnO NPs, and mixed metal oxide NPs, respectively.

Copper oxide nanoparticles (CuO NPs)

CuO nanoparticles are receiving great attention due to their distinct chemical and physical properties such as superthermal conductivity, high specific surface area, excellent stability in solutions, spin dynamics, good electrochemical activity, electron correlation effects, and proper redox potential (Verma and Kumar 2019, Cuong et al. 2022). These nanoparticles have promising applications in various areas, including catalysis, electrochemistry, biosensors, energy storage, and biocidal agents (Goel et al. 2014, Lim et al. 2015, Zampardi et al. 2018). Besides, nanoparticles of copper and its oxides have been recognized as antimicrobial materials by the US Environmental Protection Agency (EPA) (Meghana et al. 2015).

Numerous studies have been conducted using CuO NPs for the control of phytopathogenic fungi. CuO NPs were synthesized using two endophytic *Streptomyces zaomyceticus* Oc-5 and *Streptomyces pseudogriseolus* Acv-11, isolated from healthy leaves of *Oxalis corniculata*, which caused a strong inhibition of the growth of *Alternaria alternata*, *Fusarium oxysporum*, *Pythium ultimum*, and *Aspergillus niger* (Hassan et al. 2019). The CuO nanoparticles, synthesized by the culture filtrate of *Trichoderma harzianum*, significantly reduced the mycelial growth of *Alternaria alternata* and *Pyricularia oryzae* in a dose-dependent manner with the best inhibition rate at 20 ppm (Consolo et al. 2020). Also, Huang et al. (2021) obtained CuO NPs using leaf extract of *Ginkgo biloba*. At the concentration of 200 μg mL^{-1}, the suppression rate of CuO NPs on the growth of the pathogen *Bipolaris maydis* was 62.78%, and the spore germination was inhibited irrespective of suspension in a liquid or on the detached maize leaf.

Besides, CuO NPs derived from leaf extract of *Cassia fistula* via green synthesis showed fungicidal activity against *Fusarium oxysporum* f. sp. *lycopersici*, the causal agent of *Fusarium* wilt of tomato, under laboratory and greenhouse conditions (Ashraf et al. 2021). In this study, CuO NPs with particle sizes of 12–38 nm and spherical shape showed voluminous inhibition on the mycelial growth and spore germination of the fungal pathogen by 91.9% and 91.76%, respectively, at 350 μg mL^{-1}. In the pots bioassay, the disease incidence and severity were decreased to 33.3% whereas the disease was substantially reduced by 66.7% after exposure of root and foliar of infected tomato seedlings to 350 μg mL^{-1} of CuO NPs, and consequently improved the plant growth indexes and fruit quality. In addition, the photosynthetic pigments, phenolic content, and non-enzymatic antioxidant compounds such as vitamin C, lycopene, and flavonoids in tomato fruit were significantly induced with the application of CuO NPs as well as an increase in the enzymatic activity (PAL, PPO, POD, SOD, and CAT) in roots and shoots of tomato plants.

Magnesium oxide nanoparticles (MgO NPs)

MgO nanoparticles have several outstanding properties such as low cost, toxicity, catalysis, refractory materials, superconductors, and stability under harsh processing conditions (Bhattacharya et al. 2021). In this context, MgO is currently classified as a safe (GRAS) and human-food ingredient by the US Food and Drug Administration (FDA) without any harmful byproducts (Anicic et al. 2018). The antifungal and antibiofilm properties of MgO NPs synthesized using the endobacterium *Burkholderia rinojensis* were described against the fungus *Fusarium oxysporum* f. sp. *lycopersici*, which was isolated from tomato plants. These nanoparticles showed a spherical structure, with a 26.7 nm mean size, and caused severe morphological changes to the hyphal morphology and biofilm formation of the fungus with significant damage to the fungal membrane integrity (Abdel-Aziz et al. 2020). The nanoparticles of MgO NPs obtained by green synthesis with leaf extract of *Carica papaya* L. showed considerable inhibition of the fungal growth and spore germination and impede sporangium development of *Phytophthora nicotianae* and *Thielaviopsis basicola* under laboratory conditions with 36.58 and 42.35% reduction in tobacco black shank and black root rot disease, respectively, in the greenhouse experiments (Chen et al. 2020).

Titanium dioxide nanoparticles (TiO_2 NPs)

TiO_2 nanoparticles have been employed in different applications due to their powerful oxidative properties, high chemical stability, low toxicity, biocompatibility, and good corrosion resistance (Ilyas et al. 2021). Also, titanium dioxide has a long-term antimicrobial activity with non-deteriorates (Ajmal et al. 2019). TiO_2 NPs were synthesized from aqueous leaf extract of *Moringa oleifera* Lam. by the reduction of titanium dioxide salt into TiO_2 nanoparticles and observed using the analytical methods including UV-VS, FTIR, SEM, EDX, and XRD. Four concentrations of TiO_2 NPs (20, 40, 60, and 80 mg L^{-1}) were applied exogenously on wheat plants infected with *Bipolaris sorokiniana* which causes spot blotch disease. The results showed that the 40 mg L^{-1} was highly effective in reducing the disease severity with the ability to develop resistance in the wheat plants against *B. sorokiniana* (Satti et al. 2021). In this context, the leaf extract of *Luffa acutangula* was utilized as a valuable, reducing agent for the synthesis of TiO_2 NPs with high toxicity against *Aspergillus flavus*, *A. niger*, *Rhizopus oryzae*, and *Sclerotium rolfsii* at the concentration of 40 mg mL^{-1} (Anbumani et al. 2022).

Iron oxide nanoparticles (Fe_2O_3/Fe_3O_4 NPs)

Iron oxide nanoparticles have attracted attention, especially in the agricultural field, owing to their distinct chemical and physical properties like superparamagnetism, firmness in the liquid solution, low oxidation, and flexible surface chemistry, with voluminous applications in environmental regulation such as food-related processes, adsorption of dyes, antibiotic degradation, biomedical, cosmetics, bioengineering, and biosensing along with antimicrobial agents and plant growth inducers (Priya et al. 2021, Sharma et al. 2019). Recently, Akbar et al. (2022) synthesized iron oxide nanoparticles (Fe_2O_3 NPs) using the fungus *Trichoderma harzianum* to control brown rot on apple fruits caused by *Fusarium oxysporum*. The highest reduction in the fungal growth was observed at 1 mg mL^{-1} concentration by 65.4%. The same concentration significantly reduced the disease severity on the fruits by 91.5% with maintaining biochemical and organoleptic properties of fruits such as firmness, soluble solids, sugars, and ascorbic acid.

In another study, Ashraf et al. (2022) used spinach as a starting material to synthesize Fe_3O_4 nanoparticles with different extracts like pomegranate juice, white vinegar, pomegranate peel, black coffee, aloe vera peel, and aspirin as reducing/stabilizing agents to tune the properties of the Fe_3O_4 NPs. In laboratory conditions, nanoparticles significantly inhibited the mycelial growth of *F. oxysporum* by 96 and 99% at concentrations of 10 and 15 µg mL^{-1}, respectively. Under greenhouse conditions, all concentrations of Fe_3O_4 NPs caused a substantial reduction in disease severity and incidence by improving plant growth parameters as compared to the control with an increase in the enzyme activity for POD, PPO, and PAL both in the roots and shoot of tomato plants as well as enhanced the upregulation of PR-proteins and defense genes.

Zirconium oxide/dioxide nanoparticles (ZrO/ ZrO2 NPs)

ZrO nanoparticles were synthesized through the culture supernatant of the bacterium *Enterobacter* sp. RNT10 and were evaluated for its antifungal activity against the pathogen *Pestalotiopsis versicolor*, the causal agent of twig blight disease in bayberry. Biogenic ZrO NPs showed a strong inhibition zone by 25.18 mm at 20 µg mL^{-1} concentration against the fungus as well as significantly inhibited twig blight in detached leaf assay (Ahmed et al. 2021b). In another study, zirconium dioxide nanoparticles (ZrO_2 NPs) were synthesized using the leaves extract of a medicinal plant *Tinospora cordifolia* and showed antifungal activity against *Aspergillus fumigatus* and *A. niger* with the highest inhibitory effect at the concentration of 15 mg mL^{-1} (Joshi et al. 2021).

Plant pathogenic bacteria

Plant pathogenic bacteria are serious plant pathogens that infect many agriculturally important crops. The severity of bacterial diseases depends on a multiplicity of factors, including susceptible species or varieties, the genetic constitution of the host and the pathogen, and environmental conditions (Buttimer et al. 2017). Among 7100 classified bacteria, about 150 species are responsible for various plant diseases. These species mainly belong to three families - Xanthomonadaceae, Pseudomonadaceae, and Enterobacteriaceae (Tarkowski and Vereecke 2014). Plant bacterial diseases cause significant economic losses in a wide range of crops worldwide by over $1 billion dollars annually (Martins et al. 2018). The most economically important bacterial pathogens in plants are *Pseudomonas syringae* pathovars, *Ralstonia solanacearum*, *Agrobacterium tumefaciens*, *Xanthomonas oryzae* pv. *oryzae*, *Xanthomonas campestris* pathovars, *Xanthomonas axonopodis* pv. *manihotis*, *Erwinia amylovora*, *Xylella fastidiosa*, *Dickeya* (*dadantii* and *solani*), *Pectobacterium carotovorum*, and *P. atrosepticum* (Mansfield et al. 2012). The effects of biogenic metal oxide nanoparticles on some species of plant pathogenic bacteria are listed in Table 2.

Zinc oxide nanoparticles (ZnO NPs)

ZnO NPs were synthesized using plant extracts of olive leaves (*Olea europaea*), chamomile flower (*Matricaria chamomilla* L.), and red tomato fruit (*Lycopersicon esculentum* M.). At 16 μg mL^{-1}, all the biologically synthesized ZnO NPs significantly inhibited the bacterial growth, biofilm formation, swimming motility, and bacterial cell membrane of *Xanthomonas oryzae* pv. *Oryzae* (Ogunyemi et al. 2019a). Similarly, three metal oxide nanoparticles (ZnO, MnO_2, and MgO) were biologically synthesized using the cell-free supernatant of *Paenibacillus polymyxa* strain Sx3 as a reducing agent. All biogenic nanoparticles had significant inhibition effects on the growth and biofilm formation of *Xanthomonas oryzae* pv. *oryzae*, the causal agent of bacterial leaf blight disease in rice, at the concentration of 16.0 μg mL^{-1} (Ogunyemi et al. 2020c). In this context, zinc oxide nanoparticles were synthesized by *Bacillus cereus* RNT6 and significantly reduced the bacterial growth and biofilm formation of *Burkholderia glumae* and *B. gladioli*, causing bacterial panicle blight disease in rice, compared to the control treatment (Ahmed et al. 2021c).

The application of ZnO NPs, synthesized by *Matricaria chamomilla* flower extract, at the concentration of 8% on tomato plants inoculated with *Ralstonia solanacearum*, the pathogen of bacterial wilt disease, caused a great reduction in the disease severity and bacterial soil population as well as promoted the plant growth. *In vitro* conditions, ZnO NPs at the concentration of 18 μg mL^{-1} exhibited the maximum inhibition on the growth of the bacterial pathogen (Khan et al. 2021). In the same way, ZnO NPs prepared by *Withania coagulans* leaf extract showed a strong antibacterial effect against *R. solanacearum* but the combined application of leaf extract and ZnO nanoparticles gave the maximum inhibition of the pathogen both in laboratory and greenhouse conditions (Guo et al. 2022).

Table 2: Effect of biogenic metal oxide nanoparticles on plant pathogenic bacteria.

NP	Production source	Size (nm) and shape	Effective concentration	Pathogen	Host	Effect	References
ZnO	*Matricaria chamomilla* *Olea europaea* *Lycopersicon esculentum*	51.2 41 - Cubic 51.6	16 µg mL^{-1}	*Xanthomonas oryzae* pv. *oryzae*	Rice	Antibacterial activity Antibiofilm activity	Ogunyemi et al. (2019a)
ZnO TiO_2	*Citrus limon*	20–200 Polymorphic	50 µg mL^{-1}	*Dickeya dadantii*	Sweet potato	Antibacterial activity Antibiofilm activity	Hossain et al. (2019)
ZnO MgO MnO_2	*Paenibacillus polymyxa*	62.8 - Cubic 10.9 - Sheet 18.8 Spherical	16 µg mL^{-1}	*Xanthomonas oryzae* pv. *oryzae*	Rice	Antibacterial activity Antibiofilm activity	Ogunyemi et al. (2020c)
ZnO	*Bacillus cereus*	21–35 Spherical	50 µg mL^{-1}	*Burkholderia glumae* *B. gladioli*	Rice	Antibacterial activity Antibiofilm activity	Ahmed et al. (2021c)
ZnO	*Matricaria chamomilla*	8.9–32.6 Crystalline	18 µg mL^{-1} 8% - *in vivo*	*Ralstonia solanacearum*	Tomato	Reduce bacterial wilt disease severity	Khan et al. (2021)
ZnO	*Withania coagulans*	–	18 µg mL^{-1}	*Ralstonia solanacearum*	Tomato	Reduced bacterial wilt disease severity	Guo et al. (2022)
MgO MnO_2	*Matricaria chamomilla*	18.2 - Disc 16.5 Spherical	16 µg mL^{-1}	*Acidovorax oryzae*	Rice	Antibacterial activity Antibiofilm activity	Ogunyemi et al. (2019b)
MgO	*Rosmarinus officinalis*	< 20 Round	16 µg mL^{-1}	*Xanthomonas oryzae* pv. *oryzae*	Rice	Antibacterial activity Antibiofilm activity	Abdallah et al. (2019)
MgO	*Acinetobacter johnsonii*	18–45 Spherical	20 µg mL^{-1}	*Acidovorax oryzae*	Rice	Antibacterial activity	Ahmed et al. (2021a)
Fe_2O_3	*Skimmia laureola*	56–350 Cubic	6 mg mL^{-1} 6% - *in vivo*	*Ralstonia solanacearum*	Tomato	Reduced bacterial wilt disease severity	Alam et al. (2019)

Magnesium oxide nanoparticles (MgO NPs)

The MgO nanoparticles synthesized from *Rosmarinus officinalis* had an immense inhibitory effect on the growth, biofilm formation, and swimming of *Xanthomonas oryzae* pv. *oryzae*, the causal agent of bacterial blight disease in rice, after 48 h incubation (Abdallah et al. 2019). In addition, Ogunyemi et al. (2019b) described the antibacterial activity properties of magnesium oxide and manganese dioxide nanoparticles synthesized using chamomile flower extract against *Acidovorax oryzae*, the causal agent of the bacterial brown stripe of rice. MgO and MnO_2 nanoparticles at 16 μg mL^{-1} caused a remarkable reduction in bacterial pathogen growth by 62.9 and 71.3%, respectively, with a strong inhibition of the biofilm formation and swimming motility. In another study, Ahmed et al. (2021a) assessed the antibacterial activity of MgO NPs synthesized with *Acinetobacter johnsonii* strain RTN1 against rice pathogen *Acidovorax oryzae*. The maximum antibacterial effect of MgO NPs was observed at 20 μg mL^{-1} and significantly reduced the cell density of the pathogen in broth culture.

Titanium dioxide nanoparticles (TiO_2 NPs)

TiO_2 NPs and ZnO NPs were synthesized with a lemon fruit extract and investigated their antibacterial activity against *Dickeya dadantii*, which causes sweet potato stem and root rot disease. The two nanoparticles caused a great inhibition to a similar extent on the bacterial growth, swimming motility (60%), biofilm formation (64–66%), and maceration of sweet potato tuber slices (54–60%) at the concentrations of 50 μg mL^{-1} (Hossain et al. 2019).

Iron oxide nanoparticles (Fe_2O_3 NPs)

The leaf extract of *Skimmia laureola* was used to produce Fe_2O_3 NPs and evaluated their antibacterial activity against bacterial wilt pathogen *Ralstonia solanacearum in vitro* and *in planta*. These nanoparticles dramatically inhibited the bacterial growth *in vitro* at 6 mg mL^{-1}, reduced the disease severity with 6% w/v of Fe_2O_3 NPs treated root zone of tomato plants in the pot experiments and enhanced the plant shoots and root length, and fresh biomass (Alam et al. 2019).

Action mechanisms of biogenic metal oxide nanoparticles

The application of nanomaterials should achieve two vital goals of plant disease management - efficiency with minimal environmental impact and reduced toxicity to human health (Alghuthaymi et al. 2015). In general, the specific mechanism of action of nanoparticles is still somewhat unclear, but many proposed theories to explain these mechanisms have been reported (Lemire et al. 2013).

The bioactive compounds in the cellular extract of plants and microorganisms such as polyphenols, polysaccharides, polypeptides, proteins, carbohydrates, polymers, and vitamins act as reducing agents during the synthesis of biogenic metal oxide nanoparticles (Du et al. 2022). These molecules provide different functional groups to the surface of biogenic nanoparticles as well as a high degree of stability. The functional groups confer chemical bonds and active sites for physical interaction between nanoparticles and microbial cells (Xu et al. 2021).

The nanoparticles caused toxic effects on the microbial cell wall and membrane in the form of morphological alterations in the membrane structure by destroying cell membrane permeability and respiratory functions and thereby resulting in the leakage of the vital cellular components including proteins, enzymes, sugar, lipids, and nucleic acids and ultimately cell death (Hochvaldová et al. 2022). Besides, some nanoparticles lead to the formation of pits and gaps on the membrane surface of microorganisms by the release of ions that interact with disulfide or sulfhydryl groups of enzymes causing damage to the metabolic pathway and cell apoptosis (Shaikh et al. 2019). These irreversible damages on the surface of the cell wall can be observed by advanced imaging techniques such as scanning electron microscopy (SEM), transmission electron microscopy (TEM), and atomic force microscopy (AFM) (Liu et al. 2018). Therefore, this mechanism by adhesion of the nanoparticles with the cell wall is believed to be the first mechanism of antimicrobial activity (Godoy-Gallardo et al. 2021).

Using advanced chemical and biological techniques, the mechanism of action underlying the effect of biogenic nanoparticles on microorganisms might be related to the generation of reactive oxygen species (ROS) on the cell membrane surface (Patil et al. 2020, Hernández-Díaz et al. 2021). ROS are chemically reactive particles that contain oxygen, including singlet oxygen ($^1O_2$), superoxide anion ($O_2^{\bullet-}$), hydroxyl radical ($^\bullet OH$), and hydrogen peroxide (H_2O_2) (Yang et al. 2019). H_2O_2 is the major by-product of the interaction between nanoparticles and cells and works to destroy important biomolecules including proteins, lipids, and nucleic acids. The increase in the levels of lipid peroxidation (MDA content) is a key indicator of ROS generation (Yu et al. 2020).

In this context, the excess production of intracellular ROS is the main mechanism in the case of metal oxide nanoparticles via disturbing the electron transfer process, increasing the $NADP^+$/NADPH ratio, overlapping with mitochondrial function, and interfering with gene expression of oxidative stress (Wang et al. 2017, Jiang et al. 2021). The overproduction of ROS induced by biogenic nanoparticles leads to important physiological changes besides disrupting cellular structure, preventing the activity of biological macromolecular (e.g., protein, lipids, carbohydrates, enzyme), promoting DNA damage, and damage of the antioxidative system (Liao et al. 2019, Yang et al. 2019). Interestingly, the production of ROS is strongly correlated with the size, shape, concentration, and exposure period of biogenic nanoparticles (Huang et al. 2019). The enzymes such as glutathione dismutase, superoxide dismutase, and ascorbate peroxidase were upregulated or downregulated into biogenic nanoparticle treatments (Figure 2). Also, these nanoparticles can reduce adenosine triphosphate (ATP) generation, deplete glutathione, rupture lysosomes, and inhibit the ribosomal subunit from binding transfer RNA (tRNA) (Mamun et al. 2021).

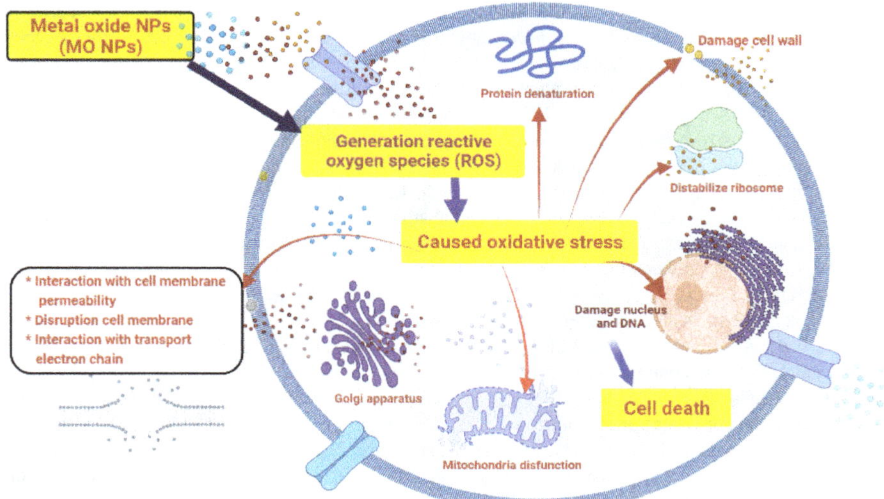

Figure 2: Action mechanisms of biogenic metal oxide nanoparticles.

Conclusions

Biosynthesis of metal oxide nanoparticles has become a promising and highly attractive area of research because of their voluminous applications, especially in the field of plant protection in economically important crops. Studies have shown that biogenic metal oxide nanoparticles possess distinct chemical and physical features compared to those prepared through other methods. Different types of natural extracts such as plants, fungi, bacteria, and algae have been employed for the synthesis of biogenic metal oxide nanoparticles without hazardous chemicals. In the longer term, new techniques depending on biosynthesized nanoparticles will be developed for the control of plant pathogens, plant disease diagnostics, and molecular biology research. Consequently, the application of novel approaches in nanobiotechnology would help in conquering the challenges to global food production.

References

Abboud, Y., Saffaj, T., Chagraoui, A., El Bouari, A., Brouzi, K., Tanane, O. and Ihssane, B. 2014. Biosynthesis, characterization and antimicrobial activity of copper oxide nanoparticles (CONPs) produced using brown alga extract (*Bifurcaria bifurcata*). Appl. Nanosci. 4: 571–576.

Abdallah, Y., Ogunyemi, S.O., Abdelazez, A., Zhang, M., Hong, X., Ibrahim, E., Hossain, A., Fouad, H. and Chen, J. 2019. The green synthesis of MgO nano-flowers using *Rosmarinus officinalis* L. (Rosemary) and the antibacterial activities against *Xanthomonas oryzae* pv. *oryzae*. BioMed. Res. Intern. 2019: 5620989.

Abdelaziz, A.M., Salem, S.S., Khalil, A.M.A., El-Wakil, D.A., Fouda, H.M. and Hashem, A.H. 2022. Potential of biosynthesized zinc oxide nanoparticles to control *Fusarium* wilt disease in eggplant (*Solanum melongena*) and promote plant growth. BioMetals. https://doi.org/10.1007/s10534-022-00391-8.

Abdel-Aziz, M.M., Emam, T.M. and Elsherbiny, E.A. 2020. Bioactivity of magnesium oxide nanoparticles synthesized from cell filtrate of endobacterium *Burkholderia rinojensis* against *Fusarium oxysporum*. Mater. Sci. Eng. C 109: 110617.

Abdo, A.M., Fouda, A., Eid, A.M., Fahmy, N.M., Elsayed, A.M., Khalil, A.M.A., Alzahrani, O.M., Ahmed, A.F. and Soliman, A.M. 2021. Green synthesis of zinc oxide nanoparticles (ZnO-NPs) by *Pseudomonas aeruginosa* and their activity against pathogenic microbes and common house mosquito, *Culex pipiens*. Materials 14: 6983.

Abel, S., Tesfaye, J.L., Shanmugam, R., Dwarampudi, L.P., Lamessa, G., Nagaprasad, N., Benti, M. and Krishnaraj, R. 2021. Green synthesis and characterizations of zinc oxide (ZnO) nanoparticles using aqueous leaf extracts of coffee (*Coffea arabica*) and its application in environmental toxicity reduction. J. Nanomater. 2021: 3413350.

Achari, G.A. and Kowshik, M. 2018. Recent developments on nanotechnology in agriculture: Plant mineral nutrition, health, and interactions with soil microflora. J. Agric. Food Chem. 66: 8647–8661.

Adebayo, E.A., Azeez, M.A., Alao, M.B., Oke, A.M. and Aina, D.A. 2021. Fungi as veritable tool in current advances in nanobiotechnology. Heliyon 7: e08480.

Agarwal, P., Gupta, R. and Agarwal, N. 2019. Advances in synthesis and applications of microalgal nanoparticles for wastewater treatment. J. Nanotechnol. 2019: 7392713.

Ağçeli, G.K., Hammachi, H., Kodal S.P., Cihangir, N. and Aksu, Z. 2020. A novel approach to synthesize TiO_2 nanoparticles: Biosynthesis by using *Streptomyces* sp. HC1. J. Inorg. Organomet. Polym. Mater. 30: 3221–3229.

Ahmad, W., Jaiswal, K.K. and Soni, S. 2020. Green synthesis of titanium dioxide (TiO_2) nanoparticles by using *Mentha arvensis* leaves extract and its antimicrobial properties. Inorg. Nano Metal. Chem. 50: 1032–1038.

Ahmed, T., Noman, M., Shahid, M., Shahid, M. and Li, B. 2021a. Antibacterial potential of green magnesium oxide nanoparticles against rice pathogen *Acidovorax oryzae*. Mater. Lett. 282: 128839.

Ahmed, T., Ren, H., Noman, M., Shahid, M., Liu, M., Ali, M.A., Zhang, J., Tian, Y., Qi, X. and Li, B. 2021b. Green synthesis and characterization of zirconium oxide nanoparticles by using a native *Enterobacter* sp. and its antifungal activity against bayberry twig blight disease pathogen *Pestalotiopsis versicolor*. NanoImpact 21: 100281.

Ahmed, T., Wu, Z., Jiang, H., Luo, J., Noman M., Shahid, M. and Li, B. 2021c. Bioinspired green synthesis of zinc oxide nanoparticles from a native *Bacillus cereus* strain RNT6: Characterization and antibacterial activity against rice panicle blight pathogens *Burkholderia glumae* and *B. gladioli*. Nanomaterials 11: 884.

Ajmal, N., Saraswat, K., Bakht, M.D., Riadi, Y., Ahsan, M.J. and Noushad, M. 2019. Cost-effective and eco-friendly synthesis of titanium dioxide (TiO_2) nanoparticles using fruit's peel agro-waste extracts: Characterization, *in vitro* antibacterial, antioxidant activities. Green Chem. Lett. Rev. 12: 244–254.

Akbar, M., Haroon, U., Ali, M., Tahir, K., Chaudhary, H.J. and Munis, M.F. 2022. Mycosynthesized Fe_2O_3 nanoparticles diminish brown rot of apple whilst maintaining composition and pertinent organoleptic properties. J. Appl. Microbiol. 132: 3735–3745.

Akinola, P.O., Lateef, A., Asafa, T.B., Beukes, L.S., Hakeem, A.S. and Irshad, H.M. 2020. Multifunctional titanium dioxide nanoparticles biofabricated via phytosynthetic route using extracts of *Cola nitida*: Antimicrobial, dye degradation, antioxidant and anticoagulant activities. Heliyon 6: e04610.

Alam, T., Khan, R.A.A., Ali, A., Sher, H., Ullah, Z. and Ali, M. 2019. Biogenic synthesis of iron oxide nanoparticles via *Skimmia laureola* and their antibacterial efficacy against bacterial wilt pathogen *Ralstonia solanacearum*. Mater. Sci. and Eng. C 98: 101–108.

Alavi, M. and Karimi, N. 2018. Characterization, antibacterial, total antioxidant, scavenging, reducing power and ion chelating activities of green synthesized silver, copper and titanium dioxide nanoparticles using *Artemisia haussknechtii* leaf extract. Artif. Cells Nanomed. Biotechnol. 46: 2066–2081.

Alghuthaymi, M.A., Almoammar, H., Rai, M., Said-Galiev, E. and Abd-Elsalam, K.A. 2015. Myconanoparticles: Synthesis and their role in phytopathogens management. Biotechnol. Biotechnol. Equip. 29: 221–236.

Alghuthaymi, M.A., Rajkuberan, C., Rajiv, P., Kalia, A., Bhardwaj, K., Bhardwaj, P., Abd-Elsalam, K.A., Valis, M. and Kuca, K. 2021. Nanohybrid antifungals for control of plant diseases: Current status and future perspectives. J. Fungi 7: 48.

Ali, M.A., Ahmed, T., Wu, W., Hossain, A., Hafeez, R., Islam Masum, M.M., Wang, Y., An, Q., Sun, G. and Li, B. 2020. Advancements in plant and microbe-based synthesis of metallic nanoparticles and their antimicrobial activity against plant pathogens. Nanomaterials 10: 1146.

Alsaggaf, M.S., Diab, A.M., ElSaied, B.E., Tayel, A.A. and Moussa, S.H. 2021. Application of ZnO nanoparticles phycosynthesized with *Ulva fasciata* extract for preserving peeled shrimp quality. Nanomaterials 11: 385.

Amin, F., Fozia, K.B., Alotaibi, A., Qasim, M., Ahmad, I., Ullah, R., Bourhia, M., Gul, A., Zahoor, S. and Ahmad, R. 2021. Green synthesis of copper oxide nanoparticles using *Aerva javanica* leaf extract and their characterization and investigation of *in vitro* antimicrobial potential and cytotoxic activities. Evid-Based Complement. Altern. Med. 2021: 5589703.

Anbumani, D., Dhandapani, K.V., Manoharan, J., Babujanarthanam, R., Bashir, A.K.H., Muthusamy, K., Alfarhan, A. and Kanimozhi, K. 2022. Green synthesis and antimicrobial efficacy of titanium dioxide nanoparticles using *Luffa acutangula* leaf extract. J. King Saud. Univ. Sci. 34: 101896.

Anicic, N., Vukomanovic, M., Koklic, T. and Suvorov, D. 2018. Fewer defects in the surface slows the hydrolysis rate, decreases the ROS generation potential, and improves the non-ROS antimicrobial activity of MgO. Small 14: 1800205.

Arciniegas-Grijalba, P.A., Patiño-Portela, M.C., Mosquera-Sánchez, L.P., Guerra Sierra, B.E., Muñoz-Florez, J.E., Erazo-Castillo, L.A. and Rodríguez-Páez, J.E. 2019. ZnO-based nanofungicides: Synthesis, characterization and their effect on the coffee fungi *Mycena citricolor* and *colletotrichum* sp. Mater. Sci. Eng. C 98: 808–825.

Ashajyothi, C., Harish, K.H., Dubey, N. and Chandrakanth, R.K. 2016. Antibiofilm activity of biogenic copper and zinc oxide nanoparticles antimicrobials collegiate against multiple drug resistant bacteria: A nanoscale approach. J. Nanostruct. Chem. 6: 329–341.

Ashraf, H., Anjum, T., Riaz, S., Ahmad, I.S., Irudayaraj, J., Javed, S., Qaiser, U. and Naseem, S. 2021. Inhibition mechanism of green-synthesized copper-oxide nanoparticles from *Cassia fistula* towards *Fusarium oxysporum* by boosting growth and defense responses in tomatoes. Environ. Sci. Nano 8: 1729–48.

Ashraf, H., Batool, T., Anjum, T., Illyas, A., Li, G., Naseem, S. and Riaz, S. 2022. Antifungal potential of green synthesized magnetite nanoparticles black coffee-magnetite nanoparticles against wilt infection by ameliorating enzymatic activity and gene expression in *Solanum lycopersicum* L. Front. Microbiol. 13: 7 54292.

Avila-Quezada, G.D., Golinska, P. and Rai, M. 2022. Engineered nanomaterials in plant diseases: Can we combat phytopathogens? Appl. Microbiol. Biotechnol. 106: 117–129.

Azizi, S., Ahmad, M.B., Namvar, F. and Mohamad, R. 2014. Green biosynthesis and characterization of zinc oxide nanoparticles using brown marine macroalga *Sargassum muticum* aqueous extract. Mater. Lett. 116: 275–277.

Balogun, S.W., James, O., Sanusi, Y. and Olayinka, O. 2020. Green synthesis and characterization of zinc oxide nanoparticles using bashful (*Mimosa pudica*), leaf extract: A precursor for organic electronics applications. SN App. Sci. 2: 504.

Batool, M., Khurshid, S., Daoush, W.M., Siddique, S.A. and Nadeem, T. 2021. Green synthesis and biomedical applications of ZnO nanoparticles: Role of PEGylated-ZnO nanoparticles as doxorubicin drug carrier against MDA-MB-231(TNBC) cells line. Crystals 11: 344.

Bhattacharya, P., Dey, A. and Neogi, S. 2021. An insight into the mechanism of antibacterial activity by magnesium oxide nanoparticles. J. Mater. Chem. B 9: 5329–5339.

Bukhari, A., Ijaz, I., Gilani, E., Nazir, A., Zain, H., Saeed, R., Alarfaji, S.S., Hussain, S., Aftab, R. and Naseer, Y. 2021. Green synthesis of metal and metal oxide nanoparticles using different plants' parts for antimicrobial activity and anticancer activity: A review article. Coatings 11: 1374.

Buttimer, C., McAuliffe, O., Ross, R.P., Hill, C., O'Mahony, J. and Coffey, A. 2017. Bacteriophages and bacterial plant diseases. Front. Microbiol. 8: 34.

Chavali, M.S. and Nikolova, M.P. 2019. Metal oxide nanoparticles and their applications in nanotechnology. SN Appl. Sci. 1: 607.

Chen, F., Xiao, Z., Yue, L., Wang, J., Feng, Y., Zhu, X., Wang, Z. and Xing, B. 2019. Algae response to engineered nanoparticles: Current understanding, mechanisms and implications. Environ. Sci.-Nano. 6: 1026–1042.

Chen, J., Wu, L., Lu, M., Lu, S., Li, Z. and Ding, W. 2020. Comparative study on the fungicidal activity of metallic MgO nanoparticles and macroscale MgO against soilborne fungal phytopathogens. Front. Microbiol. 11: 365.

Chennimalai, M., Vijayalakshmi, V., Senthil, T.S. and Sivakumar, N. 2021. One-step green synthesis of ZnO nanoparticles using *Opuntia humifusa* fruit extract and their antibacterial activities. Mater. Today: Proc. 47: 1842–1846.

Chinnaperumal, K., Govindasamy, B., Paramasivam, D., Dilipkumar, A., Dhayalan, A., Vadivel, A., Sengodan, K. and Pachiappan, P. 2018. Bio-pesticidal effects of *Trichoderma viride* formulated titanium dioxide nanoparticle and their physiological and biochemical changes on *Helicoverpa armigera* (Hub.). Pestic. Biochem. Physiol. 149: 26–36.

Consolo, V.F., Torres-Nicolini, A. and Alvarez, V.A. 2020. Mycosinthetized Ag, CuO and ZnO nanoparticles from a promising *Trichoderma harzianum* strain and their antifungal potential against important phytopathogens. Sci. Rep. 10: 20499.

Cuong, H.N., Pansambal, S., Ghotekar, S., Oza, R., Thanh Hai, N.T., Viet, N.M. and Nguyen, V.H. 2022. New frontiers in the plant extract mediated biosynthesis of copper oxide (CuO) nanoparticles and their potential applications: A review. Environ. Res. 203: 111858.

Czyżowska, A. and Barbasz, A. 2020. A review: Zinc oxide nanoparticles-friends or enemies? Int. J. Environ. Health Res. 32: 885–901.

Das, K.R., Kowshik, M., Kumar, M.P., Kerkar, S., Shyama, S.K. and Mishra, S. 2018. Native hypersaline sulphate reducing bacteria contributes to iron nanoparticle formation in saltpan sediment: A concern for aquaculture. J. Environ. Manage. 206: 556–564.

Dean, R., Van Kan, J.A., Pretorius, Z.A., Hammond-Kosack, K.E., Di Pietro, A., Spanu, P.D., Rudd, J.J., Dickman, M., Kahmann, R., Ellis, J. and Foster, G.D. 2012. The Top 10 fungal pathogens in molecular plant pathology. Mol. Plant Pathol. 13: 414–430.

Doehlemann, G., Ökmen, B., Zhu, W. and Sharon, A. 2017. Plant pathogenic fungi. Microbiol. Spectr. 5: 1–26.

Du, Z., Zhang, Y., Xu, A., Pan, S. and Zhang, Y. 2022. Biogenic metal nanoparticles with microbes and their applications in water treatment: A review. Environ. Sci. Pollut. Res. 29: 3213–3229.

Dulta, K., Agceli, G.K., Chauhan, P., Jasrotia, R. and Chauhan, P.K. 2022. Ecofriendly synthesis of zinc oxide nanoparticles by *Carica papaya* leaf extract and their applications. J. Clust. Sci. 33: 603–617.

El Sayed, M.T. and El-Sayed, A.S.A. 2020. Biocidal activity of metal nanoparticles synthesized by *Fusarium solani* against multidrug-resistant bacteria and mycotoxigenic fungi. J. Microbiol. Biotechnol. 30: 226–236.

El-Belely, E.F., Farag, M.M.S., Said, H.A., Amin, A.S., Azab, E., Gobouri, A.A. and Fouda, A. 2021. Green synthesis of zinc oxide nanoparticles (ZnO-NPs) using *Arthrospira platensis* (Class: Cyanophyceae) and evaluation of their biomedical activities. Nanomaterials 11: 95.

El-Kassas, H.Y., Aly-Eldeen, M.A. and Gharib, S.M. 2016. Green synthesis of iron oxide (Fe_3O_4) nanoparticles using two selected brown seaweeds: Characterization and application for lead bioremediation. Acta Oceanol. Sin. 35: 89–98.

El-Sheekh, M.M., El-Kassas, H.Y., El-Din, N.G.S., Eissa, D.I. and El-Sherbiny, B.A. 2021. Green synthesis, characterization applications of iron oxide nanoparticles for antialgal and wastewater bioremediation using three brown algae. Int. J. Phytoremediation 23: 1538–1552.

Elsherbiny, E.A., Taher, M.A., Abd El-Aziz, M.H. and Mohamed, S.Y. 2021. Action mechanisms and biocontrol of *Purpureocillium lilacinum* against green mold caused by *Penicillium digitatum* in orange fruit. J. Appl. Microbiol. 131: 1378–1390.

Fariq, A., Khan, T. and Yasmin, A. 2017. Microbial synthesis of nanoparticles and their potential applications in biomedicine. J. Appl. Biomed. 15: 241–248.

Fatemi, M., Mollania, N., Monemi-Moghaddam, M. and Sadeghifar, F. 2018. Extracellular biosynthesis of magnetic iron oxide nanoparticles by *Bacillus cereus* strain HMH1: Characterization and *in vitro* cytotoxicity analysis on MCF-7 and 3T3 cell lines. J. Biotechnol. 270: 1–11.

Fu, L., Wang, Z., Dhankher, O.P. and Xing, B. 2020. Nanotechnology as a new sustainable approach for controlling crop diseases and increasing agricultural production. J. Exp. Bot. 71: 507–519.

Gahlawat, G. and Choudhury, A.R. 2019. A review on the biosynthesis of metal and metal salt nanoparticles by microbes. RSC Adv. 9: 12944–12967.

Ganesan, V., Hariram, M., Vivekanandhan, S. and Muthuramkumar, S. 2020. *Periconium* sp. (endophytic fungi) extract mediated sol-gel synthesis of ZnO nanoparticles for antimicrobial and antioxidant applications. Mater. Sci. Semiconductor Process 105: 104739.

Gharpure, S., Yadwade, R. and Ankamwar, B. 2022. Non-antimicrobial and non-anticancer properties of ZnO nanoparticles biosynthesized using different plant parts of *Bixa orellana*. ACS Omega 7: 1914–1933.

Godoy-Gallardo, M., Eckhard, U., Delgado, L.M., de Roo Puente, Y.J.D., Hoyos-Nogues, M., Gil, F.J. and Perez, R.A. 2021. Antibacterial approaches in tissue engineering using metal ions and nanoparticles: From mechanisms to applications. Bioact. Mater. 6: 4470–4490.

Goel, S., Chen, F. and Cai, W. 2014. Synthesis and biomedical applications of copper sulfide nanoparticles: From sensors to theranostics. Small 10: 631–645.

Gour, A. and Jain, N.K. 2019. Advances in green synthesis of nanoparticles. Artif. Cells Nanomed. Biotechnol. 47: 844–851.

Guo, Y., Khan, R.A.A., Xiong, Y. and Fan, Z. 2022. Enhanced suppression of soil-borne phytopathogenic bacteria *Ralstonia solanacearum* in soil and promotion of tomato plant growth by synergetic effect of green synthesized nanoparticles and plant extract. J. Appl. Microbiol. 132: 3694–3704.

Gupta, K. and Chundawat, T.S. 2020. Zinc oxide nanoparticles synthesized using *Fusarium oxysporum* to enhance bioethanol production from rice-straw. Biomass Bioenergy 143: 105840.

Gur, T., Meydan, I., Seckin, H., Bekmezci, M. and Sen, F. 2022. Green synthesis, characterization and bioactivity of biogenic zinc oxide nanoparticles. Environ. Res. 204: 111897.

Hassan, S.E-D., Fouda, A., Radwan, A.A., Salem, S.S., Barghoth, M.G., Awad, M.A., Abdo, A.M. and El-Gamal, M.S. 2019. Endophytic actinomycetes *Streptomyces* spp. mediated biosynthesis of copper oxide nanoparticles as a promising tool for biotechnological applications. JBIC J. Biol. Inorg. Chem. 24: 377–393.

Hassan, S.E-D., Fouda, A., Saied, E., Farag, M.M.S., Eid, A.M., Barghoth, M.G., Awad, M.A., Hamza, M.F. and Awad, M.F. 2021. *Rhizopus oryzae*-mediated green synthesis of magnesium oxide nanoparticles (MgO-NPs): A promising tool for antimicrobial, mosquitocidal action, and tanning effluent treatment. J. Fungi 7: 372.

Hernández-Díaz, J.A., Garza-García, J.J., Zamudio-Ojeda, A., León-Morales, J.M., López-Velázquez, J.C. and García-Morales, S. 2021. Plant-mediated synthesis of nanoparticles and their antimicrobial activity against phytopathogens. J. Sci. Food Agric. 101: 1270–1287.

Hochvaldová, L., Večeřová, R., Kolář, M., Prucek, R., Kvítek, L., Lapčík, L. and Panáček, A. 2022. Antibacterial nanomaterials: Upcoming hope to overcome antibiotic resistance crisis. Nanotechnol. Rev. 11: 1115–1142.

Hossain, A., Abdallah, Y., Ali, M.A., Masum, M.M.I., Li, B., Sun, G., Meng, Y., Wang, Y. and An, Q. 2019. Lemon-fruit-based green synthesis of zinc oxide nanoparticles and titanium dioxide nanoparticles against soft rot bacterial pathogen *Dickeya dadantii*. Biomolecules 9: 863.

Huang, T., Holden, J.A., Heath, D.E., O'Brien-Simpson, N.M. and O'Connor, A.J. 2019. Engineering highly effective antimicrobial selenium nanoparticles through control of particle size. Nanoscale 11: 14937–14951.

Huang, W., Fang, H., Zhang, S. and Yu, H. 2021. Optimised green synthesis of copper oxide nanoparticles and their antifungal activity. Micro Nano Lett. 16: 374–380.

Ilyas, M., Waris, A., Khan, A.U., Zamel, D., Yar, L., Baset, A., Muhaymin, A., Khan, S., Ali, A. and Ahmad, A. 2021. Biological synthesis of titanium dioxide nanoparticles from plants and microorganisms and their potential biomedical applications. Inorg. Chem. Commun. 133: 108968.

Islam, F., Shohag, S., Uddin, M.J., Islam, M.R., Nafady, M.H., Akter, A., Mitra, S., Roy, A., Emran, T.B. and Cavalu, S. 2022. Exploring the journey of zinc oxide nanoparticles (ZnO-NPs) toward biomedical applications. Materials 15: 2160.

Jeevanandam, J., Chan, Y.S. and Danquah, M.K. 2016. Biosynthesis of metal and metal oxide nanoparticles. ChemBioEng. Rev. 3: 55–67.

Jeevanandam, J., Kiew, S.F., Boakye-Ansah, S., Lau, S.Y., Barhoum, A., Danquah, M.K. and Rodrigues, J. 2022. Green approaches for the synthesis of metal and metal oxide nanoparticles using microbial and plant extracts. Nanoscale 14: 2534–2571.

Jiang, H., Lin, Q., Yu, Z., Wang, C. and Zhang, R. 2021. Nanotechnologies for reactive oxygen species"turn-on" detection. Front. Bioeng. Biotechnol. 9: 780032.

Joshi, N.C., Chaudhary, N. and Rai, N. 2021. Medicinal plant leaves extract based synthesis, characterisations and antimicrobial activities of ZrO_2 nanoparticles (ZrO_2 NPs). BioNanoScience 11: 497505.

Kadam, V.V., Ettiyappan, J.P. and Balakrishnan, R.M. 2019. Mechanistic insight into the endophytic fungus mediated synthesis of protein capped ZnO nanoparticles. Mater. Sci. Eng. B 243: 214–221.

Kangralkar, M.V., Manjanna, J., Momin, N., Rane, K.S., Nayaka, G.P. and Kangralkar, V.A. 2021. Photocatalytic degradation of hexavalent chromium and different staining dyes by ZnO in aqueous medium under UV light. Environ. Nanotechnol. Monit. Manag. 16: 100508.

Kaur, H., Kaur, S., Singh, J., Rawat, M. and Kumar, S. 2019. Expanding horizon: Green synthesis of TiO_2 nanoparticles using *Carica papaya* leaves for photocatalysis application. Mater. Res. Express 6: 095034.

Khan, M.R., Siddiqui, Z.A. and Fang, X. 2022. Potential of metal and metal oxide nanoparticles in plant disease diagnostics and management: Recent advances and challenges. Chemosphere 297: 134114.

Khan, R. and Fulekar, M.H. 2016. Biosynthesis of titanium dioxide nanoparticles using *Bacillus amyloliquefaciens* culture and enhancement of its photocatalytic activity for the degradation of a sulfonated textile dye Reactive Red 31. J. Colloid Interface Sci. 475: 184–191.

Khan, R.A.A., Tang, Y., Naz, I., Alam, S.S., Wang, W., Ahmad, M., Najeeb, S., Rao, C., Li, Y., Xie, B. and Li, Y. 2021. Management of *Ralstonia solanacearum* in tomato using ZnO nanoparticles synthesized through *Matricaria chamomilla*. Plant Dis. 105: 3224–3230.

Kołodziejczak-Radzimska, A. and Jesionowski, T. 2014. Zinc oxide—from synthesis to application: A review. Materials 7: 2833–2881.

Koul, B., Poonia, A.K., Yadav, D. and Jin, J-O. 2021. Microbe-mediated biosynthesis of nanoparticles: Applications and future prospects. Biomolecules 11: 886.

Król, A., Railean-Plugaru, V., Pomastowski, P., Złoch, M. and Buszewski, B. 2018. Mechanism study of intracellular zinc oxide nanocomposites formation. Colloids Surf. A: Physicochemical Eng. Aspects 553: 349–358.

Kumar, A., Choudhary, A., Kaur, H., Guha, S., Mehta, S. and Husen, A. 2022. Potential applications of engineered nanoparticles in plant disease management: A critical update. Chemosphere 295: 133798.

Kumaresan, M., Anand, K.V., Govindaraju, K., Tamilselvan, S. and Kumar, V.G. 2018. Seaweed *Sargassum wightii* mediated preparation of zirconia (ZrO_2) nanoparticles and their antibacterial activity against gram positive and gram negative bacteria. Microb. Pathog. 124: 311–315.

Kutawa, A.B., Ahmad, K., Ali, A., Hussein, M.Z., Wahab, M.A.A., Adamu, A., Ismaila, A.A., Gunasena, M.T., Rahman, M.Z. and Hossain, I. 2021. Trends in nanotechnology and its potentialities to control plant pathogenic fungi: A review. Biology 10: 881.

Lakshmeesha, T.R., Kalagatur, N.K., Mudili, V., Mohan, C.D., Rangappa, S., Prasad, B.D., Ashwini, B.S., Hashem, A., Alqarawi, A.A. and Malik, J.A. 2019. Biofabrication of zinc oxide nanoparticles with *Syzygium aromaticum* flower buds extract and finding its novel application in controlling the growth and mycotoxins of *Fusarium graminearum*. Front. Microbiol. 10: 1244.

Lemire, J.A., Harrison, J.J. and Turner, R.J. 2013. Antimicrobial activity of metals: Mechanisms, molecular targets and applications. Nat. Rev. Microbiol. 11: 371–384.

Liao, F., Chen, L., Liu, Y., Zhao, D., Peng, W., Wang, W. and Feng, S. 2019. The size-dependent genotoxic potentials of titanium dioxide nanoparticles to endothelial cells. Environ. Toxicol. 34: 1199–1207.

Lim, E-K., Kim, T., Paik, S., Haam, S., Huh, Y.-M. and Lee, K. 2015. Nanomaterials for theranostics: Recent advances and future challenges. Chem. Rev. 115: 327–394.

Liu, Y., Wang, Y.-M., Sedano, S., Jiang, Q., Duan, Y., Shen, W., Jiang, J-H. and Zhong, W. 2018. Encapsulation of ionic nanoparticles produces reactive oxygen species (ROS)-responsive microgel useful for molecular detection. Chem. Commun. 54: 4329–4332.

Mamun, M.M., Sorinolu, A.J., Munir, M. and Vejerano, E.P. 2021. Nanoantibiotics: Functions and properties at the nanoscale to combat antibiotic resistance. Front. Chem. 9: 687660.

Mansfield, J., Genin, S., Magori, S., Citovsky, V., Sriariyanum, M., Ronald, P., Dow, M., Verdier, V., Beer, S.V., Machado, M.A., Toth, I., Salmond, G. and Foster, G.D. 2012. Top 10 plant pathogenic bacteria in molecular plant pathology. Mol. Plant Pathol. 13: 614–29.

Marouzi, S., Sabouri, Z. and Darroudi, M. 2021. Greener synthesis and medical applications of metal oxide nanoparticles. Ceram. Int. 47: 19632–19650.

Martins, P.M.M., Merfa, M.V., Takita, M.A. and De Souza, A.A. 2018. Persistence in phytopathogenic bacteria: do we know enough? Front. Microbiol. 9: 1099.

Mashjoor, S., Yousefzadi, M., Zolgharnain, H., Kamrani, E. and Alishahi, M. 2018. Organic and inorganic nano-Fe_3O_4: alga *Ulva flexuosa*-based synthesis, antimicrobial effects and acute toxicity to briny water rotifer *Brachionus rotundiformis*. Environ. Pollut. 237: 50–64.

Mathew, S.S., Sunny, N.E. and Shanmugam, V. 2021. Green synthesis of anatase titanium dioxide nanoparticles using *Cuminum cyminum* seed extract; effect on mung bean (*Vigna radiata*) seed germination. Inorg. Chem. Commun. 126: 108485.

Meghana, S., Kabra, P., Chakraborty, S. and Padmavathy, N. 2015. Understanding the pathway of antibacterial activity of copper oxide nanoparticles. RSC Adv. 5: 12293–12299.

Mishra, S., Keswani, C., Abhilash, P.C., Fraceto, L.F. and Singh, H.B. 2017. Integrated approach of agri-nanotechnology: Challenges and future trends. Front. Plant Sci. 8: 471.

Mohamed, A.A., Abu-Elghait, M., Ahmed, N.E. and Salem, S.S. 2021. Eco-friendly mycogenic synthesis of ZnO and CuO nanoparticles for *in vitro* antibacterial, antibiofilm, and antifungal applications. Biol. Trace Elem. Res. 199: 2788–2799.

Mohamed, A.A., Fouda, A., Abdel-Rahman, M.A., Hassan, S.E-D., El-Gamal, M.S., Salem, S.S. and Shaheen, T.I. 2019. Fungal strain impacts the shape, bioactivity and multifunctional properties of green synthesized zinc oxide nanoparticles. Biocatal. Agric. Biotechnol. 19: 101103.

Mughal, B., Zaidi, S.Z.J., Zhang, X. and Hassan, S.U. 2021. Biogenic nanoparticles: Synthesis, characterisation and applications. Appl. Sci. 11: 2598.

Nair, G.M., Sajini, T. and Mathew. B. 2022. Advanced green approaches for metal and metal oxide nanoparticles synthesis and their environmental applications. Talanta Open 5: 100080.

Nasrollahzadeh, M., Sajadi, S.M., Rostami-Vartooni, A. and Hussin, S.M. 2016. Green synthesis of CuO nanoparticles using aqueous extract of *Thymus vulgaris* L. leaves and their catalytic performance for N-arylation of indoles and amines. J. Coll. Interface Sci. 466: 113–119.

Nazir, A., Akbar, A., Baghdadi, H.B., Al-Abbad, E., Fatima, M., Iqbal, M., Tamam, N., Alwadai, N., Abbas, M. and ur Rehman, S. 2021. Zinc oxide nanoparticles fabrication using *Eriobotrya japonica* leaves extract: Photocatalytic performance and antibacterial activity evaluation. Arab. J. Chem. 14: 103251.

Nikolova, M.P. and Chavali, M.S. 2020. Metal oxide nanoparticles as biomedical materials. Biomimetics 5: 27.

Ogunyemi, S.O., Abdallah, Y., Zhang, M., Fouad, H., Hong, X., Ibrahim, E., Masum, M.M.I., Hossain, A., Mo, J. and Li, B. 2019a. Green synthesis of zinc oxide nanoparticles using different plant extracts and their antibacterial activity against *Xanthomonas oryzae* pv. *oryzae*. Artif. Cells Nanomed. Biotechnol. 47: 341–352.

Ogunyemi, S.O., Zhang, F., Abdallah, Y., Zhang, M., Wang, Y., Sun, G., Qiu, W. and Li, B. 2019b. Biosynthesis and characterization of magnesium oxide and manganese dioxide nanoparticles using *Matricaria chamomilla* L. extract and its inhibitory effect on *Acidovorax oryzae* strain RS-2. Artif. Cells Nanomed. Biotechnol. 47: 2230–2239.

Ogunyemi, S.O., Zhang, M., Abdallah, Y., Ahmed, T., Qiu, W., Ali, M. and Li, B. 2020c. The bio-synthesis of three metal oxide nanoparticles (ZnO, MnO_2, and MgO) and their antibacterial activity against the bacterial leaf blight pathogen. Front. Microbiol. 11: 3099.

Osuntokun, J., Onwudiwe, D.C. and Ebenso, E.E. 2019. Green synthesis of ZnO nanoparticles using aqueous *Brassica oleracea* L. var. *italica* and the photocatalytic activity. Green Chem. Lett. Rev. 12: 444–457.

Parthasarathy, R., Ramachandran, R., Kamaraj, Y. and Dhayalan, S. 2022. Zinc oxide nanoparticles synthesized by *Bacillus cereus* PMSS-1 induces oxidative stress-mediated apoptosis via modulating apoptotic proteins in human melanoma A375 cells. J. Clust. Sci. 33: 17–28.

Patil, S. and Chandrasekaran, R. 2020. Biogenic nanoparticles: A comprehensive perspective in synthesis, characterization, application and its challenges. J. Genet. Eng. Biotechnol. 18: 67.

Peng, Y., Li, S.J., Yan, J., Tang, Y., Cheng, J.P., Gao, A.J., Yao, X., Ruan, J.J. and Xu, B.L. 2021. Research progress on phytopathogenic fungi and their role as biocontrol agents. Front. Microbiol. 12: 670135.

Priya Naveen, Kaur, K. and Sidhu, A.K. 2021. Green synthesis: An eco-friendly route for the synthesis of iron oxide nanoparticles. Front. Nanotechnol. 3: 655062.

Raj, S.N., Anooj, E., Rajendran, K. and Vallinayagam, S.A. 2021. Comprehensive review on regulatory invention of nano pesticides in agricultural nano formulation and food system. J. Mol. Struct. 1239: 130517.

Rajwade, J.M., Chikte, R.G. and Paknikar, K.M. 2020. Nanomaterials: new weapons in a crusade against phytopathogens. Appl. Microbiol. Biotechnol. 104: 1437–1461.

Raliya, R., Biswas, P. and Tarafdar, J.C. 2015. TiO_2 nanoparticle biosynthesis and its physiological effect on mung bean (*Vigna radiata* L.). Biotechnol. Rep. 5: 22–26.

Raliya, R., Tarafdar, J.C. and Biswas, P. 2016. Enhancing the mobilization of native phosphorus in the mung bean rhizosphere using ZnO nanoparticles synthesized by soil fungi. J. Agric. Food Chem. 64: 3111–3118.

Rana, A., Yadav, K. and Jagadevan, S. 2020. A comprehensive review on green synthesis of nature-inspired metal nanoparticles: Mechanism, application and toxicity. J. Clean Prod. 272: 122880.

Rehman, S., Jermy, R., Asiri, S.M., Shah, M.A., Farooq, R., Ravinayagam, V., Ansari, M.A., Alsalem, Z., Al Jindan, R., Reshi, Z. and Khan, F.A. 2020. Using *Fomitopsis pinicola* for bioinspired synthesis of titanium dioxide and silver nanoparticles, targeting biomedical applications. RSC Adv. 10: 32137–32147.

Resmi, R., Yoonus, J. and Beena, B. 2021. A novel greener synthesis of ZnO nanoparticles from *Nilgiriantusciliantus* leaf extract and evaluation of its biomedical applications. Mater. Today: Proc. 46: 062–3068.

Ristaino, B.J., Pamela, K., Anderson, P.K., Bebber, D.P., Kate, A. and Brauman, K.A. 2021. The persistent threat of emerging plant disease pandemics to global food security. Proc. Natl. Acad. Sci. USA 118: e2022239118.

Sabzevari, S. and Hofman, J. 2022. A worldwide review of currently used pesticides' monitoring in agricultural soils. Sci. Total Environ. 812: 152344.

Salem, D.M., Ismail, M.M. and Tadros, H.R. 2020. Evaluation of the antibiofilm activity of three seaweed species and their biosynthesized iron oxide nanoparticles (Fe_3O_4-NPs). Egypt. J. Aquat. Res. 46: 333–339.

Salem, D.M.S.A., Ismail, M.M. and Aly-Eldeen, M.A. 2019. Biogenic synthesis and antimicrobial potency of iron oxide (Fe_3O_4) nanoparticles using algae harvested from the Mediterranean Sea, Egypt. Egypt. J. Aquat. Res. 45: 197–204.

Saravanakumar, K., Shanmugam, S., Varukattu, N.B., MubarakAli, D., Kathiresan, K. and Wang, M-H. 2019. Biosynthesis and characterization of copper oxide nanoparticles from indigenous fungi and its effect of photothermolysis on human lung carcinoma. J. Photochem. Photobiol. B 190: 103–109.

Sardar, M., Ahmed, W., Al Ayoubi, S., Nisa, S., Bibi, Y., Sabir, M., Khan, M.M., Ahmed, W. and Qayyum, A. 2022. Fungicidal synergistic effect of biogenically synthesized zinc oxide and copper oxide nanoparticles against *Alternaria citri* causing citrus black rot disease. Saudi J. Biol. Sci. 29: 88–95.

Sarkar, J., Mollick, M.M.R., Chattopadhyay, D. and Acharya, K. 2017. An eco-friendly route of γ-Fe_2O_3 nanoparticles formation and investigation of the mechanical properties of the HPMC-γ- Fe_2O_3 nanocomposites. Bioprocess Biosyst. Eng. 40: 351–359.

Sathiyavimal, S., Veeramani, S.V., Saravanan, M., Rajalakshmi, G., Kaliannan, T., Al-Misned, F.A. and Pugazhendhi, A. 2021. Green chemistry route of biosynthesized copper oxide nanoparticles using

Psidium guajava leaf extract and their antibacterial activity and effective removal of industrial dyes. J. Environ. Chem. Eng. 9: 105033.

Satti, S.H., Raja, N.I., Javed, B., Akram, A., Mashwani, Z.-u.-R., Ahmad, M.S. and Ikram, M. 2021. Titanium dioxide nanoparticles elicited agro-morphological and physicochemical modifications in wheat plants to control *Bipolaris sorokiniana*. PLoS One 16: e0246880.

Savary, S., McRoberts, N., Esker, P.D., Willocquet, L. and Teng, P.S. 2017. Production situations as drivers of crop health: Evidence and implications. Plant Pathol. 66: 867–876.

Savary, S., Willocquet, L., Pethybridge, S.J., Esker, P., McRoberts, N. and Nelson, A. 2019. The global burden of pathogens and pests on major food crops. Nat. Ecol. Evol. 3: 430–439.

Selvam, K., Albasher, G., Alamri, O., Sudhakar, C., Selvankumar, T., Vijayalakshmi, S. and Vennila, L. 2022. Enhanced photocatalytic activity of novel *Canthium coromandelicum* leaves based copper oxide nanoparticles for the degradation of textile dyes. Environ. Res. 211: 113046.

Shaikh, S., Nazam, N., Rizvi, S.M.D., Ahmad, K., Baig, M.H., Lee, E.J. and Choi, I. 2019. Mechanistic insights into the antimicrobial actions of metallic nanoparticles and their implications for multidrug resistance. Int. J. Mol. Sci. 20: 2468.

Sharma, A., Sharma, S., Sharma, K., Chetri, S.P.K., Vashishtha, A., Singh, P., Kumar, R., Rathi, B. and Agrawal, V. 2016. Algae as crucial organisms in advancing nanotechnology: A systematic review. J. Appl. Phycol. 28: 1759–1774.

Sharma, D., Kanchi, S. and Bisetty, K. 2019. Biogenic synthesis of nanoparticles: A review. Arabian J. Chem. 12: 3576–3600.

Singh, P., Singh, K.R., Singh, J., Das, S.N. and Singh, R.P. 2021. Tunable electrochemistry and efficient antibacterial activity of plant-mediated copper oxide nanoparticles synthesized by *Annona squamosa* seed extract for agricultural utility. RSC Adv. 11: 18050–18060.

Strange, R.N. and Scott, P.R. 2005. Plant disease: A threat to global food security. Ann. Rev. Phytopathol. 43: 83–116.

Suarez, O.J. and León-Molina, H.B. 2022. Theoretical and experimental approach to the production of ZnO nanoparticles by controlled precipitation method in methanol. ChemistrySelect 7: e20210345.

Suriyaraj, S.P., Ramadoss, G., Chandraraj, K. and Selvakumar, R. 2019. One pot facile green synthesis of crystalline bio-ZrO_2 nanoparticles using *Acinetobacter* sp. KCSI1 under room temperature. Mater. Sci. Eng. C 105: 110021.

Tarkowski, P. and Vereecke, D. 2014. Threats and opportunities of plant pathogenic bacteria. Biotechnol. Adv. 32: 215–229.

Thakur, B., Kumar, A. and Kumar, D. 2019. Green synthesis of titanium dioxide nanoparticles using *Azadirachta indica* leaf extract and evaluation of their antibacterial activity. South Afr. J. Bot. 124: 223–227.

Verma, N. and Kumar, N. 2019. Synthesis and biomedical applications of copper oxide nanoparticles: An expanding horizon. ACS Biomater. Sci. Eng. 5: 1170–1188.

Vidovix, T.B., Quesada, H.B., Januário, E.F.D., Bergamasco, R. and Vieira, A.M.S. 2019. Green synthesis of copper oxide nanoparticles using *Punica granatum* leaf extract applied to the removal of methylene blue. Mater. Lett. 257: 126685.

Vijayanandan, A.S. and Balakrishnan, R.M. 2018. Biosynthesis of cobalt oxide nanoparticles using endophytic fungus *Aspergillus nidulans*. J. Environ. Manage. 218: 442–450.

Wang, G., Jin, W., Qasim, A.M., Gao, A., Peng, X., Li, W., Feng, H. and Chu, P.K. 2017. Antibacterial effects of titanium embedded with silver nanoparticles based on electron-transfer-induced reactive oxygen species. Biomaterials 124: 25–34.

Worrall, E.A., Hamid, A., Mody, K.T., Mitter, N. and Pappu, H.R. 2018. Nanotechnology for plant disease management. Agronomy 8: 285.

Xu, L., Zhu, Z. and Sun, D.-W. 2021. Bioinspired nanomodification strategies: moving from chemical-based agrosystems to sustainable agriculture. ACS Nano 15: 12655–12686.

Yang, B., Chen, Y. and Shi, J. 2019. Reactive oxygen species (ROS)-based nanomedicine. Chem. Rev. 119: 4881–4985.

Yu, Z., Li, Q., Wang, J., Yu, Y., Wang, Y., Zhou, Q. and Li, P. 2020. Reactive oxygen species-related nanoparticle toxicity in the biomedical field. Nanoscale Res. Lett. 15: 115.

Yusof, H.M., Mohamad, R., Zaidan, U.H. and Rahman, N.A. 2020. Sustainable microbial cell nanofactory for zinc oxide nanoparticles production by zinc-tolerant probiotic *Lactobacillus plantarum* strain TA4. Microb. Cell Factories 19: 10.

Zaki, S.A., Ouf, S.A., Albarakaty, F.M., Habeb, M.M., Aly, A.A. and Abd-Elsalam, K.A. 2021. *Trichoderma harzianum*-mediated ZnO nanoparticles: Aa green tool for controlling soil-borne pathogens in cotton. J. Fungi 7: 952.

Zampardi, G., Thöming, J., Naatz, H., Amin, H.M., Pokhrel, S., Mädler, L. and Compton, R.G. 2018. Electrochemical behavior of single CuO nanoparticles: Implications for the assessment of their environmental fate. Small 14: 1801765.

Zhu, W., Hu, C., Ren, Y., Lu, Y., Song, Y., Ji, Y., Han, C. and He, J. 2021. Green synthesis of zinc oxide nanoparticles using *Cinnamomum camphora* (L.) Presl leaf extracts and its antifungal activity. J. Environ. Chem. Eng. 9: 106659.

Chapter 8

Plant Virus Disease Management by Nanobiotechnology

Taswar Ahsan,[1,*] *Adnan Khalil*,[2] *Ansar Javeed*[3] and *Esha Basharat*[4]

Introduction

Currently, research indicates that worldwide interest in food production shall double by 2050 (Tilman et al. 2011). Climate change is disturbingly predicted to interrupt food production in many agriculturally delicate regions by increasing dry spell occurrences and raising normal day-to-day temperatures (Elmer and White 2018). The irritants and pathogens found on plants cause 20% to 40% of yields to be lost annually (Worrall et al. 2018). The pathogens caused plant diseases that seriously impacted global rural economies. It is their ability to transform themselves frequently through genetic recombination to overcome the terrible conditions that cause such a great deal of devastation. By comparing and methodically infecting, both susceptible and tolerant plants are tainted; however, tolerant plants are able to fight infection spread and reduce viral replication (Sharma and Prasad 2017). The difficulties that plant pathologists and agriculture scientists face are enormous (Ahsan and Yuanhua 2021). A majority of the world's crops have been destroyed by plant pathogens, specifically viruses. In terms of efficiency and environment safety, genetically modified plants and chemicals do not deliver the results advertised (Elmer and White 2018). At the moment, chemical pesticides remain the most effective means of controlling crop diseases. Unfortunately, these pesticides are usually harmful, whether they affect humans or the environment directly or indirectly (Khamis et al. 2017). Despite the

[1] Department of Resources and Environmental Microbiology, Shenyang Agricultural University, China.
[2] Department of Physics, Khawaja Fareed University of Science and Technology, Raheem Yar Khan, Pakistan.
[3] College of Bioscience and Biotechnology, Shenyang Agricultural University, Shenyang China.
[4] Faculty of Sciences, University of Central Punjab, Pakistan.
* Corresponding author: taswarahsan@163.com

fact that pesticides are not highly specific, they can also negatively impact other organisms, thus disrupting ecological balances when working to kill pathogens. Aside from pathogenic microbes, pesticide resistance may become increasingly problematic, which demands the constant development and application of new pesticides (Hussien et al. 2018). To avoid their negative environmental impacts, especially in developing countries, it is imperative to reduce the use of chemical pesticides in food production (Ruffo Roberto et al. 2019). Conventional ways of managing viral diseases target the vector with pesticides, natural predators, or physical barriers (reflective mulches and UV-absorbing sheets), among others (Spadoni et al. 2014). Along with these culturing approaches, disease control also involves sowing early, suppressing weeds, establishing crop-free periods, propagating virus-free crops, and discarding affected crops. A potent approach to disease management has been very difficult to achieve in the past, due to the complicated epidemiological parameters surrounding viral disease outbreaks, such as vector migration dynamics, evolving viruses, and unknown host ranges of the viruses (Tiemann et al. 2015). Agricultural pest control may be possible in the near future, thanks to recent developments in nanotechnology. By using nanotechnology in agriculture, we can promote eco-friendly agriculture and reduce the cost of production by reducing the use of chemical pesticides and fertilizers, delivering chemical pesticides and fertilizers in smart and targeted ways, and monitoring insect pest infestation, soil conditions, and regulating plant growth (Rikta and Rajiv 2021). The RNA interference (RNAi)-mediated response, for example, consists of silencing the target gene via double-stranded RNA. Using RdRP (RNA-dependent RNA polymerase), excess RNA is converted into dsRNA, allowing for RNA silencing. This technique is also used to control insect vectors. Nonetheless, studies are needed in order to evaluate resistance efficiency in the field (Chauhan et al. 2019).

A perspective on plant viral disease management

In the world, viruses cause an average 20% loss of cultivation. Nevertheless, there are some infections that can cause crop losses of up to 100%, mostly in developing countries, at a cost of 60 billion US dollars per year (Cai et al. 2019). By changing climate and cultural practices, the virus has caused severe damage to crops (Jones 2016). As virus infections increase, methods for controlling them must be developed that are effective. In greenhouses, chemical and biological methods are used to control insect vector populations (Sarwar 2020), mainly because viruses depend on insect vectors for their survival, transmission, and propagation (Ziegler-Graff 2020). There are other practices such as reducing virus sources (infected plants or plant remains), altering the color of the environment in order to alter the attraction of insects to colors, and using mineral oils to interfere with the transmission process (Sarwar 2020). Other strategies used in the laboratory for the treatment of viral infections in crops include traditional and sophisticated practices (Chauhan et al. 2019). Within conventional strategies, we find meristem tip culture, cryotherapy, thermotherapy, and chemotherapy. Nevertheless, because they are primarily utilized in tissue cultures and take into account the decrease of the original inoculum, these approaches have significant limitations. RNA silencing, cross-protection, transgenic plants, gene pyramiding, and protein-protein interaction are examples of sophisticated approaches (Chauhan et al. 2019). The virus can be removed in plant tissues (rather than merely preventing or postponing infection) in the latter case, although this method interferes with vectors, requires the use of suppressed strains and resistant or tolerant types, and is highly costly (Bragard et al. 2013). Chemotherapy is the use of antivirals to treat certain illnesses in animals. It is also beneficial for treating viruses in plant tissue cultures (Wang et al. 2018). Acycloguanosine (Acyclovir), 5-azacytidine, cytarabine, 5-bromouracil, ribavirin (Virazole), 2-thiouracil azidothymidine (Zidovudine), and 2, 4-dioxohexahydro-2, 5-triazine are antivirals utilized in plant tissue culture settings (DHT). Potato leaf-roll virus is one of the diseases that has been treated (PLRV) (Singh 2015), and grapevine vitivirus (Panattoni et al. 2007), sugarcane mosaic virus (SCMV) (Vargas-Hermandez et al. 2020), and Indian citrus ringspot virus (ICRSV) (Zhao et al. 2020) were also treated respectively. Meanwhile, modern approaches use dsRNA to silence cells, such as the RNA interference (RNAi)-mediated response. RdRP (RNA-dependent RNA polymerase) converts excess RNA to dsRNA, allowing for RNA silencing; this approach is also used to regulate vector insects. However, research is required to assess the effectiveness of resistance in the agricultural field (Chauhan et al. 2019). The adoption of resistant cultivars or a cross-protection approach, which consists of a systematic infection with a second virus to promote resistance to the target virus, is also used to manage viral infections in plants. The adoption of resistant varieties is one of the key ways employed to decrease crop losses since they are the most affordable and beneficial; nonetheless, the creation of highly efficient and persistent virus-resistant/immune crop types is necessary (Zhao et al. 2020).

Nanoparticles' synthesis and activity

Nanoparticles from plant extracts

Plants are well recognized for their capacity to decrease metal ions on the surface as well as in different organs or tissues located at a distance from ion penetration sites. The research of metal ion accumulation in plants revealed that metals may be converted to NPs (Makarov et al. 2014). For example, *Medicago sativa* and *Brassica juncea*, were treated with 50 nm size AgNPs and in results 12.4, and 13.6 weight % of AgNPs were detected, respectively (Harris and Bali 2008). In additional investigations, 4 nm gold icosahedra were found in *M. sativa* (Gardea-Torresdey et al. 2002) and 2 nm semi-spherical copper particles were found in Iris pseudacorus (Manceau 2008) grown on metal salt-containing substrates. Several *in vitro* techniques using plant extracts as reducing agents for the production of NPs have been explored in recent years (Makarov et al. 2014). Extracts of many plant species, as well as a range of acids and metal salts, such as copper, gold, silver, platinum, and iron, were utilized in this green synthesis of NPs (Ghosh et al. 2012). Plant materials utilized for NPs production are more beneficial than microbial or chemical approaches since there are no microbe effects or toxic chemical contamination. Furthermore, it consumes less energy and has simple and wide ramifications (Masum et al. 2019). Because of the inclusion of biomolecules such as phenols, terpenoids, ketones, carboxylic acids, aldehydes, enzymes, amides, and flavonoids (Prabhu and Poulose 2012, Masum et al. 2019, Abdallah et al. 2019), the green synthesis of NPs mediated by plant extracts requires the alleviation of metal ions (Rai and Yadav 2013). Extracts derived from various sections of the plant, including roots, stems, barks, leaves, flowers, fruits, and seeds, have been employed in the green preparation of nanoparticles (Raju et al. 2015, Rautela and Rani 2019). Plant extracts have the potential to act as reducing and stabilizing agents in the green production of NPs (Hossain et al. 2019). In recent years, a wide range of green synthesized NPs from various plant species has been described. AgNPs, for example, have been produced from *Phyllanthus emblica* fruit extract (Masum et al. 2019) and extract from the leaf of *Citrus limon* (Vankar and Shukla 2012), green tea (*Camellia sinensis*) (Nakhjavani et al. 2017), *Coffea Arabica* (Dhand et al. 2016), neem (*Azadirachta indica*) (Ahmed et al. 2016), *Acalypha indica* (Krishnaraj et al. 2012), phytochemical extract of *Aloe vera* (Tippayawat et al. 2016), latex of *Jatropha gossypifolia* (Borase et al. 2014), *Morinda citrifolia* root extract (Suman et al. 2013), *Phoenix dactylifera* (Oves et al. 2018), and extract of inflorescence of *Mangifera indica* (Qayyum et al. 2017). Antibacterial activity was demonstrated with zinc oxide nanoparticles (ZnONPs) and titanium dioxide nanoparticles (TiO_2NPs) produced from extracts of fresh lemon fruits (*Citrus limon*) (Hossain et al. 2019). Antibacterial activity was shown by zinc oxide NPs (ZnONPs) produced from extracts of chamomile flower (*Matricaria chamomilla*), olive leaf (Olea europaea), and red tomato fruit (*Lycopersicon esculentum*). MgO nano-flowers (MgONFs) with antibacterial potential were synthesized using aqueous rosemary extract (Abdallah et al. 2019).

Plant extract proportions and compositions, metal salt concentrations, reaction pH, and reaction temperature are all parameters that influence the quality, size, and form of these green synthesized NPs (Shah et al. 2015).

Microbe-based synthesis of NPs

Microbe-mediated green synthesis has been proposed as an alternate technique of NP design and development (Singh et al. 2018, Ahmed et al. 2020). Microbes may be utilized to synthesize metallic NPs such as gold, silver, copper, zinc, titanium, palladium, and nickel in a safe and cost-effective manner. Microbes can be used to synthesize NPs in both extracellular and intracellular environments (Singh et al. 2016). For extracellular formation of NPs, the culture filtrate is collected by centrifugation and combined with an aqueous metallic salt solution for extracellular production. The color change of the combined solution is used to monitor NP synthesis. The pale yellow to dark brown tint, for example, is a sign of silver NP synthesis (AgNPs) (Ibrahim et al. 2020).

For intracellular synthesis, after growing microorganisms under optimal growth conditions, the biomass is rinsed completely with sterile water and incubated with a metal ion solution. As previously stated, the color shift indicates the synthesis of NPs. After that, ultrasonication, centrifugation, and washing are used to capture NPs (Ali et al. 2020).

Virucidal activities of NPs

Biogenic AgNPs are intriguing environmental friendly instruments capable of combating harmful bacteria and viruses (Singh et al. 2020). Although substantial progress has been made by employing biogenic produced AgNPs as antibacterial agents, the antiviral activity of these NPs has received less attention than AgNPs manufactured using traditional approaches (chemical and/or physical procedures). However, few significant studies have shown the capacity of biogenic AgNPs to inhibit harmful viruses (El-Sheekh et al. 2022). Silver nanoparticles (AgNPs), for example, exhibit a broad spectrum of antiviral action against viruses of many families by interacting with gp120, competing for virus-cell attachment, inactivating the virus prior to entrance, interacting with double-stranded DNA, and connecting with virus (Galdiero et al. 2011).

Recently, AgNPs were prepared from the *Pseudomonas fluorescens* bacteria to control the tobacco mosaic virus. The size of silver nanoparticles was 8–10 nm (Ahsan 2020). A norovirus research found that gold/copper sulfide core-layer NPs (Au/CuSNP) destroy the viral capsid protein (Broglie et al. 2015). NPs' antiviral activity has been proven *in vitro* and *in vivo* with several plants. NPs are commonly characterized as working on the virus's surface. For example, a study showed *in vitro* antiviral efficacy of ZnONP and SiO_2NP against tobacco mosaic virus (TMV) (Cai et al. 2019). Envelope glycoproteins appear to interact with NPs, causing direct damage to the viral capsid. Controlling vectors is another method of preventing plant virus transmission, as insect vectors transmit 70% of plant viruses (Cai et al. 2019). A common action mechanism of NPs is given in the Figure 1.

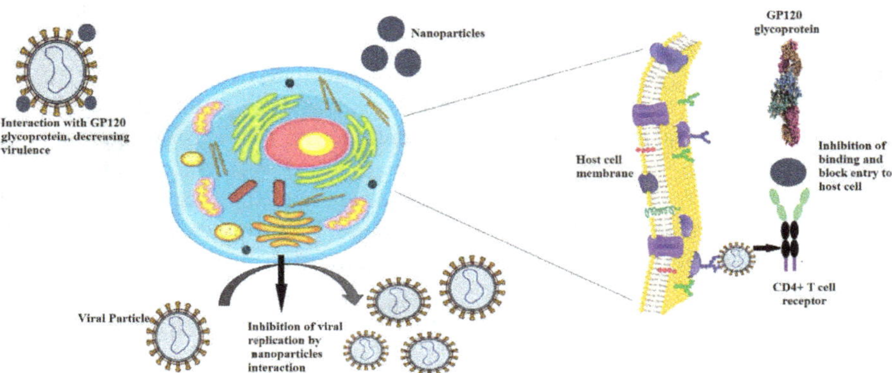

Figure 1: A general action mechanism of NPs against the virus. The picture adopted from the (Tortella et al. 2021).

Nanoclay sheets

Double-stranded RNA-induced RNAi has recently been proposed as a viable approach for providing viral resistance in crops. The topical application of dsRNA/siRNA SIGS (Spray Induced Gene Silencing), for example, has garnered interest due to its practicality and inexpensive cost when compared to transgenic plants. Once on the leaf surface, dsRNAs can either move directly to pest target cells (insects or diseases) or be picked up indirectly by plant cells and transported to pest cells (Worrall et al. 2018, Cagliari et al. 2019).

RNA molecules are directly taken up by the plant when they come into contact with its surface or feed on its tissues. RNA molecules are also indirectly up taken by insects and pathogens in indirect uptake. This occurs when the molecules first enter the vascular system of the plant. A transmembrane channel protein such as sid-1 can facilitate dsRNA uptake into cells (Zotti et al. 2018) or endocytosis (Pinheiro et al. 2018, Vélez and Fishilevich 2018). Gene silencing by RNAi relies on the release of dsRNA or siRNA molecules at cellular levels (Zotti and Smagghe 2015). Dicer 2 (DCR-2) is an enzyme that breaks up dsRNAs into siRNA fragments when they are unloaded into the cytoplasm (Tomari et al. 2007). Upon inclusion in the RISC complex (RNA-induced silencing complex), which is composed of the Argonaute 2 (AGO-2) protein (Ketting 2011), siRNA fragments quandary to a complementary messenger RNA (mRNA), cleave it, and prevent protein synthesis, eventually affecting the susceptibility or demise of the target organism (Huvenne and Smagghe 2010). A method which uses exogenous double-stranded RNA (dsRNA) to inhibit disease in plants seems like an extremely promising alternative to focusing on infections via transgenic products and pesticides. However, the shakiness of dsRNA splashed onto plants is an important test, as showering stripped dsRNA onto plants affords a guarantee against homologous infections for just 5 days. Innovative approaches, for example, the use of nanoparticles as dsRNA carriers for increased security and supported discharge, are emerging as critical challenging developments. Discovering about the component of section, transport, and preparation of exogenously connected dsRNA in plants is still limited. The cost of dsRNA and the administrative structure will have a significant impact on the innovation's functional selection (Mitter et al. 2017a). Nanoparticles can be used to protect plants in two ways: (a) as pesticides or other active components, such as double-stranded RNA (dsRNA), or (b) as carriers for existing pesticides or other actives, such as dsRNA, and can be applied via spray application or drenching/soaking onto seeds, foliar tissue, or roots. As carriers, nanoparticles can offer various advantages, including (i) increased shelf life, (ii) better solubility of pesticides that are weakly water soluble, (iii) decreased toxicity, and (iv) increased site-specific absorption into the target pest (Mitter et al. 2017a). A general action mechanism of dsRNA application is given in Figure 2.

Recent progress of nano-claysheet of DsRNA for plant protection

Plants contain double-stranded RNA molecules that are followed by single-stranded interfering RNA molecules, which diminish RNA homologous to the soybean mosaic virus's P3 coding region (Yang et al. 2018). However, it is possible that genetic modification of some leguminous plants is impossible. Embrapa 5.1 transgenic beans have been reported only to be able to guard the Golden Mosaic virus from the geminivirus (Bonfim et al. 2007). RNAi-inducing dsRNA molecules are alternative techniques for generating RNAi antivirus, specifically in transformation or regenerative resistant plants. The induction of RNAi by dsRNA in nanoparticles of double-layer (LDH) (BioClay) and spraying the particles continuously on plants'

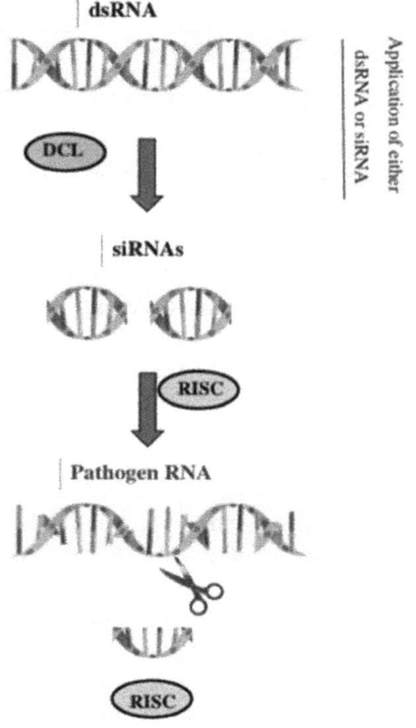

Figure 2: A general action mechanism of dsRNA applications against the plant virus (Ahsan and Wu 2021).

surfaces has been discovered recently (Mitter et al. 2017a). By comparison with recent techniques of antivirus RNAi to plants that are non-transgenic, this technique offered protection from the virus for twenty days rather than five when compared to naked dsRNA (Mitter et al. 2017a). While, most of the viruses through the vectors (insects) in the field grown plants (Baulcombe 2004, Groen et al. 2017). The study of RNAi has led to the advancement of innovative disease management strategies. A RNAi action is an action which occurs in eukaryotes and is capable of eliciting symptoms such as resistance against viral pathogens, growth, and development, even against insects, viruses, fungi, and weeds (Baulcombe 2004, Borges and Martienssen 2015). By using Dicer-like (DCL) enzymes, dsRNA induced RNAi in plants, which led to small interfering RNAs (siRNA). Following the RNA-induced silencing complex (RISC) is the siRNA, which guides the RISC by coupling at the base to cause the pathogenic RNA to no longer function as a translation template (Baulcombe 2004). An innovative approach to plant disease management evolved by modifying genetics with the RNAi mechanism (Robinson et al. 2014). There are few reports on dsRNA applications against plant virus management as shown in Table 1.

Table 1: Application of dsRNA in plant virus disease management.

Experimental application of nanoclay	Target of interest	References
dsRNA applications	Nib and CP genes of BCMV	(Worrall et al. 2019)
dsRNA applications	CP, P126, RP of TMV TMV Tobacco virus resistance for 7–20 d	(Konakalla et al. 2016)
dsRNA applications	CP, P126, RP of TMV TMV Tobacco virus	(Niehl et al. 2018)
Applying naked dsRNA	Induce antiviral RNAi in non-transgenic plants	(Mitter et al. 2017b)
BioClay carrying BCMV-specific dsRNA molecules	Aphid-mediated transmission of BCMV	(Vickers 2017) (Carr et al. 2018)
dsRNA	mRNA, pepper mottle virus	(Yoon et al. 2021)
dsRNA	2b and CP genes of *Cucumber mosaic virus*	(Holeva et al. 2021)

Conclusion

Nanobiotechnology is an emerging technology, specifically in nanophytopathology. Mostly, synthetic formulation of chemicals is in practice to control plant virus diseases. Rarely genetically modified crops have been introduced to combat the plant viruses. However, both practices are unhealthy and laborious, respectively. Nanotechnology found the solution to this issue; however, metallic nanoparticle also has some environmental pollution-related concerns. However, nanobiotechnology has a significant impact to combat plant viruses. In plant virus disease management by the direct applications of nanoparticle, only a few studies have been reported. Not much work has been done in this direction. There is a need to explore the action mechanism. However, dsRNA nanoclay sheets have an efficient role in resistance induction against the plant virus and, to some extent, against the virus vectors.

References

Abdallah, Y., Ogunyemi, S.O., Abdelazez, A., Zhang, M., Hong, X., Ibrahim, E., Hossain, A., Fouad, H., Li, B. and Chen, J. 2019. The green synthesis of MgO nano-flowers using *Rosmarinus officinalis* L.(Rosemary) and the antibacterial activities against *Xanthomonas oryzae pv*. oryzae. Biomed. Res. Int. 2019.

Ahmed, S., Saifullah Ahmad, M., Swami, B.L. and Ikram, S. 2016. Green synthesis of silver nanoparticles using *Azadirachta indica* aqueous leaf extract. J. Radi. Res. Appl. Sci. J. 9: 1–7.

Ahmed, T., Shahid, M., Noman, M., Niazi, M.B.K., Mahmood, F., Manzoor, I., Zhang, Y., Li, B., Yang, Y. and Yan, C. 2020. Silver nanoparticles synthesized by using *Bacillus cereus* SZT1 ameliorated the damage of bacterial leaf blight pathogen in rice. Pathogens 9: 160.

Ahsan, T. 2020. Biofabrication of silver nanoparticles from *Pseudomonas fluorescens* to control tobacco mosaic virus. Egypt. J. Biol. Pest Control. 30: 1–4.

Ahsan, T. and Yuanhua, W. 2021. Plant virus disease management by two modern applications (dsRNA nano-clay sheet and CRISPR/Cas). Arch. Phytopathol. Plant Prot. 54: 1292–1304.

Ali, M., Ahmed, T., Wu, W., Hossain, A., Hafeez, R., Islam Masum, M., Wang, Y., An, Q., Sun, G. and Li, B. 2020. Advancements in plant and microbe-based synthesis of metallic nanoparticles and their antimicrobial activity against plant pathogens. J. Nanomater. 10: 1146.

Baulcombe, D. 2004. RNA silencing in plants. Nature 431: 356–363.

Bonfim, K., Faria, J.C., Nogueira, E.O., Mendes, É.A. and Aragão, F.J. 2007. RNAi-mediated resistance to Bean golden mosaic virus in genetically engineered common bean (*Phaseolus vulgaris*). Mol. Plant Microbe Interact. 20: 717–726.

Borase, H.P., Patil, C.D., Salunkhe, R.B., Suryawanshi, R.K., Salunke, B.K. and Patil, S.V. 2014. Transformation of aromatic dyes using green synthesized silver nanoparticles. Bioprocess Biosyst. Eng. 37: 1695–1705.

Borges, F. and Martienssen, R.A. 2015. The expanding world of small RNAs in plants. Nat. Rev. Mol. Cell Biol. 16: 727–741.

Bragard, C., Caciagli, P., Lemaire, O., Lopez-Moya, J., MacFarlane, S., Peters, D., Susi, P. and Torrance, L. 2013. Status and prospects of plant virus control through interference with vector transmission. Annu. Rev. Phytopathol. 51: 177–201.

Broglie, J.J., Alston, B., Yang, C., Ma, L., Adcock, A.F., Chen, W. and Yang, L. 2015. Antiviral activity of gold/copper sulfide core/shell nanoparticles against human norovirus virus-like particles. Plos One 10: e0141050.

Cagliari, D., Dias, N.P., Galdeano, D.M., Dos Santos, E.Á., Smagghe, G. and Zotti, M.J. 2019. Management of pest insects and plant diseases by non-transformative RNAi. Front. Plant Sci. 1319.

Cai, L., Liu, C., Fan, G., Liu, C. and Sun, X. 2019. Preventing viral disease by ZnONPs through directly deactivating TMV and activating plant immunity in *Nicotiana benthamiana*. Environ. Sci. Nano. 6: 3653–3669.

Carr, J.P., Donnelly, R., Tungadi, T., Murphy, A.M., Jiang, S., Bravo-Cazar, A., Yoon, J.-Y., Cunniffe, N.J., Glover, B.J. and Gilligan, C.A. 2018. Viral manipulation of plant stress responses and host interactions with insects. Adv. Virus Res. 102: 177–197.

Chauhan, P., Singla, K., Rajbhar, M., Singh, A., Das, N. and Kumar, K. 2019. A systematic review of conventional and advanced approaches for the control of plant viruses. J. Appl. Biol. Biotech. 7: 8–8.

Dhand, V., Soumya, L., Bharadwaj, S., Chakra, S., Bhatt, D. and Sreedhar, B. 2016. Green synthesis of silver nanoparticles using Coffea arabica seed extract and its antibacterial activity. Mater. Sci. Eng: C 58: 36–43.

El-Sheekh, M.M., Shabaan, M.T., Hassan, L. and Morsi, H.H. 2022. Antiviral activity of algae biosynthesized silver and gold nanoparticles against Herps Simplex (HSV-1) virus *in vitro* using cell-line culture technique. Int. J. Environ. Health Res. 32: 616–627.

Elmer, W. and White, J.C. 2018. The future of nanotechnology in plant pathology. Annu. Rev. Phytopathol. 56: 111–133.

Galdiero, S., Falanga, A., Vitiello, M., Cantisani, M., Marra, V. and Galdiero, M. 2011. Silver nanoparticles as potential antiviral agents. Molecules 16: 8894–8918.

Gardea-Torresdey, J.L., Parsons, J., Gomez, E., Peralta-Videa, J., Troiani, H., Santiago, P. and Yacaman, M.J. 2002. Formation and growth of Au nanoparticles inside live alfalfa plants. Nano Lett. 2: 397–401.

Ghosh, S., Patil, S., Ahire, M., Kitture, R., Gurav, D.D., Jabgunde, A.M., Kale, S., Pardesi, K., Shinde, V. and Bellare, J. 2012. Gnidia glauca flower extract mediated synthesis of gold nanoparticles and evaluation of its chemocatalytic potential. J. Nanobiotech. 10: 1–9.

Groen, S.C., Wamonje, F.O., Murphy, A.M. and Carr, J.P. 2017. Engineering resistance to virus transmission. Curr. Opin. Virol. 26: 20–27.

Harris, A.T. and Bali, R. 2008. On the formation and extent of uptake of silver nanoparticles by live plants. J. Nanopart. Res. 10: 691–695.

Holeva, M.C., Sklavounos, A., Rajeswaran, R., Pooggin, M.M. and Voloudakis, A.E. 2021. Topical application of double-stranded RNA targeting 2b and CP genes of Cucumber mosaic virus protects plants against local and systemic viral infection. Plants 10: 963.

Hossain, A., Abdallah, Y., Ali, M., Masum, M., Islam, M., Li, B., Sun, G., Meng, Y., Wang, Y. and An, Q. 2019. Lemon-fruit-based green synthesis of zinc oxide nanoparticles and titanium dioxide nanoparticles against soft rot bacterial pathogen *Dickeya dadantii*. Biomolecules 9: 863.

Hussien, A., Ahmed, Y., Al-Essawy, A.-H. and Youssef, K. 2018. Evaluation of different salt-amended electrolysed water to control postharvest moulds of citrus. Trop. Plant Pathol. 43: 10–20.

Huvenne, H. and Smagghe, G. 2010. Mechanisms of dsRNA uptake in insects and potential of RNAi for pest control: A review. J. Insect Physiol. 56: 227–235.

Ibrahim, E., Zhang, M., Zhang, Y., Hossain, A., Qiu, W., Chen, Y., Wang, Y., Wu, W., Sun, G. and Li, B. 2020. Green-synthesization of silver nanoparticles using endophytic bacteria isolated from garlic and its antifungal activity against wheat *Fusarium* head blight pathogen *Fusarium graminearum*. Nanomat. 10: 219.

Jones, R. 2016. Future scenarios for plant virus pathogens as climate change progresses. Adv. Virus Res. 95: 87–147.

Ketting, R.F. 2011. The many faces of RNAi. Dev. Cell. 20: 148–161.

Khamis, Y., Hashim, A.F., Margarita, R., Alghuthaymi, M.A. and Abd-Elsalam, K.A. 2017. Fungicidal efficacy of chemically-produced copper nanoparticles against *Penincillium digitatum* and *Fusarium solani* on citrus fruit. Philipp. Agric. Sci. 100: 69–78.

Konakalla, N.C., Kaldis, A., Berbati, M., Masarapu, H. and Voloudakis, A.E. 2016. Exogenous application of double-stranded RNA molecules from TMV p126 and CP genes confers resistance against TMV in tobacco. Planta 244: 961–969.

Krishnaraj, C., Ramachandran, R., Mohan, K. and Kalaichelvan, P. 2012. Optimization for rapid synthesis of silver nanoparticles and its effect on phytopathogenic fungi. Spectrochimica Acta Part A: Mol. Biom. Spect. 93: 95–99.

Makarov, V., Love, A., Sinitsyna, O., Makarova, S., Yaminsky, I., Taliansky, M. and Kalinina, N. 2014. "Green" nanotechnologies: Synthesis of metal nanoparticles using plants. Acta Naturae 6: 35–44.

Manceau, A., Nagy, K.L., Marcus, M.A., Lanson, M., Geoffroy, N., Jacquet, T. and Kirpichtchikova, T. 2008. Formation of metallic copper nanoparticles at the soil-root interface. Environ. Sci. Technol. 42: 1766–1772.

Masum, M., Islam, M., Siddiqa, M., Ali, K.A., Zhang, Y., Abdallah, Y., Ibrahim, E., Qiu, W., Yan, C. and Li, B. 2019. Biogenic synthesis of silver nanoparticles using *Phyllanthus emblica* fruit extract and its inhibitory action against the pathogen *Acidovorax oryzae* strain RS-2 of rice bacterial brown stripe. Front. Microbiol. 10: 820.

Mitter, N., Worrall, E.A., Robinson, K.E., Li, P., Jain, R.G., Taochy, C., Fletcher, S.J., Carroll, B.J., Lu, G. and Xu, Z.P. 2017a. Clay nanosheets for topical delivery of RNAi for sustained protection against plant viruses. Nat. Plants 3: 1–10.

Mitter, N., Worrall, E.A., Robinson, K.E., Xu, Z.P. and Carroll, B.J. 2017b. Induction of virus resistance by exogenous application of double-stranded RNA. Curr. Opin. Virol. 26: 49–55.

Nakhjavani, M., Nikkhah, V., Sarafraz, M., Shoja, S. and Sarafraz, M. 2017. Green synthesis of silver nanoparticles using green tea leaves: Experimental study on the morphological, rheological and antibacterial behaviour. Heat & Mass Trans. 53: 3201–3209.

Niehl, A., Soininen, M., Poranen, M.M. and Heinlein, M. 2018. Synthetic biology approach for plant protection using ds RNA. Plant Biotechnol. J. 16: 1679–1687.

Oves, M., Aslam, M., Rauf, M.A., Qayyum, S., Qari, H.A., Khan, M.S., Alam, M.Z., Tabrez, S., Pugazhendhi, A. and Ismail, I.M. 2018. Antimicrobial and anticancer activities of silver nanoparticles synthesized from the root hair extract of *Phoenix dactylifera*. Mater. Sci. Eng.: C 89: 429–443.

Panattoni, A., D'Anna, F., Cristani, C. and Triolo, E. 2007. Grapevine vitivirus A eradication in *Vitis vinifera* explants by antiviral drugs and thermotherapy. J. Virolo. Meth. 146: 129–135.

Pinheiro, D.H., Velez, A.M., Fishilevich, E., Wang, H., Carneiro, N.P., Valencia-Jimenez, A., Valicente, F.H., Narva, K.E. and Siegfried, B.D. 2018. Clathrin-dependent endocytosis is associated with RNAi response in the western corn rootworm, *Diabrotica virgifera virgifera LeConte*. Plos One 13: e0201849.

Prabhu, S. and Poulose, E.K. 2012. Silver nanoparticles: Mechanism of antimicrobial action, synthesis, medical applications, and toxicity effects. Int. Nano Lett. 2: 1–10.

Qayyum, S., Oves, M. and Khan, A.U. 2017. Obliteration of bacterial growth and biofilm through ROS generation by facilely synthesized green silver nanoparticles. Plos One 12: e0181363.

Rai, M. and Yadav, A. 2013. Plants as potential synthesiser of precious metal nanoparticles: Progress and prospects. IET Nanobiotechnol. 7: 117–124.

Raju, D., Mehta, U.J. and Ahmad, A. 2015. Simple recovery of intracellular gold nanoparticles from peanut seedling roots. J. Nanosci. Nanotechnol. 15: 1575–1581.

Rautela, A. and Rani, J. 2019. Green synthesis of silver nanoparticles from Tectona grandis seeds extract: Characterization and mechanism of antimicrobial action on different microorganisms. J. Ana. Sci. Tech. 10: 1–10.

Rikta, S.Y. and Rajiv, P. 2021. Applications of silver nanomaterial in agricultural pest control. In Silver Nanomaterials for Agri-Food Applications (Elsevier), pp. 453–470.

Robinson, K.E., Worrall, E.A. and Mitter, N. 2014. Double stranded RNA expression and its topical application for non-transgenic resistance to plant viruses. J. Plant Biochem. Biotechnol. 23: 231–237.

Ruffo Roberto, S., Youssef, K., Hashim, A.F. and Ippolito, A. 2019. Nanomaterials as alternative control means against postharvest diseases in fruit crops. Nanomat. 9: 1752.

Sarwar, M. 2020. Insects as transport devices of plant viruses. In Appl. P. Vir. (Elsevier), pp. 381–402.

Shah, M., Fawcett, D., Sharma, S., Tripathy, S.K. and Poinern, G.E.J. 2015. Green synthesis of metallic nanoparticles via biological entities. Mat. 8: 7278–7308.

Sharma, N. and Prasad, M. 2017. An insight into plant–Tomato leaf curl New Delhi virus interaction. The Nucleus 60: 335–348.

Singh, A., Gautam, P.K., Verma, A., Singh, V., Shivapriya, P.M., Shivalkar, S., Sahoo, A.K. and Samanta, S.K. 2020. Green synthesis of metallic nanoparticles as effective alternatives to treat antibiotics resistant bacterial infections: A review. Biotechol. Rep. 25: e00427.

Singh, B. 2015. Effect of antiviral chemicals on *in vitro* regeneration response and production of PLRV-free plants of potato. J. Crop Sci. Biotech. 18: 341–348.

Singh, J., Dutta, T., Kim, K.-H., Rawat, M., Samddar, P. and Kumar, P. 2018. 'Green' synthesis of metals and their oxide nanoparticles: Applications for environmental remediation. J. Nanobiotechol. 16: 1–24.

Singh, P., Kim, Y.-J., Zhang, D. and Yang, D.-C. 2016. Biological synthesis of nanoparticles from plants and microorganisms. Trends Biotechnol. 34: 588–599.

Spadoni, A., Guidarelli, M., Sanzani, S.M., Ippolito, A. and Mari, M. 2014. Influence of hot water treatment on brown rot of peach and rapid fruit response to heat stress. Post. Bio. Tech. 94: 66–73.

Suman, T., Rajasree, S.R., Kanchana, A. and Elizabeth, S.B. 2013. Biosynthesis, characterization and cytotoxic effect of plant mediated silver nanoparticles using Morinda citrifolia root extract. Colloids Surf. B; Biointerfaces 106: 74–78.

Tiemann, L., Grandy, A., Atkinson, E., Marin-Spiotta, E. and McDaniel, M. 2015. Crop rotational diversity enhances below ground communities and functions in an agroecosystem. Ecol. Lett. 18: 761–771.

Tilman, D., Balzer, C., Hill, J. and Befort, B.L. 2011. Global food demand and the sustainable intensification of agriculture. PNAS 108: 20260–20264.

Tippayawat, P., Phromviyo, N., Boueroy, P. and Chompoosor, A. 2016. Green synthesis of silver nanoparticles in aloe vera plant extract prepared by a hydrothermal method and their synergistic antibacterial activity. Peer J. 4: e2589.

Tomari, Y., Du, T. and Zamore, P.D. 2007. Sorting of Drosophila small silencing RNAs. Cell 130: 299–308.

Tortella, G., Rubilar, O., Fincheira, P., Pieretti, J.C., Duran, P., Lourenço, I.M. and Seabra, A.B. 2021. Bactericidal and virucidal activities of biogenic metal-based nanoparticles: Advances and perspectives. Antibiotics 10: 783.

Vankar, P.S. and Shukla, D. 2012. Biosynthesis of silver nanoparticles using lemon leaves extract and its application for antimicrobial finish on fabric. Appl. Nanosci. 2: 163–168.

Vargas-Hernandez, M., Macias-Bobadilla, I., Guevara-Gonzalez, R.G., Rico-Garcia, E., Ocampo-Velazquez, R.V., Avila-Juarez, L. and Torres-Pacheco, I. 2020. Nanoparticles as potential antivirals in agriculture. Agri. 10: 444.

Vélez, A.M. and Fishilevich, E. 2018. The mysteries of insect RNAi: A focus on dsRNA uptake and transport. Pestic. Biochem. Phys. 151: 25–31.

Vickers, N.J. 2017. Animal communication: When i'm calling you, will you answer too? Curr. Biol. 27: R713–R715.

Wang, M.-R., Cui, Z.-H., Li, J.-W., Hao, X.-Y., Zhao, L. and Wang, Q.-C. 2018. *In vitro* thermotherapy-based methods for plant virus eradication. Plant Meth. 14: 1–18.

Worrall, E.A., Bravo-Cazar, A., Nilon, A.T., Fletcher, S.J., Robinson, K.E., Carr, J.P. and Mitter, N. 2019. Exogenous application of RNAi-inducing double-stranded RNA inhibits aphid-mediated transmission of a plant virus. Front. Plant Sci. 10: 265.

Worrall, E.A., Hamid, A., Mody, K.T., Mitter, N. and Pappu, H.R. 2018. Nanotechnology for plant disease management. Agron. 8: 285.

Yang, X., Niu, L., Zhang, W., Yang, J., Xing, G., He, H., Guo, D., Du, Q., Qian, X. and Yao, Y. 2018. RNAi-mediated SMV P3 cistron silencing confers significantly enhanced resistance to multiple Potyvirus strains and isolates in transgenic soybean. P. Cell. Rep. 37: 103–114.

Yoon, J., Fang, M., Lee, D., Park, M., Kim, K.-H. and Shin, C. 2021. Double-stranded RNA confers resistance to pepper mottle virus in *Nicotiana benthamiana*. Appl. Bio. Chem. 64: 1–8.

Zhao, Y., Yang, X., Zhou, G. and Zhang, T. 2020. Engineering plant virus resistance: From RNA silencing to genome editing strategies. P. Biotech. J. 18: 328–336.

Ziegler-Graff, V. 2020. Molecular insights into host and vector manipulation by plant viruses. Viruses 12: 263.

Zotti, M. and Smagghe, G. 2015. RNAi technology for insect management and protection of beneficial insects from diseases: lessons, challenges and risk assessments. Neotrop. Entom. 44: 197–213.

Zotti, M., Dos Santos, E.A., Cagliari, D., Christiaens, O., Taning, C.N.T. and Smagghe, G. 2018. RNA interference technology in crop protection against arthropod pests, pathogens and nematodes. Pest Manag. Sci. 74: 1239–1250.

Chapter 9

Nano Clay Materials and their Role in Plant Disease Management

Oluwatoyin Adenike Fabiyi

Introduction

The dependence of the world on agriculture has enabled the production of requisite chemical nutrients for soil augmentation to increase plant growth and development. Herewith, the goal of increasing food production, and minimizing hunger through maximization of yield was realized for demographic growth reasons. Howbeit, the world population is anticipated to increase. Instinctively, the need for food will double and will warrant enlargement in land area, which will equally imply a surge in tonnes of fertilizer and pesticide application worldwide (FAO 2008, Jahnavi et al. 2015).

Owing to the needless abuse of fertilizer and pesticide in crop production in order to manage the high population in a dependable manner, fertilizer and pesticide production is envisioned to shoot up. The formation process of nitrogen fertilizers like urea and N.P.K is energy exhaustive and intense. With the quantity of application for most crops, it is considered expensive when compared with what is actually taken up by the plant which is remarkably and outstandingly low. Ascribable to leaching, volatilization and erosion, a greater amount of macro and micronutrients added to the soil is lost (De Rosa et al. 2010). It is estimated that close to 90% of pesticide is lost due to several factors during application (Ghormade et al. 2011). There is a great need for efficiency of nutrient use for dependable agriculture (Suppan 2017, Mahanta et al. 2019).

To polish up nutrient uptake by plants, scientists are looking inwards and the application of nanotechnology in agriculture is needed for precision farming (De Rosa et al. 2010, Sekhon 2010, Dimkpa and Bindraban 2018) (Figure 1). The obligation to

Department of Crop Protection, Faculty of Agriculture, University of Ilorin, Ilorin, Nigeria.
Email: fabiyitoyinike@hotmail.com

cut down fertilizer adoption in agriculture is of high noteworthiness because of the troubles associated with fertilizer use, which ranges from environmental to economic and technical difficulties (Rouse-Miller et al. 2020). In order to reduce leaching (nitrates from nitrogen fertilizers), increase nutrient efficiency uptake of plants and reduce the volatilization of fertilizers as nitrous oxide, it is highly important to intensify efforts and diversify into several other approaches of fertilizer application (Bindraban et al. 2015). Operational molecules are necessary to ameliorate agricultural production systems and boost yield. Encapsulation and immobilization of functional nutrients, amino acids, bio-stimulants, and phytochemicals, in order to preserve the soil for a worthwhile biological action for crop development, is of paramount importance (Tarafdar et al. 2012). With this method, the soil's active ingredients will be less exposed to unfavorable environmental factors. Their viability will be enhanced and they will equally be stable in storage.

Materials that could be employed for such a goal are liposomes, polymers, and clays. Clays have the capacity to convey diverse substances, owing to their cation exchange capacity and high surface area. They are naturally occurring, inexpensive, and organic in nature with flexibility in chemical composition depending on their sources (Naderi and Danesh-Shahraki 2013). The binding of nutrients into pellets with the polymer of nano-composite origin is a good approach that will certainly go a long way in lessening leaches on agricultural fields and reducing environmental protection costs (Chinnamuthu and Boopathi 2009). Similarly, embedded nano-biosensor in biopolymer-coated fertilizers, which are delineated to supply nutrients in reciprocation to signals from microorganisms in the plant rhizosphere, will be of great significance to crop growth (Monreal 2012). Multiple benefits could be derived from the intercalation of fertilizers, pesticides, and macro and micronutrients into nano clays.

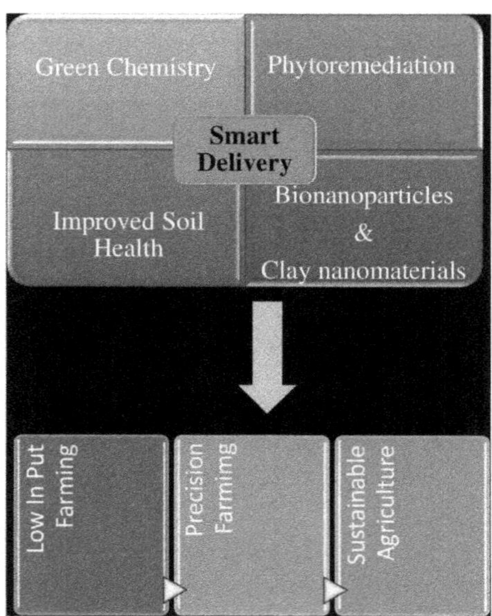

Figure 1: The benefits of nanotechnology to agriculture.

Through this, the speedy breakdown of molecules by environmental factors is prevented. Dependable release of agro-inputs is guaranteed with prolonged treatment while utilizing the same quantity of active principle (Bansiwal et al. 2006, Kashyap et al. 2015, Fabiyi et al. 2020a). Explicitly, modern studies and opportunities abound in the next few decades with the application of controlled release of pesticides, fertilizers, and phytohormones (Liu et al. 2006a). Production systems based on nano clays are expected to bring forth immense satisfaction and profit in agriculture. The intent of this chapter is to appraise contemporary developments in the interposing of agro-inputs into nano clays with remarkable output to achieve sustainability in modern agriculture.

Plant disease management and nanotechnology

Plants experience interruptions and impairments which alter their normal function. All classes and species of plants, be it cultivated or wild, are predisposed to disease infection. Some type of disease characteristically affects a group of plant species (Leach et al. 2014, Ul Haq et al. 2020, Schmid et al. 2021). The disease may be pervasive or cyclical; it is usually influenced by the variety, crop type, environmental situations, the presence of the causative organism, and change in climate (McDonald and Linde 2002, Hovmoller et al. 2011, Lobell and Gourdji 2012, Newton et al. 2012). Resistance and predisposition vary, depending on the varieties and species of the crop (Jones and Dangl 2006, Anderson et al. 2010, McNutt et al. 2013, Pelczar et al. 2021).

Anomalous and irregular weather situations brought about by climate change bring forth differences in communities of pathogens with reference to each ecosystem (Chakraborty et al. 2000, Garrett et al. 2006) culminating in increasing communities of disease-causing pathogens (Coakley et al. 1999, Mondal and Timmer 2002). Historically, the effect of diseases on plants dates back to roughly 250 million years as evidenced by fossils. Other documents like the Bible equally described plant diseases such as blights, mildews and rusts. All of these are recorded to have incited severe changes and famine in some nations' economies (Augustyn et al. 2021). A short time ago, outbreaks of diseases with high magnitude in history includes blight infection of potato in Ireland, which occurred in the year 1845 and lasted up till 1860 (Ó Gráda 1995, Mokyr and Ó Gráda 2002); this was followed by the epidemic of downy and powdery mildews, which infested grapes in France 1861 and 1878 (Johnson 1935, Augustyn et al. 2021, Boland et al. 2021). The rust of coffee in Sri Lanka (Ceylon) which emanated in the 1870s and extended till the 1900s was witnessed, while the Fusarium wilt damaging flax and cotton, plus the southern bacterial wilt infection of tobacco all came up in the same country at the same time (Berkeley 1869, Waller 1982, Kushalappa and Eskes 1989, Peiris and Arasaratnam 2021). In Central America, precisely Panama, Sigatoka leaf spot of banana was reported; this started from 1900 and extended till 1965 (Mourichon et al. 1997). For the wheat crop, black stem rust occurred in 1916, 1935, and again from 1953–1954 in the US coupled with southern corn leaf blight in 1970 (Steven 1919).

Central and South America experienced coffee rust in 1960, with a repeat in 2012 up until today (Griffin et al. 2021, Augustyn et al. 2021, Baker 2014, McCook and Vandermeer 2015). The Panama disease of bananas came up in Africa, Asia, and Australia from 1990 to the present day (Vala 1996, Castelan et al. 2013, Divya and Venethaji 2020). Significant economic losses with a huge reduction in income for the farmers were witnessed; this extends to the distributors and the consumers who are at the mercy of high prices (ICO 2014, ICO 2015). In developing economies, the annual loss of about 50% is sometimes recorded on important crops owing to a lack of access to disease management strategies (Baker 2014, Alivelino et al. 2015).

Invariably, this culminates in starvation, hunger, and indigence in society. It is often catastrophic because of the large dependence on crops for food in major economies. In history, a considerable outbreak of diseases has ended up with mass migration of the citizenry due to extensive famine (Mokyr and Ó Gráda 2002). For instance, America witnessed the mass migration of the Irish population in response to the ruinous outburst of water mould *Phytophthora infestans* infestation of potatoes in 1845, which climaxed into 12.5% death from starvation and illnesses associated with famine, while roughly about 19% of the population became refugees in the United States (Mokyr 2021, Royde-Smith et al. 2021). The water mould had monumental sway on the cultural, political, and economic growth of the United States and Europe. Diseases take place among weather pitfalls, innumerable species of cultivated plants are vulnerable to infection and would hardly survive without human involvement (Webb et al. 2010).

Wild plants often evade disease attacks but cultivated ones are highly predisposed to infection (Stewart 2018, Pelczar et al. 2021). This is so, in view of the fact that cultivated plants on a large and wide expanse of land are of the same genetic composition and planted together closely, there is the tendency for a pathogen or pest to spread disease under such conditions (McDonald and Linde 2002). By and large, the plant exhibits symptoms of disease when it is disturbed constantly by the causative agents which brings about changes and dysfunction in the biochemical and physiological functions of the plant (Vanderplank 1963, Webb et al. 2010, Royde-Smith et al. 2021). The various organisms and environments are capable of altering the plant system and thus producing disease in the process. Biotechnology and sundry approaches are of high significance in the control of pests, pathogens, and plant diseases (Atolani and Fabiyi 2020, Fabiyi et al. 2021a, Fabiyi and Olatunji 2021a, b, Fabiyi 2021a, b, c, d, Fabiyi 2022, Fabiyi et al. 2022).

A smart and new method of addressing plant diseases lately is employing nanotechnology, and a number of avenues for the administration of this technology have evolved (Fabiyi et al. 2020b). A handful of positive outcomes on plant disease management with nanotechnology have been documented in the literature. Ranging from fungi, bacteria, nematodes and insect pests and diseases (Ramasubburayan et al. 2017, Fabiyi et al. 2021b). Green synthesis of nanoparticles is specified to be potent against plant-parasitic nematodes (Khan et al. 2013). Fabiyi and Olatunji (2018) presented the importance of nanoparticles in nematode control on groundnuts. Similarly, nanoparticles have been stated to be effective on juveniles and eggs of plant-parasitic nematodes, while application on other crops such as rice and carrots has brought forth outstanding results (Fabiyi et al. 2018, 2020c, 2021). Mineral nutrition hiccups are linked with accessibility to crops; to correct this, inorganic materials of nano origin have been put into use. Among these are synthetic or natural zeolite, nanoporous silica, SiO2 nanoparticles and clay montmorillonite-urea (Ditta and Arshad 2016).

Nano clay composites

The composition of nano clay is of various types. Some are devised through the perforation of polymer chains into clay, and an organic and polymer layer is then created. The layers occur repeatedly in turns in an interspersed manner (Weiss et al. 2006). In some other formations, the clay layers are irregularly dispersed in the polymer array. The polymer probes extensively into the clay layers; with this arrangement, there is an excellent interface betwixt polymer and clay, thus the composite exhibits good properties (Mukhopadhyay and De 2014). That said, there are hydroxyl cations of polymeric nature where cation charge equilibrium is interchanged.

These belong to a new class of materials, clays of ratio 2:1 lamellae supported with small pillars of nanosize made up of organic cations. In some setups, organophilic clay materials are prepared by a replacement of the customary inorganic cations with organic cations, while some unimportant groups are modified with acid and salts for enhanced catalytic ability. A change in the chemical and mineral composition of some nano clay types may be brought about by thermal or mechanical modification (Mukhopadhyay and De 2014). The inclusion of clay materials in nano preparations has always upgraded the water absorption characteristics of such substances. Supplementation of nano polymer composite with 7% montmorillonite clay has been proved to boost the water-absorbing ability of the material than ordinary incorporation of monomer acrylic acid (Liu et al. 2006b). Sepiolite at 15% was also known to improve absorbency of liquid (Zhang et al. 2005). Analogously, the water releasing pace in soils amended with nano clays is very low, meaning that the augmentation of soils with nano clay composites retains water for plant development. The clay nanocomposites had the ability to conserve moisture, safe-guard water loss, and a superb water-absorbing ability.

These properties translate into the good capability of plants in such amended soil to withstand water stress in drought situations (Jatav et al. 2013). Several clay types exhibit different release sequences. As demonstrated by Sarkar et al. (2012), kaolinitic, micaceous, and montmorillonite clays were synthesized into polymer composite separately. They observed that under 48 hours, montmorillonite composite clay had 70% nutrient discharge while 90% was documented in kaolinitic clay composite. However, with a blend of amorphous aluminosilicate, the pace of nutrient release was not dissimilar among the different clay composites.

Essentials of nano clay materials in agriculture

Nano polymers of biodegradable nature are used to correct desertification in agriculture (Vundavali et al. 2015). Polyacrylamides cross-linked with nano clay composite plated with calcium have been used largely in ameliorating drought pressure impacts, through which plant yield enhancement has been attained with realization in the stability of agricultural production (Ali et al. 2014). The delivery manner of nanoporous utilization has risen to an advanced level. Zinc oxide nanoparticles are designed to penetrate the roots of plants, an example is seen in ryegrass (Lin and Xing 2008); likewise, nanotubes of carbon have infiltrated tomato roots for nutrient delivery purposes (Khodakovskaya et al. 2009).

Nano fertilizers are used in the supply of vital crop nutrients through elements of nanostructured features as carriers; it is incorporated by absorption into matrices of zeolite or clay. Nano zeolite-based fertilizers have the ability to discharge nutrients steadily to crops; this makes nutrients accessible all through the plants' growing period, thus guiding against nutrient loss via leaching, volatilization, denitrification or by way of fixing nitrogen (Mahanta et al. 2019). Bio-compatible polymeric nanoparticles and kaolin enhance regular nutrient demand for crops and soil (Wilson et al. 2008). Under field conditions, application mode of fertilizers is a function of their ability to be effective.

Encapsulated nano clay fertilizers have imparted skill into spraying such that the bottlenecks pertaining to losses from splashes and drifts have been subdued (Matthews and Thomas 2000). Nano clay encapsulations are also used in controlling the limited availability of certain soil nutrients like nickel, copper, and iron; these may be packed in form of organic nanomaterials or emulsions. Occasionally also, the crop growth stage may be used to match the quantity of fertilizer, nutrients or pesticides needed and could be applied in a controlled manner (Mahanta et al. 2019). Mesoporous silica nanoparticles are absorbed swiftly by plants through the symplastic and apoplastic routes via the epidermis to the vascular bundles where it circulates around the plant (Le et al. 2014). Superior dissolution, rapid absorption, and assimilation are seen in the nano clay fertilizers contrary to what is achieved in the old customary method of fertilizer application.

The nano clay technique could be employed for important plant nutrients like molybdenum, phosphorus, zinc, copper, magnesium, nitrogen, calcium, and potassium (Ditta and Arshad 2016, Mahanta et al. 2019). An appealing substitute process for agro input application is the nano clay option. This lowers pollution risk to the environment and the cost of fertilizer application and other inputs in crop production is lessened. Application of necessary nutrients at an optimum level is indispensable for the best result. Exceeding the ideal level could have a negative outcome on crop growth; hence, dose optimization is of vital importance for different crops in relation to the manufacturer's specifications.

Montmorillonite clay seems to be the most efficacious, dependable and the foremost slow-release material for plant nutrients. Kottegoda et al. (2011), in their research on nanoparticles of hydroxyapatite (HA) composition hinged on hybrid nanostructures encapsulated with extremely soluble urea as the origin of nitrogen fertilizer, established that encapsulation ability of the surface of urea molecules was increased by the nanorod structures, thus enhancing the dependable release of fertilizer for agricultural use. Also, Wu and Liu (2008) examined the polymerization of methacrylic acid into chitosan nanoparticles for practicable inclusion of water-soluble micro and macronutrients like phosphorus, potassium, and nitrogen. The interior was coated with CTS (chitosan) while the surface coat had a polymer which is a super absorbent material [poly (acrylic acid-co-acrylamide) (P(AA-co-AM))].

The material was stated to have good water holding and properly controlled delivery ability, coupled with exceptionally degradable characteristics, which makes it dependable for horticultural and agricultural purposes.

Correspondingly, Corradini et al. (2010) investigated chitosan as a slow-release material for fertilizers as well as an anti-bacterial substance. Additionally, chitosan and alginate were made as carriers and encapsulated with imidacloprid by Guan et al. (2008); they achieved 82% efficiency in the composite, and the release of imidacloprid was extended eight folds more than the regular method of pesticide application. The mixture of chitosan with glutaraldehyde modeled with paramagnetic Fe_3O_4 has been employed as nanosensor in agricultural fields to monitor plant diseases, detect pesticides and heavy metal pollution. Such composites are designed to remove heavy metals from the environment after detection (Ahmed and Fekry 2013, Kashyap et al. 2015). Description of the form of release was influenced by the concentration of materials used. Calcium carbonate nanocomposite is reported to have increased the growth parameters of groundnut at low concentrations against the effect of organic fertilizer and humic acid (Liu et al. 2009). Layers of kaolin clay were again tested in form of coated material to be used as controlled release in fertilizer (Liu et al. 2009).

The clay mineral halloysite made into nanotubes is a systematic well-planned means of monitoring the quantity of pesticides discharged into the environment, while being an eco-friendly approach to pesticide application, it equally lessens pesticide residue build-up in the environment (Du et al. 2010). Moderation of herbicide toxicity in the environment was evaluated by Grillo et al. (2014). They employed tripolyphosphate with chitosan as carriers for paraquat and minimal environmental toxicity of the herbicide was achieved. Nanoceramics are deployed as filters in irrigation waters. The application is hinged on the principle of like charges (Argonide 2005). Several micro-organisms, endotoxins and organic materials are negatively charged and are removed by the ceramic filters (Karn et al. 2009). Owing to the mechanical properties of clay nanofibers, they are regarded as useful materials in soil and water purification (Sehaqui et al. 2014).

Bentonite nano clay is of significance in the remediation of agricultural soils from heavy metal pollution. Merrikhpour and Sobhan Ardakani (2017) extracted cobalt and chromium with adsorbent of nano clay nature. They opined that bentonite has a high degree of activity in heavy metal removal. Nano clay montmorillonite was confirmed by Baghaie and Keshavarzi (2018) to be a good material for heavy metal removal from water. They noted a reduction in soil cadmium concentration with a corresponding rise in soil cation exchange capacity. The respiratory activity of soil microbes also increased. This observation was corroborated by Monajjem et al. (2016), who reiterated that nano clays improve soil CEC and invariably reduce soil cadmium load. The application of clay composite polymers in rain-fed agriculture was emphasized by Guo et al. (2005), who remarked that owing to the ability of clay composites to release nutrients and water slowly, they could be the good candidates in rain-fed agriculture since there will be no anxiety regarding nutrient wash off. It will also be useful in arid lands because of the water absorbency characteristics. It will serve as a nutrient and water bag such that the frequency of irrigation is reduced.

Conclusion

Changing climate, population increase, and low outturn of agricultural produce call for nanotechnology embracement. Polymer composites of nano clay present a distinctive opportunity to improve agricultural productivity, evade obstacles, plus ebb and flow presented by the trials of climate change. Projections from the econometric school of thought express that food quantity production must rise by the year 2050, though this has been highly contentious and negated. Whether this is unambiguous or not, it is essential to tweak agricultural production. Nanotechnology advocates the enlargement of food production, while minimizing the negative consequences of pesticides and fertilizer application in the environment. The public health footprints of nitrate flow from agricultural fields and eventual contamination of municipal water are sufficient to adopt nanotechnology in fertilizer application as a pressing duty. A policy should be instituted to create awareness for the general public on the need for nano technology-empowered agriculture.

References

Ahmed, R.A. and Fekry, A.M. 2013. Preparation and characterization of a nanoparticles modified chitosan sensor and its application for the determination of heavy metals from different aqueous media. Int. J. Electrochem. Sci. 8: 6692–6708.

Ali, M.A., Rehman, I., Iqbal, A., Din, S., Rao, A.Q., Latif, A., Samiullah, T.R., Azam, S. and Husnain, T. 2014. Nanotecnology: A new frontier in agriculture. Adv. Life Sci. 1(3): 129–138.

Anderson, J.P., Gleason, C.A., Foley, R.C., Thrall, P.H., Burdon, J.B. and Singh, K.B. 2010. Plants versus pathogens: An evolutionary arms race. Functional Plant Biology 37(6): 499–512.

Argonide Nanoceram filters Argonide Corp. 2005. http://sbir.nasa.gov/SBIR/successes/ss/9-072text.html.

Atolani, O. and Fabiyi, O.A. 2020. Plant parasitic nematodes management through natural products: Current progress and challenges. pp. 297–315. *In*: Ansari, R.A., Rizvi, R. and Mahmood, I. (eds.). In Management of Phytonematodes: Recent Advances and Future Challenges. Singapore.

Augustyn, A., Zeidan, A., Zelazko, A., Eldridge, A., McKenna, A. and Tikkanen, A. 2021. Britannica, The Editors of Encyclopaedia. Panama disease. *Encyclopedia Britannica*, 15 Aug. 2019. https://www.britannica.com/science/Panama-disease. Accessed 23 September 2021.

Avelino, J., Cristancho, M., Georgiou, S., Imbach, P., Aguilar, L., Bornemann, G., Läderach, P., Anzueto, F., Hruska, A.J. and Morales, C. 2015. The coffee rust crises in Colombia and Central America (2008–2013): Impacts, plausible causes and proposed solutions. Food Sec. 7: 303–321.

Baker, P. 2014. The "Big Rust": An update on the coffee leaf rust situation. Coffee Cocoa Int. 40: 37–39.

Bansiwal, A.K., Rayalu, S.S., Labhasetwar, N.K. and Devotta, S. 2006. Surfactant-modified zeolite as a slow release fertilizer for phosphorous. J. Food Chem. 54: 4773–9.

Berkeley, M.J. 1874. The gardeners chronicle: A weekly illustrated. Journal of Horticulture and Allied Subjects. Gardeners Chronicle, London, 139.1.

Bindraban, P.S., Dimkpa, C., Nagarajan, L., Roy, A. and Rabbing, R. 2015. Revisiting fertilisers and fertilisation strategies for improved nutrient uptake by plants. Biol. Fertil. Soils 51(5): 897–911.

Boland, Frederick Henry, Fanning, Ronan, Ranelagh, John O'Beirne, Kay, Sean and Edwards, Robert Walter Dudley. Ireland. *Encyclopedia Britannica*, 22 Sep. 2021. https://www.britannica.com/place/Ireland. Accessed 23 September 2021.

Castelan, F.P., Abadie, C., Hubert, O., Chilin-Charles, Y., De Bellaire, L.L. and Chillet, M. 2013. Relation between the severity of Sigatoka disease and banana quality characterized by pomological traits and fruit green life. Crop Protection 50: 61–65.

Chakraborty, S., Tiedemann, A.V. and Teng, P.S. 2000. Climate change: Potential impact on plant diseases. Environmental Pollution 108: 317–326.

Chinnamuthu, C.R. and Boopathi, P.M. 2009. Nanotechnology and agroecosystem. Madras. Agric. J. 96: 17–31.

Coakley, S.M., Scherm, H. and Chakraborty, S. 1999. Climate change and plant disease management. Annual Rev. Phytopathol. 37: 399–426.

Corradini, E., De Moura, M.R. and Mattoso, L.H.C. 2010. A preliminary study of the incorporation of NPK fertilizer into chitosan nanoparticles. Express Polym. Lett. 4(8): 509–515.

Current world fertiliser trends and outlook to 2011/12. 2008. Food and Agriculture Organization of the United Nations, Rome.

De Rosa, M.R., Monreal, C., Schnitzer, M., Walsh, R. and Sultan, Y. 2010. Nanotechnology in fertilizers. Nat. Nanotechnol. J. 5: 91.

Dimkpa, C.O. and Bindraban, P.S. 2018. Nano-fertilizers: New products for the Industry? J. Agric. Food Chem. 66(26): 6462–6473. http://pubs.acs.org/doi/pdf/10.1021/acs.jafc.7b02150.

Ditta, A. and Arshad, M. 2016. Applications and perspectives of using nanomaterials for sustainable plant nutrition. Nano Technol. Rev. 5: 209.

Divya, S. and Venethaji, M. 2020. Efficacy of mineral oil with fungicides in management of sigatoka leaf spot of banana. J. Krishi Vigyan 9(1): 355–357. Doi: 10.5958/2349-4433.2020.00191.9.

Du, M., Guo, B. and Jia, D. 2010. Newly emerging applications of halloysite nanotubes: A review. Polym. Int. 59(5): 574–82.

Fabiyi, O.A. 2021a. Sustainable management of *Meloidogyne incognita* infecting carrot: Green synthesis of silver nanoparticles with *Cnidoscolus aconitifolius*: (*Daucus carota*). Vegetos 34(2): 277–285.

Fabiyi, O.A. 2021b. Application of furfural in sugarcane nematode pest management. Pak. J. Nematol. 39(2): 151–155. Doi | https://dx.doi.org/10.17582/journal.pjn/2021.39.2.151.155.

Fabiyi, O.A. 2021c. Evaluation of Nematicidal activity of *Terminalia glaucescens* fractions against *Meloidogyne incognita* on *Capsicum chinense*. Hort Res. 29(1): 67–74. https://doi.org/10.2478/johr-2021-0006.

Fabiyi, O.A. 2021d. Application of furfural in sugarcane nematode pest management. Pak. J. Nematol. 39(2): 151–155. doi.org/10.17582/journal. PJN /2021.39.2.151.155.

Fabiyi, O.A. 2022. Fractions from *Mangifera indica* as an alternative in *Meloidogyne incognita* management. Pak. J. Nematol. 40(1): 65–74.

Fabiyi, O.A. and Olatunji, G.A. 2018. Application of green synthesis in nano particles preparation: *Ficus mucoso* extracts in the management of *Meloidogyne incognita* parasitizing groundnut *Arachis hypogea*. Indian J. Nematol. 48(1): 13–17.

Fabiyi, O.A. and Olatunji, G.A. 2021a. Environmental sustainability: Bioactivity of *Leucaena leucocephala* leaves and pesticide residue analysis in tomato fruits. Acta Univ. Agric. et Silvic. Mendelianae Brun. 69(4): 473–480.

Fabiyi, O.A. and Olatunji, G.A. 2021b. Toxicity of derivatized citrulline and extracts of water melon rind (*Citrullus lanatus*) on root-knot nematode (*Meloidogyne incognita*). Trop. Agric. 98(4): 347–355.

Fabiyi, O.A., Alabi, R.O. and Ansari, R.A. 2020b. Nanoparticles' synthesis and their application in the management of phytonematodes: An overview. pp. 125–140. *In*: Ansari, R.A., Rizvi, R. and Mahmood, I. (eds.). In Management of Phytonematodes: Recent Advances and Future Challenges. Singapore.

Fabiyi, O.A., Olatunji, G.A. and Saadu, A.O. 2018. Suppression of *Heterodera sacchari* in rice with agricultural waste-silver nano particles. J. Solid Waste Technol. Manag. 44(2): 87–91.

Fabiyi, O.A., Olatunji, G.A., Atolani, O. and Olawuyi, R.O. 2020c. Preparation of bio-nematicidal nanoparticles of *Eucalyptus officinalis* for the control of cyst nematode (*Heterodera sacchari*). J. Anim. Plant. Sci. 30(5): 1172–1177.

Fabiyi, O.A., Baker, M.T., Claudius-Cole, A.O., Alabi, R.O and Olatunji, G.A. 2022. Application of cyclic imides in the management of root-knot nematode (*Meloidogyne incognita*) on cabbage. Indian Phytopathol. 75(2): 457–466.

Fabiyi, O.A., Claudius-Cole, A.O., Olatunji, G.A., Abubakar, D.O. and Adejumo, O.A. 2021a. Evaluation of the *in vitro* response of *Meloidogyne incognita* to silver nano particle liquid from agricultural wastes. Agrivita J. Agricultural Sci. 43(3): 524–534.

Fabiyi, O.A., Claudius-Cole, A.O. and Olatunji, G.A. 2021b. *In vitro* assessment of n-phenyl imides in the management of *Meloidogyne incognita*. Sci. Agric. Bohem. 52(3): 60–65.

Fabiyi, O.A., Saliu, O.D., Claudius-Cole, A.O., Olaniyi, I.O., Oguntebi, O.V. and Olatunji, G.A. 2020a. Porous starch citrate biopolymer for controlled release of carbofuran in the management of root knot nematode *Meloidogyne incognita*. Biotechnol. Rep. 25(e00428).

Garrett, K., Dendy, S., Frank, E., Rouse, M. and Travers, S. 2006. Climate change effects on plant disease: Genomes to ecosystems. Annual Rev. Phytopathol. 44: 489–509.

Ghormade, V., Deshpande, M.V. and Paknikar, K.M. 2011. Perspectives for nano-biotechnology enabled protection and nutrition of plants. Biotechnol. Adv. 29: 792–803.

Griffin, Ernst C., Avila, Héctor Fernando., Minkel, C.W., Gade, Daniel W., Knapp, Gregory W., Germani, Gino., Dorst, Jean P. and Ramos, Victor A. 2021. South America. *Encyclopedia Britannica*, 10 Sep. 2021. https://www.britannica.com/place/South-America. Accessed 23 September 2021.

Grillo, R., Pereira, A.E.S., Nishisaka, C.S., de Lima, R., Oehlke, K., Greiner, R. and Fraceto L.F. 2014. Chitosan/tripolyphosphate nanoparticles loaded with paraquat herbicide: an environmentally safer alternative for weed control. J. Hazard. Mater. 278: 163–171.

Guan, H., Chi, D., Yu, J. and Li, X. 2008. A novel photodegradable insecticide: Preparation, characterization and properties evaluation of nano-imidacloprid. Pestic. Biochem. Physiol. 92: 83–91.

Guo, M., Liu, M., Zhan, F. and Wu, L. 2005. Preparation and properties of a slow-release membrane encapsulated urea fertilizer with superabsorbent and moisture preservation. Ind. Eng. Chem. Res. 44: 4206–4211.

Hovmoller, M.S., Sorensen, C.K., Walter, S. and Justesen, A.F. 2011. Diversity of *Puccinia striiformis* on cereals and grasses. Annual Rev. Phytopathol. 49: 197–217.

International Coffee Organisation. 2014. Annual Rev. 2013–2014. International Coffee Organisation, London.

International Coffee Organisation. 2015. ICO Composite and Group Indicator Prices (annual monthly averages). International Coffee Organisation, London.

Jahnavi, S., Pravin, P. and Nirmal, D. 2015. Nano-clay composite and phyto-nanotechnology a new horizon to food security issue in Indian agriculture. J. Global Bio. Sci. 4(5): 2187–2198.

Jatav, G.K., Mukhopadhyay, R. and De, N. 2013. Characterization of swelling behaviour of nano clay composite. Int. J. Innovative Res in Sci. Eng. and Tech. 2: 1560–1563.

Johnson, G.F. 1935. The early history of copper fungicides. Agricultural History 9(2): 67–79. https://www.jstor.org/stable/3739659.

Jones, J.D. and Dangl, J.L. 2006. The plant immune system. Nature 444(7117): 323–329.

Karn, B., Kuiken, T. and Otto, M. 2009. Nanotechnology and *in situ* remediation: A review of benefits and potential risks. Environ. Hlth. Perspect. 117(12): 1823–1831.

Kashyap, P.L., Xiang, X. and Heiden, P. 2015. Chitosan nanoparticle based delivery systems for sustainable agriculture. Int. J. Biol. Macromol. 77: 36.

Khan, M., Khan, M., Adil, S.F., Tahir, M.N., Tremel, W., Alkhathlan, H.Z. and Siddiqui, M.R.H. 2013. Green synthesis of silver nanoparticles mediated by *Pulicaria glutinosa* extract. Inter. J. Nanomed. 8: 1507–1516.

Khodakovskaya, M., Dervishi, E., Mahmood, M., Xu, Y., Li, Z., Watanabe, F. and Biris, A.S. 2009. Carbon nanotubes are able to penetrate plant seed coat and dramatically affect seed germination and plant growth. ACS Nano 3: 3221–3227.

Kottegoda, N., Munaweera, I., Madusanka, N. and Karunaratne, V. 2011. A green slow release fertiliser composition based on urea modified hydroxyapatite nanoparticles encapsulated wood. Current Sci. 101(3): 73–78.

Kushalappa, A.C. and Eskes, A.B. 1989. Advances in coffee rust research. Annual Rev. Phytopathol. 27: 503–31.

Le, V., Rui, Y., Gui, X., Li, X., Liu, S. and Han, Y. 2014. Uptake, transport, distribution and Bio-effects of SiO_2 nanoparticles in Bt-transgenic cotton. J. Nano Biotechnol. 12: 50.

Leach, J.E., Leung, H. and Tisserat, N.A. 2014. Plant disease and resistance. Encyclopaedia of Agric. and Food Sys. 4: 360–374. Doi: 10.1016/B978-0-444-52512-3.00165-0.

Lin, D.H. and Xing, B.S. 2008. Root uptake and phytotoxicity of ZnO nanoparticles. Environ. Sci. and Technol. 42(15): 5580–5585.

Liu, P., Li, L., Zhou, N., Zhang, J., Wei, S. and Shen, J. 2006a. Synthesis and properties of a poly (acrylic acid) montmorillonite superabsorbent nanocomposite. J. Appl. Polym. Sci. 102: 5725–5730.

Liu, Q.L., Chen, B., Wang, Q.L., Fang, X.H. and Lin, J.X. 2009. Carbon nanotubes as molecular transporters for walled plant cells. Nano Lett. 9(3): 1007–1010.

Liu, X., Feng, Z., Zhang, S., Zhang, J., Xiao, Q. and Wang, Y. 2006b. Preparation and testing of cementing nano-subnano composites of slow or controlled release of fertilizers. Sci. Agric. Sinica 39: 1598–1604.

Lobell, D.B. and Gourdji, S.M. 2012. The influence of climate change on global crop productivity. Plant Physiol. 160(4): 1686–1697.

Mahanta, N., Ashok Dambale, A. and Rajkhowa, M. 2019. Nutrient use efficiency through nano fertilizers. Inter. J. Chemical Studies 7(3): 2839–2842.

Matthews, G.A. and Thomas, N. 2000. Working towards more efficient application of Pesticides. Pest. Manag. Sci. 56: 974–976.

McCook, S. and Vandermeer, J. 2015. The big rust and the red queen: Long term perspectives on coffee rust research. Phytopathol. 105(9): 1164–73. doi.org/10.1094/PHYTO-04-15-0085-RVW.

McNutt, Louise-Anne and Krug, Allison. 2013. Prevalence. *Encyclopedia Britannica*, 16 Dec. 2013. https://www.britannica.com/science/prevalence. Accessed 23 September 2021.

Merrikhpour, H. and Sobhan Ardakani, S. 2017. Investigation of the efficiency of bentonite nanoclay as an adsorbent for removal of Cr and Co from aqueous solutions: Adsorption isotherm and kinetics studies. J. Neyshabur University of Med. Sci. 5(3): 9–24.

Mokyr, J. and Ó Gráda, C. 2002. What do people die of during famines: The Great Irish Famine in comparative perspective. European Rev. Econ. History 6(3): 339–363. Cambridge University Press. http:// dx.doi.org /10.1017/ S1361491 602000163.

Mokyr, Joel. 2021. Great Famine. *Encyclopedia Britannica*, 27 Jul. 2021. https://www.britannica.com/event/Great-Famine-Irish-history. Accessed 22 September 2021.

Monajjem, M.A., Heidari, A. and Bagheri Marandi, G.H. 2016. Comparison the ability of nano-clays and clays extracted from different soils in retention of some heavy metals. J. Water Soil Conserv. 23(3): 189–205. Doi: 10.22069/jwfst.2016.3193.

Mondal, S.N. and Timmer, L.W. 2002. Environmental factors affecting pseudothecial development and ascospore production of *Mycosphaerella citri*, the cause of citrus greasy spot. Phytopathol. 92: 1267–1275.

Monreal, C. 2012. Development of Intelligent Nano-fertilizers. Slide 13. February 2012. https://www.scribd.com/document/240227289/Development-of Intelligent-Nano-Fertilizers-Monreal#.

Mouri chon, X., Carlier, J. and Foure, E. 1997. Sigatoka leaf spot diseases. INIBAP Muse disease fact sheet No. 8. 4 pp.

Mukhopadhyay, R. and De, N. 2014. Nano clay polymer composite: Synthesis, characterization, properties and application in rain fed agriculture. Global J. Bio Science and Biotechnol. 3(2): 133–138.

Naderi, M.R. and Danesh-Shahraki, A. 2013. Nano fertilizers and their roles in sustainable agriculture. Inter. J. Agric. Crop Sci. 5(19): 2229.

Newton, A.C., Torrance, L., Holden, N., Toth, I.K., Cooke, D.E.L., Block, V.C. and Gilroy E.M. 2012. Climate change and defence against pathogens in plants. Adv. Appl. Microbiol. 81: 89–132.

Ó Gráda, C. 1995. The Great Irish Famine. Research Repository. University College Dublin. Cambridge University Press, 1–26 pp. http://hdl.handle.net/10197/363 ogradac_bookchap_pub_075.pdf (ucd.ie).

Peiris, Gerald Hubert and Arasaratnam, Sinnappah. 2021. Sri Lanka. *Encyclopedia Britannica*, 19 Sep. 2021. https://www.britannica.com/place/Sri-Lanka. Accessed 23 September 2021.

Pelczar, Michael J., Shurtleff, Malcolm C., Kelman, Arthur P. and Rita, M. 2021. Plant disease. *Encyclopedia Britannica*, 31 Aug. 2021. https://www.britannica.com/science/plant-disease. Accessed 22 September 2021.

Ramasubburayan, R., Sumathi, S., Prakash, S., Ramkumar, V.S., Titus, S., Immanuel, G. and Palavesam, A. 2017. Synthesis of nano silver by a marine epibiotic bacterium *Bacillus vallismortis* and its potent eco-friendly antifouling properties. Environ. Nanotechnol. Monitoring & Manag. 8: 112–120.

Rouse-Miller, J., Bartholomew, E.S., Martin, C.C.G and Vilpigue, P. 2020. Bioprospecting compost for long-term control of plant parasitic nematodes. pp. 35–50. *In*: Ansari, R.A., Rizvi, R. and Mahmood, I. (eds.). In Management of Phytonematodes: Recent Advances and Future Challenges. Singapore.

Royde-Smith, John Graham and Showalter, Dennis E. 2021. World War I. *Encyclopedia Britannica*, 21 Jul. 2021. https://www.britannica.com/event/World-War-I. Accessed 23 September 2021.

Sarkar, S., Datta, S.C. and Biswas, D.R. 2012. Synthesis and characterization of nanoclay-polymer composites from soil clay with respect to their water-holding capacities and nutrient - release behaviour. J. Appl. Polym. Sci. 131(6): 39951. https://doi.org/10.1002/app.39951.

Schmid, Rudolf, Lambers, Hans, Woodwell, George M., Rothwell, Gar W., Dickison, William C. and Yopp, John H. 2021. Plant. *Encyclopedia Britannica*, 15 Jun. 2021. https://www.britannica.com/plant/plant. Accessed 23 September 2021.

Sehaqui, H., de Larraya, U.P., Liu, P., Pfenninger, N., Mathew, A.P., Zimmermann, T. and Philippe Tingaut, P. 2014. Enhancing adsorption of heavy metal ions onto bio-based nanofibers from waste pulp residues for application in wastewater treatment. Cellulose 21: 2831–2844. https://doi.org/10.1007/s10570-014-0310-7.

Sekhon, B.S. 2010. Food nanotechnology—An overview. Nanotechnol. Sci. & Applications 3: 1–15.

Stevens, F.L. 1919. Foot-rot disease of wheat historical and bibliographic. State of Illinois department of registration and education division of the natural history survey. Bulletin. Vol XIII. Article IX. Bulletin (illinois.edu).

Stewart, R.E. 2018. Agricultural Technology. *Encyclopedia Britannica*, 17 May. 2018. https://www.britannica.com/technology/agricultural-technology. Accessed 23 September 2021.

Suppan, S. 2017. Applying nanotechnology to fertilizer: Rationales, research, risks and regulatory challenges. The Institute for Agriculture and Trade Policy. 2–31 pp. www.iatp.org. A governança dos riscos socioambientais da nanotecnologiae o marco legal de ciência, tecnologia e inovação do Brasil. https://editorakarywa.wordpress.com/2017/03/29/a-governanca-dos-riscossocioambientais/#more-236.

Tarafdar, J.C., Raliya, R. and Tathore, I. 2012. Microbial synthesis of phosphorous nanoparticle from tri-calcium phosphate using *Aspergillus tubingensis* TFR-5. J. Bio Nanosci. 6: 84–89.

Ul Haq, Imran and Ijaz, Siddra. 2020. History and recent trends in plant disease control. pp. 1–13. *In*: Ul Haq, Imran and Ijaz, Siddra (eds.). Plant Disease Management Strategies for Sustainable Agriculture through Traditional and Modern Approaches.

Vala, D.G. 1996. Fungal and bacterial disease of banana. pp. 37–46. *In*: Singh, S.J. (ed.). Advances in Disease of Fruit Crops in India. Kalyani Publishers, Ludhiana.

Vanderplank, J.E. 1963. Plant Diseases: Epidemics and Control. New York: Academic Press, 349 pp.

Vundavalli, R., Vundavalli, S., Nakka, M. and Rao, S. 2015. Biodegradable nano-hydrogels in agricultural farming-alternative source for water resources. Procedia Mat. Sci. 10: 548–554.

Waller, J.W. 1982. Coffee rust-epidemiology and control. Crop Protection 1(4): 383–404. https://doi.org/10.1016/0261-2194(82)90022-9.

Webb, K.M., Oña, I., Bai, J., Garret, K.A., Mew, T., Vera-Cruz, C.M. and Leach, J.E. 2010. A benefit of high temperature: Increased effectiveness of a rice bacterial blight resistance gene. New Phytol. 185(2): 568–576.

Weiss, J., Takhistov, P. and McClements, D.J. 2006. Functional materials in food nanotechnology. J. Food Sci. 71: R107–R116.

Wilson, Andrew W. 2018. Saprotroph. *Encyclopedia Britannica*, 21 Nov. 2018. https://www.britannica.com/science/saprotroph. Accessed 23 September 2021.

Wilson, M.A., Tran, N.H., Milev, A.S., Kannangara, G.S.K., Volk, H. and Lu, G.H.M. 2008. Nanomaterials in soils. Geoderma 146: 291–302.

Wu, L. and Liu, M. 2008. Preparation and properties of chitosan–coated NPK compound fertiliser with controlled-release and water retention. Carbohydrate Polymers 72: 240–247.

Zhang, F., Guo, Z., Gao, H., Li, Y., Ren, L., Shi, L. and Wang, L. 2005. Synthesis and properties of sepiolite/poly (acrylic acid-co-acrylamide) nanocomposites. Polymer Bulletin 55: 419–428.

Chapter 10

Advances in Agro-Nanotechnologies for Pest Management

Maicon S.N. dos Santos,[1] *Carolina E.D. Oro*,[2] *João H.C. Wancura*,[3] *Giovani L. Zabot*[1] and *Marcus V. Tres*[1,*]

Introduction

Agricultural pests cause 20 to 40% of global agricultural production loss each year. In this way, pest control in agriculture is fundamental to ensure a good harvest with high productivity and quality of agricultural products. Good planning and monitoring of plantations to quickly identify any type of pest that can reach plantations are essential to avoid losses in the field. For this, modern agriculture presents technological tools and biological and chemical control strategies, which can be used in a combined way to ensure a good harvest for the farmer, minimizing agricultural losses, costs, and environmental impacts (Worrall et al. 2018).

Damage caused by insects and pests is among the most crucial factors in the productivity reduction of cultures. Agricultural losses may occur in pre-harvest and during storage (post-harvest). The main factors that influence the appearance of pests in different agricultural crops are the lack of crop rotation in agro-ecosystems, planting in regions or seasons favorable to pest attack, unbalanced fertilization, and inappropriate use of pesticides.

At no further moment of history, agri-food systems have been facing such a new and unprecedented set of hazards, including extreme weather conditions, mega-fires, desert grasshoppers, and emerging biological threats, such as Covid-19 pandemic.

[1] Laboratory of Agroindustrial Processes Engineering (LAPE), Federal University of Santa Maria, 1040, Sete de Setembro St., Center DC, Cachoeira do Sul, Brazil.

[2] Department of Food Engineering, Integrated Regional University of Alto Uruguay and Missions, 1621, Sete de Setembro Av., Fátima, Erechim, Brazil.

[3] Department of Teaching, Research, and Development, Sul-Rio-Grandense Federal Institute (IFSul), 81, General Balbão St., Center DC, Charqueadas, Brazil.

* Corresponding author: marcus.tres@ufsm.br

The pandemic of Covid-19 amplified the risks to agri-food systems and produced negative effects on human lives, livelihoods, and economies worldwide. Drought is the first factor of agricultural production losses, followed by floods, pests, diseases, and fires.

The concept of agricultural pest is applicable to any species, race, or biotype of plant, animal, or pathogenic agent that damages plants or plant products. Specific control actions are needed for each type of pest, due to their characteristics of reproduction, survival, and so on. Pest control methods must be selected based on technical parameters of effectiveness, economy, and environmental preservation. The main methods used in pest control are chemical, biological, physical, and mechanical control methods (Rai and Ingle 2012).

Within this context, pesticides act in the prevention and control of insects and pests responsible for large losses in crops. The main modes of action of pesticides are by direct contact of the product on the target, ingestion of leaves that contain the active ingredient, repulsion, or attraction, and subsequent death by the active ingredient. Figure 1 presents a summary of the main modes of action of pesticides.

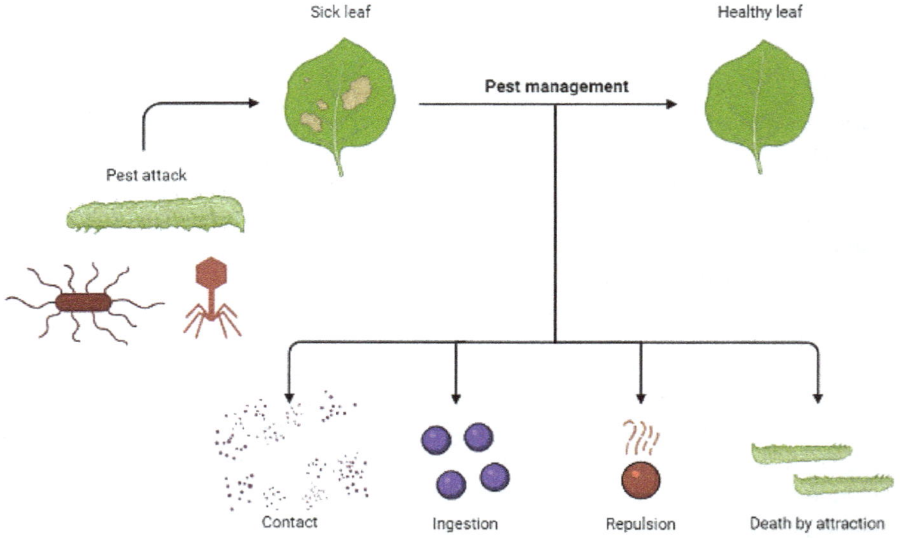

Figure 1: Main modes of action of pesticides for pest management approach.

The main technique for the control of agricultural pests used in the crops is the use of chemical defenses, which prevent the growth of weeds and the emergence of insects and other parasites. New technologies have emerged to reduce the use of these chemical defenses, reducing costs and impacts to the environment. These new strategies can be combined with plant monitoring technologies, which apply defensive controlled and on-demand, in the so-called precision agriculture, capable of reducing the quantities of defensive use.

Another method used is biological control. Biological control uses natural predators of certain types of pests as an alternative to control pests in the crop. It can be used in a complementary way to chemical methods for reducing environmental impacts and costs. In general, biological control consists of the use of living organisms, such as other beneficial insects, predators, parasitoids, and microorganisms, released in the plantation to control a particular pest.

Mechanical control in pest control is based on the use of techniques that allow the direct elimination of pests and physical control consists of the use of methods such as fire, drainage, flooding, temperature, and electromagnetic radiation. In addition to these traditional methods, the use of nanomaterials has gained prominence (Sinha et al. 2017). Nanomaterials have several advantages and applications that will be addressed in this chapter. Essential oils and extracts containing the active ingredients of interest can be encapsulated in nanomaterials and improve field application with the controlled release (Hussain et al. 2017, Maia et al. 2019, Šeregelj et al. 2019, Taban et al. 2020, Hasan et al. 2021).

Accordingly, the objective of this chapter is to describe the application of nanomaterials as potential strategies to control a number of harmful pests considering the agricultural approach. Additionally, the chapter presents information on the main properties and application strategies of agronanomaterials, the interactions of these products with invasive organisms, and technological alternatives that enhance the promising prospect of these materials.

Nanomaterials: Characteristics and functions

In comparison to others compounds, nanomaterials have interesting characteristics, mainly in terms of high surface area and high ability of adsorption in both liquid and gas phases that permits a good efficacy to their employment in distinct applications such as energy, transport, medicine, food safety, environmental science, information technology and homeland security (Saleh 2016).

There are various types of nanomaterials, and they are projected with a diversity of innovative forms that will arise in the future. Nevertheless, based on their construction, nanomaterials are currently categorized as carbon-based, metal-based, dendrimers, and composites (Kabir et al. 2018).

Composite nanomaterials are projected according to properties of interest, focusing on applications or requirements for choosing the matrix, configuration (form), orientation, and curing phase (Sahay et al. 2014). By the way, the adding of nanoparticles to a polymer matrix has been the most typical technique implemented for the preparation of polymer nanocomposites. Baigorria and Fraceto (2022) investigated the carbendazim removal for the decontamination of an aqueous medium employing nanostructured composite hydrogels as an adsorption method. The systems were designed from polymeric networks composed of novel chitosan-based hydrogels with organoclay. According to the results presented by the authors, the adsorption assays returned a kinetic model of pseudo-first-order and Langmuir-type isotherms, with a maximum adsorption capacity observed of $0.4618\ mg\cdot g^{-1}$ to the nanocomposite hydrogels. Also, a total of three utilizations of the nanocomposite were evaluated, where a loss of 60% of the adsorption capacity was observed after the first cycle. Baigorria and Fraceto (2022) described that the results indicated that the use of organoclay as an active element of the hydrogels is crucial to the carbendazim removal process from an aqueous medium, showing the potential of these novel nanocomposite materials for application in the treatment of contaminated water with fungicides.

Carbon-based nanomaterials have been intensively investigated by the scientific and engineering public due to their exclusive and interesting physical-chemical, mechanical, thermal, and optical properties (Mazari et al. 2021). Appropriately, materials composed mainly of carbon, commonly present the form of hollow spheres, ellipsoids or tubes (Saleh 2016). Carbon nanotubes, for example, are strong and hard materials with respect to stiffness. Lu et al. (2022) verified that the mixture of 0.1 wt% of unzipped carbon nanotubes with high specific surface area enhanced the compressive and flexural strengths of a cement paste in 22 % and 51%, respectively. The results demonstrated that this class of nanomaterials present intriguing potential to improve the mechanical properties of cementitious composites. On the other hand, graphene is a one-atom-thick layer of carbon organized in a two-dimensional hexagonal array with notable heat and electrical conductivity, and with optical transparency in the range of infrared and visible light (Kabir et al. 2018). In an interesting research (Kappagantula et al. 2022), analyzed the effect of graphene content (ranging from 0 to 250 ppm) and density of defect on electrical properties of macro-scale 3D composites wires of copper/graphene. The highest enhancement of performance was reached with 15 ppm of graphene (of low defect density), resulting in a 2.7% increase in the electrical conductivity compared with samples containing only Cu. Also, according to results presented by the authors, the highest improvement in relation to the high-defect density graphene was at a concentration of 100 ppm, with a 1.6% of rising in electrical conductivity over the control sample.

Metal-based nanomaterials consist of quantum dots (fluorescent semiconductors with 2 to 10 nm), nanogold, nanosilver, and monometallic oxides such as titanium dioxide, zinc oxide, and iron oxide (Mazari et al. 2021). Kapinder et al. (2021) describe that nanoparticles of gold are inert at bulk scale and become chemically active when their size is reduced due to the wide spacing between the atomic coordinates, which make them suitable for their use in catalytic systems. Zhao et al. (2021) designed a novel multi-residue electrochemical biosensor of nanogold/mercaptomethamidophos to detect simultaneously eleven types of organophosphorus pesticides and the total amount of these compounds, applied to the detection of pesticides in samples of apple and cabbage. The biosensor showed good electrochemical properties, stability, and repeatability, with a wide linear range of 0.1 to 1,500 $ng \cdot mL^{-1}$ and a low detection limit of 0.019 to 0.077 $ng \cdot mL^{-1}$, performing a fast detection of the total quantity of pesticides in the samples analyzed. Similarly, nanoparticles of silver have an elevated surface area that makes them a potential antimicrobial agent as compared to bulk silver (Kapinder et al. 2021). However, their potential to manage agricultural pests still is little explored, enhancing the growth of the scientific community to applied nanoparticles of silver on pest control (Kabir et al. 2018).

Dendrimers are polymers extremely ramified with a structure rigorously defined. They are symmetrical molecules that can reach the size of a human, where their monodisperse structure consists of branches similar to trees, built on a linear polymer core (Kabir et al. 2018). The structural characteristics of dendrimers are well defined and highly uniform, with interesting functional properties for the application of biopesticides (Mlynarczyk et al. 2021). Dendrimers are highly effective carriers with the applied product, mainly due to their size accuracy, low polydispersion, and structures that change functionally and improve their role in the environment (Liu et al. 2015).

Pest-nanomaterial mechanisms of action

Recent advances in environmental and human health issues have directed to the research for alternative strategies that involve pest control and provide minimal influence on the environment (Athanassiou et al. 2018). The application of nanopesticides has been considered for the control of a number of agricultural pests (Kitherian et al. 2021, Pittarate et al. 2021, Shekhar et al. 2021).

Exposure of nanopesticides can be performed via inhalation, dermal contact, and ingestion (Bhan et al. 2018). Inhalation has the strongest potential for control, as it can cause high levels of stress that lead to an increase in cytokines, reactive oxygen species (ROS), and pro-inflammatory mediators, as well as alterations in the membrane and mitrochondrial respiratory chain (Shahzad and Manzoor 2021). At the cellular level, the nanopesticide comes into contact with sulfur in proteins or phosphorus in DNA, leading to rapid denaturation of organelles and enzymes, causing loss of function and, consequently, cell death (Benelli 2018). Ag nanopesticides, for example, promote significant reduction of the enzyme acetylcholinesterase, establishing oxidative stress and cellular breakdown by affecting antioxidants, and detoxifying enzymes (Yadav et al. 2021). Additionally, protein synthesis and gonadotropin release are drastically affected, leading to immunological imbalance in gene regulation. Appropriately, the total disruption of the protective barrier of insects occurs and leads to their death (Deka et al. 2021). On the other hand, TiO_2-based nanoproducts promoted increased insulin signaling pathway and increased fat, carbohydrate and protein metabolism (Tian et al. 2016). ZnO-based products directly influenced the configuration of the thorax, affected the midgut, antennae, anal gills, brush mouth and anal papilla (Banumathi et al. 2017). Considering a series of damages caused by a series of nanoformulations, the main effects on insects are represented in Figure 2.

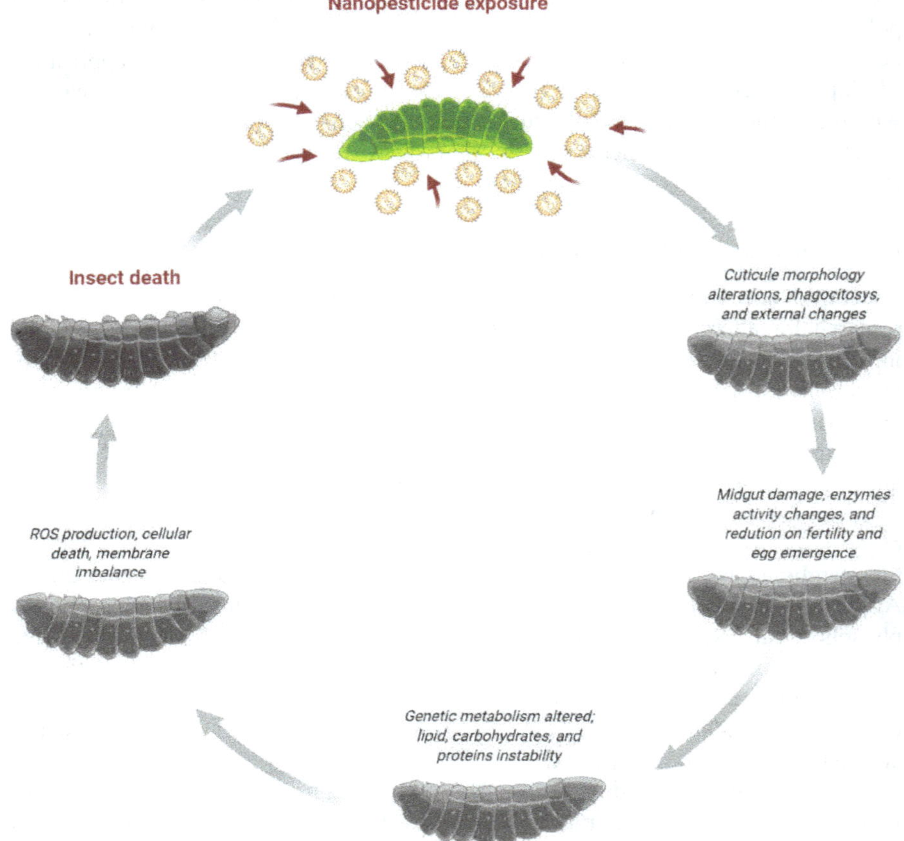

Figure 2: Representation of the toxic mode of action of nanopesticides against insects.

Current applications of nanomaterials for pests control

Insects, larvae and eggs

The African cotton leafworm (*Spodoptera littoralis*) has attracted the attention of researchers in recent years due to its economic impact on cotton cultivation. Besides feeding on the plant, the insects cause the leaves' skeletonization and damage to the roots, causing losses to farmers with the harm to crops. In this scenario, Saad et al. (2021) investigated the larvicidal activity of silver nanoparticles produced from peels extracts of watermelon and pomegranate, residues rich in polyphenols. According to results presented by the authors, the pomegranate-Ag nanoparticles were most efficient larvicidal effect against the first instar larvae instar, followed by the watermelon-Ag nanoparticles, the pomegranate extract, whereas the peel extracts of watermelon were the least efficacious compound.

Malaikozhundan and Vinodhini (2018) investigated the synthesis, physical characterization, and insecticidal effect of nanoparticles of zinc oxide coated with leaf extracts of *Pongamia pinnata* against pulse beetle (*C. maculatus*). *C. maculatus* is the main pest associated with the storage of cowpea worldwide. The insecticidal effect of the nanoparticles evaluated against the cowpea bruchid was established according to the mortality and modifications in the digestive activities of the enzyme that form the insects' midgut. According to the results presented by the authors, the *P. pinnata*-ZnO nanoparticles diminished the hatchability and fecundity of the *C. maculatus* eggs significantly. Applying 25 $\mu g \cdot mL^{-1}$ of *P. pinnata*-ZnO nanoparticles, the authors reported a substantial delay in the larval, pupal, and total development period of the *C. maculatus*. Moreover, *P. pinnata*-ZnO nanoparticles caused total mortality of the insects at treatment with 25 $\mu g \cdot mL^{-1}$ of the bioproduct (the LC_{50} parameter was estimated to be 10.85 $\mu g \cdot mL^{-1}$).

In research with a similar proposal, Jafir et al. (2021) prepared nanoparticles of silver from extracts of *Ocimum basilicum* and analyzed their pesticide action against distinct instar of tobacco cutworm (*Spodoptera litura*). Also, the authors compared the insecticidal potential of the silver nanoparticles with typical synthetic insecticides applied in diverse instar of *S. litura* on Pakistan (Coragen®, Proclaim®, Talstar®, and Tracer®). Assays accomplished by the authors showed that *O. basilicum*-Ag nanoparticles produced the maximum mortality (21.7–96.7%) in the second instar larvae at a rate of 100–1,500 $mg \cdot L^{-1}$ followed by Coragen® (18.3–91.7%), Proclaim® (13.3–78.3%), and Talstar® (13.3–68.3%). These results demonstrated that the *O. basilicum*-Ag nanoparticles are efficient as biogenic composites for the combat against the population of *S. litura* compared to some popular synthetic chemicals that may bring risks for human health as well as the environment.

Nematodes

Nematodes (plant parasites) are among the most dangerous pests with the potential to attack a variety of plants in agrosystems. In this context, Heflish et al. (2021) reported results to the nematicidal action – against nematodes of root-knot *Meloidogyne incognita* – of silver nanoparticles prepared from aqueous extracts of *Acalypha wilkesiana* leaves as a stabilizing agent. The *A. wilkesiana*-Ag nanoparticles were analyzed employing X-ray diffraction, where the crystalline structure of the compound with face-center cubic and 20 nm of size was confirmed. The authors also investigated the nematicidal activity of *A. wilkesiana*-Ag nanoparticles applied *in vitro* against the nematode species *M. incognita* at different concentrations (25, 50, and 100 $mg \cdot mL^{-1}$). The hatching eggs were observed and the larvae movement after 24 and 48 h was evaluated. Results reported that the application of 100 $mg \cdot mL^{-1}$ of Ag-nanoparticles was the most effective treatment (after 48 h), returning 53.3% of nematode mortality and, in general, reducing egg hatching and movement of the larvae.

In an interesting work, Tauseef et al. (2021) proposed the synthesis of copper oxide nanoparticles from the extract of walnut and their employment at different concentrations (50, 100, and 200 ppm) as a nematicidal agent against second-stage juveniles of *Meloidogyne incognita*. Under *in vitro* and *in planta* assays, the authors evaluate the proper timing of application of the CuO nanoparticles to diminish the biotic stress induced by the nematode in cowpea. The use of 100 ppm of CuO nanoparticles three days before the nematode inoculation returned the maximum enhancement in growth, yield, and biochemical attributes to the cowpea plants in comparison to inoculated control samples. However, the study reported that higher concentrations of nanoparticles ($\geq$ 200 ppm) also caused high toxicity to the cowpea plants.

Mites

Wang et al. (2019) evaluated the synergy between nanoparticles of graphene oxide and three usual pesticides used in pest management (beta-cyfluthrin, pyridaben, and chlorpyrifos) as acaricidal against two spider mites (*Tetranychus truncatus* and *T. urticae* Koch). The data presented by the authors showed that the graphene oxide nanoparticles can improve the activity of the pesticides considered in the study. In comparison with the pesticides, the combination between beta-cyfluthrin-graphene oxide, pyridaben-graphene oxide, and chlorpyrifos-graphene oxide presented elevation in the toxicity 1.77-, 1.56-, and 1.55-fold, respectively, against *Tetranychus truncatus*, and 1.50-, 1.75-, and 1.78-fold upper contact toxicity against *T. urticae.* Wang et al. (2019) discussed that the synergistic mechanism is associated with the large surface area of the graphene oxide nanoparticles, that work as a carrier of the pesticides, facilitating their adsorption on the mites' cuticle, reducing pesticide losses, enhancing the dispersibility, and effectiveness of the compounds.

Ticks are associated with several diseases in humans and animals, being responsible for economic losses associated with treatment, prophylaxis, and prevalence, achieving US$573.6 million and US$3.2 billion per year in countries such as Mexico and Brazil (Rodríguez-Vivas et al. 2017). In this scenario, Zaheer et al. (2021) investigated the acaricidal activity of nanoparticles of zinc oxide prepared from lemongrass (*Cymbopogon citratus*) and neem (*Azadirachta indica*), via ethanol extraction, against *Hyalomma* ticks (*in vitro*). Data reported by the authors showed high acaricidal action of the nanoparticles synthesized: the LC_{50} of neem-ZnO nanoparticles for adult tick was 4.76 $mg{\cdot}L^{-1}$ *versus* 4.92 $mg{\cdot}L^{-1}$ to the lemon grass-ZnO nanoparticles; while the LC_{90} of neem-ZnO nanoparticles was 8.87 $mg{\cdot}L^{-1}$ against 9.10 $mg{\cdot}L^{-1}$ to the nanoparticles of lemon grass-ZnO nanoparticles (the mortality of the adult ticks was noted 24 to 48 h after exposure).

Termites

Chotikhun et al. (2018) investigated the resistance of particle-boards panels during 12 weeks, fabricated from Eastern red-cedar (*Juniperus virginiana* L.) employing nanoparticles of SiO_2 and adding modified starch as a binder, against subterranean termites (*Reticulitermes flavipes* Kollar). Results reported by the authors showed that the panels of nanoparticles had only 9.92 to 14.07 wt% of loss in comparison to a conventional radiata pine control sample that had 87.65 wt% of loss. Regarding the damage rating index for a choice-feeding assay, the samples of radiata pine also had the highest value of 4.0 against the lowest value observed for panels with 0.60 $g{\cdot}cm^{-3}$ of density and 1 wt% of SiO_2 nanoparticles (2.25). Panels with

0.60 g·cm^{-3} of density and 3 wt% of SiO_2 nanoparticles presented the lowest average value of loss (9.25 wt%) against 14.08 wt% for the control panels, representing promising results in terms of experimental particle-boards panels fabricated using nano-SiO_2 aiming resistance against subterranean termites.

Mishra et al. (2021) analyzed a greener preparation of nanoparticles of silver utilizing leaf extracts of *Glochidion eriocarpum*, a shrub largely distributed in Asia, and their action against wood termites. The authors tested the proposal that the nanoparticles of Ag reduce the digestive enzymes in the termite gut. The *G. eriocarpum*-Ag nanoparticles synthesis formed spherical materials ranging from 4.0 to 44.5 nm and demonstrated high thermal stability with insignificant loss of weight at 700°C. Results obtained by the authors revealed powerful repellent effectiveness of the *G. eriocarpum*-Ag nanoparticles regarding the termites (80.97%). In the same way, the antifeedant influence of the *G. eriocarpum*-Ag nanoparticles was also found to be interesting, indicating an antifeedant index of 53.49 after one week of tests. Furthermore, the *G. eriocarpum*-Ag nanoparticles contact produced evident morphological alterations in termites where a simulation of molecular docking showed probable attenuation of the xylanase of bacterial origin and endoglucanase, digestive enzymes from termite gut, *via* partial "obstruction" of the catalytic site by the *G. eriocarpum*-Ag nanoparticles.

Agro-nanotechnologies and environmental impacts

Excessive use of products and misuse in management can cause contamination both in agricultural production and in water sources close to the application area. Due to these risks, there is an inspection by the public agency of each country and by the pesticide companies. Accordingly, three important issues must be considered when applying crop protection products: what to apply (product quality), how to apply (application quality), and when to apply (time of application) (Zhang et al. 2019). To define what to apply, a correct analysis of the problem must be conducted to control the economic and environmental impact that this problem may cause. It is considered a pest when a certain species that is harmful to agriculture and its development interferes with the ecosystem and causes epidemic diseases.

Generally, when an agricultural pest is considered, it is possible to associate it to an overpopulation of a certain living being such as mites, insects, bacteria, viruses, and fungi. It causes an imbalance such as devastation of the plantation, extinction of another species, infectious and parasitic diseases. In the global hydrological cycle, nanopesticides from agricultural systems and industrial processing come into direct contact with the soil, mainly due to the lixiviation process. The directing of these residues in the soil to other regions can affect water sources and affect the water supply, minimizing the quality of the resource and exposing humans and animals to direct contact with these products. Accordingly, Figure 3 indicated specific scenarios of nanopesticide-environment association and applications that can generate aggravations for natural resources and human and animal health, mainly. This factor is a result of the strong distribution capacity of nanoparticles and the high potential for bioaccumulation in soil, aquatic environments, food, and animals (Grillo et al. 2021). Furthermore, wastewater treatment plants act directly in the release of nanomaterials, mainly in the disposal of concentrated sludge or specific effluent (Rajput et al. 2020).

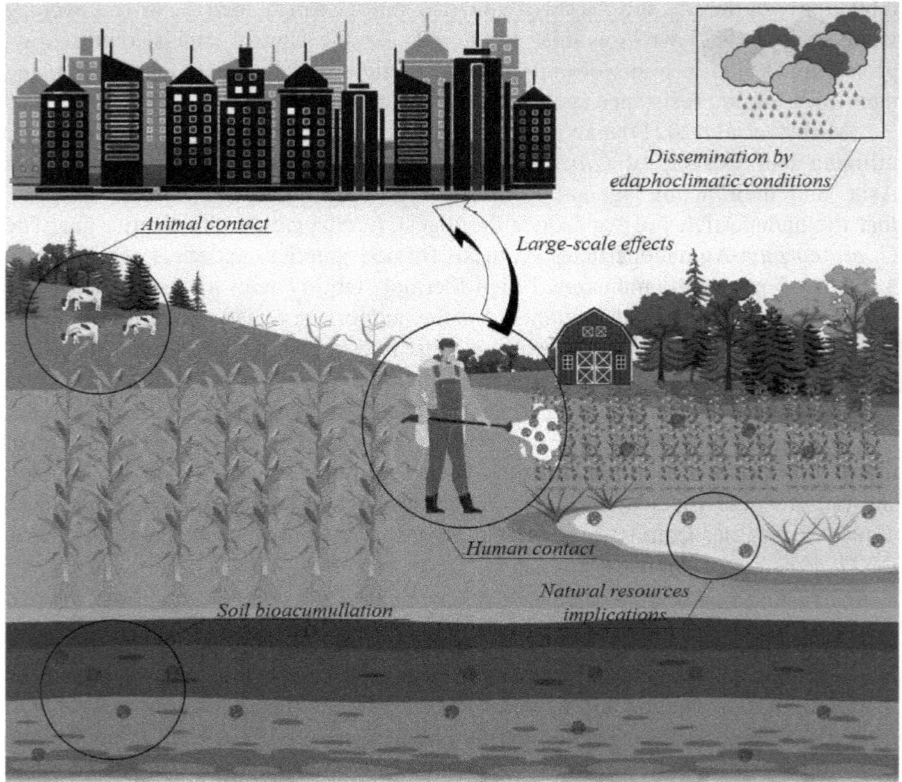

Figure 3: Ecotoxicological performance and dissemination sources of nanopesticides in the environment.

Future perspectives and promising overview

The application of nanomaterials in the management of agricultural pests has reported extremely promising results in recent years. The use of these products in chemical and/or biological action perspectives motivated the adoption of nanoformulations based on nanoparticles, nanoemulsions, etc., that set up a secure and highly effective control mechanism. Nevertheless, chemical pesticides are still the key factor driving the pest control market. Conventional pesticides are associated to the extinction of a diversity of species and harm a significant number of endangered species. However, more than 4 million tons of synthetic-based pesticides are employed annually (Gea et al. 2022). The global market for these products generates a flow of more than US$ 45 billion every year (Pretty and Bharucha 2015). This worrying scenario promotes assertive strategies as management alternatives. Recent innovations in agricultural systems raise pertinent questions regarding the continued use and scale-up of nanopesticides. One of the biggest challenges on the validity of nanopesticide applications involves regulatory and guideline configuration (Deka et al. 2021). Regulatory bodies show the gaps regarding the use of nanopesticides in application programs and projects. Considering the significant range of properties, nanopesticides require risk assessments, especially if instability of environmental conditions and uncertainties of product efficacy are considered when directed in adverse circumstances. Studies report the use of nanopesticides associated to a series of obstacles, from the imbalance of the mechanisms of action of the products, even affecting the biochemical and metabolic profile of plant species, to the insecurity of soil communities, which can lead to serious damage to the environment (Zhang et al. 2019, Grillo et al. 2021). Correspondingly, the optimization of the nanopesticide action under different abiotic conditions, involving conditions of light, humidity, pH, and temperature, can improve the action potential of the product, minimizing risks and impacts to the environment and non-target communities (Chaud et al. 2021). Since approximately 60% of nanoproducts do not act directly in the control of target organisms, the indiscriminate adoption of chemical solvents as action enhancers can drastically affect the chain of non-targeted organisms. Moreover, high levels of toxicity in soil, natural resources, and accumulation in the environment lead to a wide asymmetry in the plant, animal, and human chain (Sun et al. 2019). Therefore, studies of environmental perspectives are one of the main emerging topics for the coming years. Figure 4 illustrates important gaps to be explored in the coming years in the entire context that involves the development and application of nanomaterials in pest management.

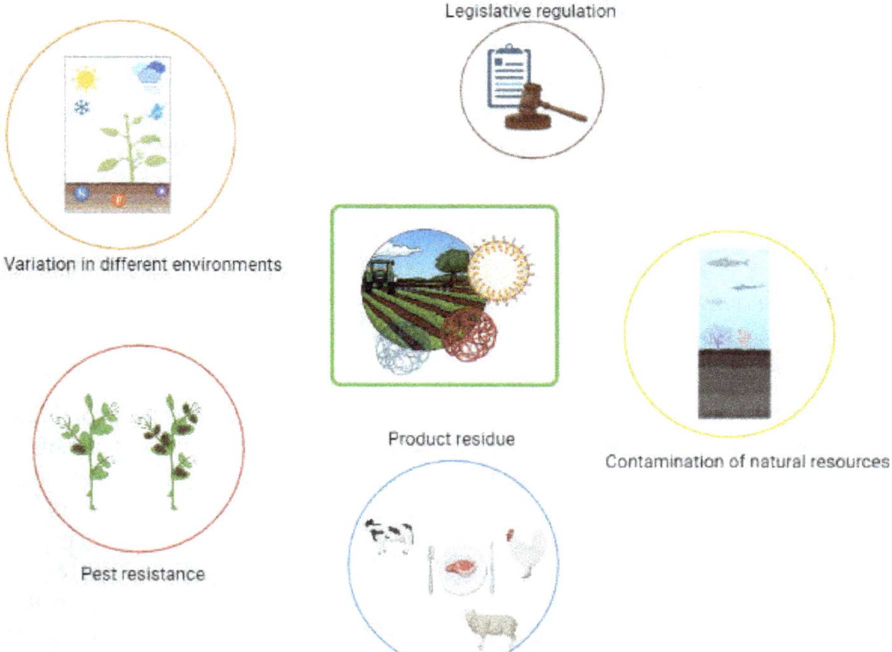

Figure 4: Major challenges of the application of nanopesticides for future perspectives.

Additionally, studies have advertised the importance of economic parameterization of nanopesticides such as cost/benefit efficiency characterization, scale-up feasibility, commercial profile, availability and distribution dynamics, acceptance of smaller-scale farmers, etc. (Gahukar and Das 2020). The safety in the use of nanopesticides has been considered in the scientific literature and on an industrial scale, and the cost of converting synthetic pesticides into nanopesticides is circumvented by the many benefits that these products offer (Zobir et al. 2021). This panorama is verified by the emergence of advanced nanotechnologies, such as the emergence of products that combine plant protection with pest management. Contextually, the complexity of biopesticides that integrate different protection actions in agroecosystems has gained light.

Appropriately, contemporary perspectives on nanotechnology in pest control include (i) strategies to enhance nanoinsecticide action in the field, (ii) micro and macroeconomic analysis of the use of nanopesticides in different environments and based on different technological configurations, (iii) analysis of economic risks based on the direct and successive application of nanoproducts, (iv) quantitative and qualitative estimates of damage to ecosystem biodiversity and toxicological residues in the environment, and (v) regulatory requirements and specifications and legislative exploration of government agencies and organizations that direct guidelines on the use of nanopesticides.

Finally, one of the trending topics in the agricultural field is the expansion of production with minimal impacts on the ecosystem. Based on innovative actions that transcend the frontiers of knowledge, it is safe to provide an assertive vision regarding the success and promising future of nanopesticides. Encouraging results have been obtained established on nanotechnology aimed at pesticides, biopesticides, and plant biodefensives. Accordingly, nanotechnology is considered an effective and positive strategy for agricultural production in the coming years. Considering the population expansion and the pressure exerted on agricultural systems to meet the growing food demand, nanotechnology emerges as an effective and sustainable alternative. Appropriately, the paths of knowledge for the coming years involve the adoption of strategies that explore technology-based tools and exert strong sustainable potential with minimal impacts on biodiversity.

Conclusions

The 2030 Agenda, established by the United Nations (UN), proposes the exploration of sustainable assertions in a variety of aspects of society. Agriculture is one of the most important fields in which the projection of sustainability is a top priority for the coming years. In the contemporary agricultural perspective, the current incidence of pests has resulted in billions of dollars lost annually. This adversity generates numerous other problems that involve significant costs in the management of invasive species and in the excessive use of agrochemicals as a decisive element in the fight against these species. Nonetheless, in addition to economic losses, the direct contact of humans and animals and the exacerbated incidence of synthetic pesticides in the environment and in natural resources pose substantial dangers to society and concerns that aspire to remedy the recurrent ecological impacts. Appropriately, nanotechnology has been an ideal strategy to overcome environmental impacts in pest management processes.

The emergence of practices that involve compensating the strong agricultural demand and establishing sustainable parameters has been gradually explored globally. The use of nanomaterials in pest management is an emerging alternative and outlines the adoption of nanoformulations with economic and environmental benefits. A series of applications of nanopesticides based on different elements has resulted in highly promising results. Among the main benefits of these products, the requirement for lower dosages, the ability to biocontrol, and the degradation of a series of chemical molecules resulting from synthetic applications are among those that most expose the direct use of nanomaterials. This scenario shows nanotechnology as an alternative to remediate contamination and environmental pollution and a range of applications, from remote sensing to crop analysis and quantification of damage caused by pests and diseases. Furthermore, nanomaterials are characterized by an extremely promising area in the context of modern agriculture. Based on the results published so far, nanopesticides lead a new domain of research to be explored on a large scale in the coming years and provide a positive and optimistic scenario regarding the use of environmentally-friendly technologies for the coming decades.

References

Athanassiou, C.G., Kavallieratos, N.G., Benelli, G., Losic, D., Usha Rani, P. and Desneux, N. 2018. Nanoparticles for pest control: Current status and future perspectives. J. Pest Sci. (2004): 91. https://doi.org/10.1007/s10340-017-0898-0.

Baigorria, E. and Fraceto, L.F. 2022. Novel nanostructured materials based on polymer/organic-clay composite networks for the removal of carbendazim from waters. J. Clean Prod. 331: 129867. https://doi.org/10.1016/j.jclepro.2021.129867.

Banumathi, B., Vaseeharan, B., Ishwarya, R., Govindarajan, M., Alharbi, N.S., Kadaikunnan, S., Khaled, J.M. and Benelli, G. 2017. Toxicity of herbal extracts used in ethno-veterinary medicine and green-encapsulated ZnO nanoparticles against *Aedes aegypti* and microbial pathogens. Parasitol. Res. 116: 1637–1651. https://doi.org/10.1007/s00436-017-5438-6.

Benelli, G. 2018. Mode of action of nanoparticles against insects. Environ. Sci. Pollut. Res. 25: 12329–12341. https://doi.org/10.1007/s11356-018-1850-4.

Bhan, S., Mohan, L. and Srivastava, C.N. 2018. Nanopesticides: A recent novel ecofriendly approach in insect pest management. J. Entomol. Res. 42: 263–270. https://doi.org/10.5958/0974-4576.2018.00044.0.

Chaud, M., Souto, E.B., Zielinska, A., Severino, P., Batain, F., Oliveira-Junior, J. and Alves, T. 2021. Nanopesticides in agriculture: Benefits and challenge in agricultural productivity, toxicological risks to human health and environment. Toxics 9(6). https://doi.org/10.3390/toxics9060131.

Chotikhun, A., Hiziroglu, S., Kard, B., Konemann, C., Buser, M. and Frazier, S. 2018. Measurement of termite resistance of particleboard panels made from Eastern redcedar using nano particle added modified starch as binder. Measurement 120: 169–174. https://doi.org/10.1016/j.measurement.2018.02.028.

Deka, B., Babu, A., Baruah, C. and Barthakur, M. 2021. Nanopesticides: A systematic review of their prospects with special reference to tea pest management. Front. Nutr. 8. https://doi.org/10.3389/fnut.2021.686131.

Gahukar, R.T. and Das, R.K. 2020. Plant-derived nanopesticides for agricultural pest control: Challenges and prospects. Nanotechnol. Environ. Eng. 5: 1–9. https://doi.org/10.1007/s41204-020-0066-2.

Gea, M., Zhang, C., Tota, R., Gilardi, G., Di Nardo, G. and Schilirò, T. 2022. Assessment of five pesticides as endocrine-disrupting chemicals: Effects on estrogen receptors and aromatase. Int. J. Environ. Res. Public Health 19(4). https://doi.org/10.3390/ijerph19041959.

Grillo, R., Fraceto, L.F., Amorim, M.J.B., Scott-Fordsmand, J.J., Schoonjans, R. and Chaudhry, Q. 2021. Ecotoxicological and regulatory aspects of environmental sustainability of nanopesticides. J. Hazard Mater. 404. https://doi.org/10.1016/j.jhazmat.2020.124148.

Hasan, M., Ahmad-Hamdani, M.S., Rosli, A.M. and Hamdan, H. 2021. Bioherbicides: An eco-friendly tool for sustainable weed management. Plants 10: 1–21. https://doi.org/10.3390/plants10061212.

Heflish, A.A., Hanfy, A.E., Ansari, M.J., Dessoky, E.S., Attia, A.O., Elshaer, M.M., Gaber, M.K., Kordy, A., Doma, A.S., Abdelkhalek, A. and Behiry, S.I. 2021. Green biosynthesized silver nanoparticles using *Acalypha wilkesiana* extract control root-knot nematode. J. King Saud. Univ.-Sci. 33: 101516. https://doi.org/10.1016/j.jksus.2021.101516.

Hussain, I., Singh, N.B., Singh, A. and Singh, H. 2017. Allelopathic potential of sesame plant leachate against *Cyperus rotundus* L. Ann. Agrar. Sci. 15: 141–147. https://doi.org/10.1016/j.aasci.2016.10.003.

Jafir, M., Ahmad, J.N., Arif, M.J., Ali, S. and Ahmad, S.J.N. 2021. Characterization of *Ocimum basilicum* synthesized silver nanoparticles and its relative toxicity to some insecticides against tobacco cutworm, *Spodoptera litura* Feb. (Lepidoptera; Noctuidae). Ecotoxicol. Environ. Saf. 218: 112278. https://doi.org/10.1016/j.ecoenv.2021.112278.

Kabir, E., Kumar, V., Kim, K.-H., Yip, A.C.K. and Sohn, J.R. 2018. Environmental impacts of nanomaterials. J. Environ. Manage. 225: 261–271. https://doi.org/10.1016/j.jenvman.2018.07.087.

Kapinder, Dangi K. and Verma, A.K. 2021. Efficient & eco-friendly smart nano-pesticides: Emerging prospects for agriculture. Mater. Today Proc. 45: 3819–3824. https://doi.org/10.1016/j.matpr.2021.03.211.

Kappagantula, K.S., Smith, J.A., Nittala, A.K. and Kraft, F.F. 2022. Macro copper-graphene composites with enhanced electrical conductivity. J. Alloys Compd. 894: 162477. https://doi.org/10.1016/j.jallcom.2021.162477.

Kitherian, S., Thangapandi, V. and Jesu Antony, M.R. 2021. Seaweed Lobophora variegata-based silver nanopesticide for environmental friendly management of economically important pest, *Spodoptera litura*. Environ. Nanotechnology, Monit. Manag. 16: 100531. https://doi.org/10.1016/j.enmm.2021.100531.

Lu, D., Shi, X. and Zhong, J. 2022. Understanding the role of unzipped carbon nanotubes in cement pastes. Cem. Concr. Compos. 126: 104366. https://doi.org/10.1016/j.cemconcomp.2021.104366.

Maia, P.D.D.S., dos Santos Baião, D., da Silva, V.P.F., Calado, V.M. de, A., Queiroz, C., Pedrosa, C., Valente-Mesquita, V.L. and Pierucci, A.P.T.R. 2019. Highly stable microparticles of cashew apple (*Anacardium occidentale* L.) juice with maltodextrin and chemically modified starch. Food Bioprocess Technol. 12: 2107–2119. https://doi.org/10.1007/s11947-019-02376-x.

Malaikozhundan, B. and Vinodhini, J. 2018. Nanopesticidal effects of *Pongamia pinnata* leaf extract coated zinc oxide nanoparticle against the Pulse beetle, *Callosobruchus maculatus*. Mater. Today Commun. 14: 106–115. https://doi.org/10.1016/j.mtcomm.2017.12.015.

Mazari, S.A., Ali, E., Abro, R., Khan, F.S.A., Ahmed, I., Ahmed, M., Nizamuddin, S., Siddiqui, T.H., Hossain, N., Mubarak, N.M. and Shah, A. 2021. Nanomaterials: Applications, waste-handling, environmental toxicities, and future challenges—A review. J. Environ. Chem. Eng. 9: 105028. https://doi.org/10.1016/j.jece.2021.105028.

Mishra, S., Wang, W., de Oliveira, I.P., Atapattu, A.J., Xia, S.-W., Grillo, R., Lescano, C.H. and Yang, X. 2021. Interaction mechanism of plant-based nanoarchitectured materials with digestive enzymes of termites as target for pest control: Evidence from molecular docking simulation and *in vitro* studies. J. Hazard. Mater. 403: 123840. https://doi.org/10.1016/j.jhazmat.2020.123840.

Mlynarczyk, D.T., Dlugaszewska, J., Kaluzna-Mlynarczyk, A. and Goslinski, T. 2021. Dendrimers against fungi—A state of the art review. J. Control Release 330: 599–617. https://doi.org/10.1016/j.jconrel.2020.12.021.

Pittarate, S., Rajula, J., Rahman, A., Vivekanandhan, P., Thungrabeab, M., Mekchay, S. and Krutmuang, P. 2021. Insecticidal effect of zinc oxide nanoparticles against *Spodoptera frugiperda* under laboratory conditions. Insects 12: 1–11. https://doi.org/10.3390/insects12111017.

Pretty, J. and Bharucha, Z.P. 2015. Integrated pest management for sustainable intensification of agriculture in Asia and Africa. Insects 6: 152–182. https://doi.org/10.3390/insects6010152.

Rai, M. and Ingle, A. 2012. Role of nanotechnology in agriculture with special reference to management of insect pests. Appl. Microbiol. Biotechnol. 94: 287–293. https://doi.org/10.1007/s00253-012-3969-4.

Rajput, V., Minkina, T., Mazarji, M., Shende, S., Sushkova, S., Mandzhieva, S., Burachevskaya, M., Chaplygin, V., Singh, A. and Jatav, H. 2020. Accumulation of nanoparticles in the soil-plant systems and their effects on human health. Ann. Agric. Sci. 65: 137–143. https://doi.org/10.1016/j.aoas.2020.08.001.

Rodríguez-Vivas, R.I., Grisi, L., Pérez de León, A.A., Villela, H.S., Torres-Acosta, J.F. de, J., Sánchez, H.F., Salas, D.R., Cruz, R.R., Saldierna, F. and Carrasco, D.G. 2017. Potential economic impact assessment for cattle parasites in Mexico. Review. Rev. Mex. Ciencias Pecu 8: 61. https://doi.org/10.22319/rmcp.v8i1.4305.

Saad, A.M., El-Saadony, M.T., El-Tahan, A.M., Sayed, S., Moustafa, M.A.M., Taha, T.F. and Ramadan, M.M. 2021. Polyphenolic extracts from pomegranate and watermelon wastes as substrate to fabricate sustainable silver nanoparticles with larvicidal effect against *Spodoptera littoralis*. Saudi J. Biol. Sci. 28: 5674–5683. https://doi.org/10.1016/j.sjbs.2021.06.011.

Sahay, R., Reddy, V.J. and Ramakrishna, S. 2014. Synthesis and applications of multifunctional composite nanomaterials. Int. J. Mech. Mater. Eng. 9: 25. https://doi.org/10.1186/s40712-014-0025-4.

Saleh, T.A. 2016. Nanomaterials for pharmaceuticals determination. Bioenerg. Open Access 5. https://doi.org/10.4172/2167-7662.1000226.

Šeregelj, V., Tumbas Šaponjac, V., Lević, S., Kalušević, A., Ćetković, G., Čanadanović-Brunet, J., Nedović, V., Stajčić, S., Vulić, J. and Vidaković, A. 2019. Application of encapsulated natural bioactive compounds from red pepper waste in yogurt. J. Microencapsul. 36: 704–714. https://doi.org/10.1080/02652048.2019.1668488.

Shahzad, K. and Manzoor, F. 2021. Nanoformulations and their mode of action in insects: A review of biological interactions. Drug Chem. Toxicol. 44: 1–11. https://doi.org/10.1080/01480545.2018.1525393.

Shekhar, S., Sharma, S., Kumar, A., Taneja, A. and Sharma, B. 2021. The framework of nanopesticides: A paradigm in biodiversity. Mater. Adv. 2: 6569–6588. https://doi.org/10.1039/d1ma00329a.

Sinha, K., Ghosh, J. and Sil, P.C. 2017. New pesticides: A cutting-edge view of contributions from nanotechnology for the development of sustainable agricultural pest control. New Pesticides and and Soil Sensors 47–49. https://doi.org/10.1016/B978-0-12-804299-1.00003-5.

Sun, Y., Liang, J., Tang, L., Li, H., Zhu, Y., Jiang, D., Song, B., Chen, M. and Zeng, G. 2019. Nano-pesticides: A great challenge for biodiversity? Nano Today 28: 100757. https://doi.org/10.1016/j.nantod.2019.06.003.

Taban, A., Saharkhiz, M.J. and Khorram, M. 2020. Formulation and assessment of nano-encapsulated bioherbicides based on biopolymers and essential oil. Ind. Crops Prod. 149: 112348. https://doi.org/10.1016/j.indcrop.2020.112348.

Tauseef, A., Hisamuddin, Gupta J., Rehman, A. and Uddin, I. 2021. Differential response of cowpea towards the CuO nanoparticles under *Meloidogyne incognita* stress. South African J. Bot. 139: 175–182. https://doi.org/10.1016/j.sajb.2021.02.017.

Tian, J.H., Hu, J.S., Li, F.C., Ni, M., Li, Y.Y., Wang, B.B., Xu, K.Z., Shen, W.D. and Li, B. 2016. Effects of TiO_2 nanoparticles on nutrition metabolism in silkworm fat body. Biol. Open 5: 764–769. https://doi.org/10.1242/bio.015610.

Wang, X., Xie, H., Wang, Z. and He, K. 2019. Graphene oxide as a pesticide delivery vector for enhancing acaricidal activity against spider mites. Colloids Surfaces B Biointerfaces 173: 632–638. https://doi.org/10.1016/j.colsurfb.2018.10.010.

Worrall, E.A., Hamid, A., Mody, K.T., Mitter, N. and Pappu, H.R. 2018. Nanotechnology for plant disease management. Agronomy 8: 1–24. https://doi.org/10.3390/agronomy8120285.

Yadav, J., Jasrotia, P., Bhardwaj, A.K., Kumar, S., Singh, M. and Singh, G.P. 2021. Nanopesticides: Current status and scope for their application in agriculture. Plant Prot. Sci. 58: 1–17. https://doi.org/10.17221/102/2020-pps.

Zaheer, T., Imran, M., Pal, K., Sajid, M.S., Abbas, R.Z., Aqib, A.I., Hanif, M.A., Khan, S.R., Khan, M.K., Sindhu, Z.D. and Rahman, S. 2021. Synthesis, characterization and acaricidal activity of green-mediated ZnO nanoparticles against *Hyalomma* ticks. J. Mol. Struct. 1227: 129652. https://doi.org/10.1016/j.molstruc.2020.129652.

Zhang, X., Xu, Z., Wu, M., Qian, X., Lin, D., Zhang, H., Tang, J., Zeng, T., Yao, W., Filser, J., Li, L. and Sharma, V.K. 2019. Potential environmental risks of nanopesticides: Application of $Cu(OH)_2$

nanopesticides to soil mitigates the degradation of neonicotinoid thiacloprid. Environ. Int. 129: 42–50. https://doi.org/10.1016/j.envint.2019.05.022.

Zhao, G., Zhou, B., Wang, X., Shen, J. and Zhao, B. 2021. Detection of organophosphorus pesticides by nanogold/mercaptomethamidophos multi-residue electrochemical biosensor. Food Chem. 354: 129511. https://doi.org/10.1016/j.foodchem.2021.129511.

Zhou, Q., Wu, Y., Sun, Y., Sheng, X., Tong, Y., Guo, J., Zhou, B. and Zhao, J. 2021. Magnetic polyamidoamine dendrimers for magnetic separation and sensitive determination of organochlorine pesticides from water samples by high-performance liquid chromatography. J. Environ. Sci. 102: 64–73. https://doi.org/10.1016/j.jes.2020.09.005.

Zobir, S.A.M., Ali, A., Adzmi, F., Sulaiman, M.R. and Ahmad, K. 2021. A review on nanopesticides for plant protection synthesized using the supramolecular chemistry of layered hydroxide hosts. Biology (Basel) 10. https://doi.org/10.3390/biology10111077.

CHAPTER 11

Botanicals-Based Nanoformulations for the Management of Insect-Pests

Ngangom Bidyarani,[1,#] *Jyoti Jaiswal*[1] and *Umesh Kumar*[2,*]

Introduction

Major factors responsible for causing food damage are insects, pests, pathogens, microorganisms, weeds, drought, temperature, humidity, and salinity (Pandey et al. 2017). Among the mentioned factors, insect pest significantly leads to the damage of plant from the initial stage of seedling to throughout the growth stages (Shukla et al. 2020). From pre to the post-harvest procedure of the grains, the plant growth used to be harmed at different stages depending upon the kinds and life cycle of the insects-pests. The farmers have to spend a lot of money to protect the plant from such losses. Farmer uses extensive synthetic agrochemicals and pesticides to control the crop/plant damages ignoring their complexity (Damalas 2009), but these are costly and pose threat to the non-target beneficial organisms besides causing harm to both human health and environment. This non-regulated extensive use of agrochemicals and pesticides has led to the development of ever-increasing resistance among insects and pests. To minimize the threat caused by synthetic agrochemicals and pesticides, an effective strategy is needed to control the crop losses without harming the non-target beneficial organisms. Management of the current agricultural problems using modern nanotechnology-based methods needs to be implemented so that the overall crop losses can be checked. The most promising area of nanotechnology is site-targeted drug delivery and sustained release of drugs (Patra et al. 2018) which has emerged as a successful strategy in pharmaceutics to deliver engineered nanomedicines (Mu et al. 2018). Through agro-nanotechnology, site-specific

[1] School of Nano Sciences, Central University of Gujarat, Gandhinagar-382030, Gujarat, India.
[2] Nutrition Biology department, School of Interdisciplinary and Applied Sciences, Central University of Haryana, Mahendragarh-123031.
* Corresponding author: umesh.kumar@cug.ac.in, umeshkumar@cuh.ac.in
# ICAR-Central Institute for Cotton Research, Nagpur-440010.

targeted delivery, and controlled release of the agrochemicals and fertilizers (plant nutrients) can be achieved which can lead to a decrease in resistance development among insects, pests, and phytopathogens.

Botanicals are plant or plant-derived products having medicinal or therapeutical properties. This also includes flavour, or aroma which are used as pharmaceutical drugs or as pesticides, insecticides or as antimicrobial agents and are biologically active. They have always been preferred over commercial ones for medicinal and agricultural purposes because of their fewer side effects and environment-friendly nature. Several plant-based materials (plant extract, plant-derived biomolecules) have been reported in the literature for the biosynthesis or green synthesis of different nanomaterials (Peralta-Videa et al. 2016). However, there are very few reports of the synthesis of plant-derived protein nanoparticles for crop protection. Plants used to get damaged by various pests and microorganisms whose control requires hazardous synthetic pesticides and chemicals. Various measures are required to diminish environmental effects and control the number of insect-pests-infected crops. Nowadays, the application of plant-based botanicals (active ingredients) is emerging to control plant diseases (Table 1). Alkaloids, isoflavone, terpenoids, and phenylpropanoids, etc., are substances extracted from plants in the form of secondary metabolites, and they have enormous biological functions (Oliveira et al. 2018).

Control measures for the management of insect's pests

Various practices adopted for the management of insects/pests are:

1. Cultural control (Dhaliwal et al. 2004), which includes altering agricultural practices to affect the life cycle of the insects/pests.

2. Mechanical and physical control, which includes the use of different physical and mechanical devices to block entry and even killing the insect pests (Vincent et al. 2003).

3. Biological control employs the different predators, parasites, pathogens, and competitors (biological factors) to check the population of insects and pests (Divya and Sankar 2009). Also, a method named sterile insect techniques (SIT) to control the population of many insects leading to a reduction in their number (Hendrichs and Robinson 2009, Dunn and Follett 2017, Scott et al. 2017). Another method called incompatible insect technique (IIT) is quite efficient to control different insects/pests and disease vectors (Zhang et al. 2015, Nikolouli et al. 2018).

4. Chemical control is a chemically formulated method for the management of pest population which causes a negative effect on soil structure, fertility, mineral cycles, and soil micro-flora (García-García et al. 2016), affecting nontarget insect/pests and even human beings (Le Goff and Giraudo 2019). But this method is unadvisable unless the population of insects/pests is uncontrollable as it causes several deadly diseases in humans, including endocrine disruption, breast cancer, cytotoxicity, and reproductive toxicity (Gangemi et al. 2016).

Table 1: Different botanicals possessing insecticide and pesticidal properties.

Botanicals	**Plant species**	**Mode/site of action**	**References**
Azadirachtin	*Azadirachta indica*	Inhibits the activity of acetylcholinesterase	Nathan et al. 2008
Carvacrol	*Origanum vulgare*	Binds to nicotinic acetylcholine receptors	Tong et al. 2013
Eugenol	*Syzygium aramaticum*	Stimulates octopamine receptors	Price and Berry 2006
Thymol	*Thymus vulgaris*	Stimulates GABAA receptors	Enan 2005
Linalool	*Aniba rosaeodora*	Competitive and reversible inhibition of acetylcholinesterase activity	Perry et al. 2000, Ryan and Byrne 1988
Alpha terpineol	*Citrus sinensis*	Reduces intracellular levels of second messenger cAMP (tyramine receptors)	Enan 2001
Rotenone	*Derris* sp.	Mitochondrial cytotoxin	Copping and Duke 2007, Isman and Paluch 2011
Pyrethrin	*Chrysanthemum cinerarifolium*	Agonists of voltage-sensitive sodium channels (GABAergic system)	Wakeling et al. 2012, Soderlund and Bloomquist 1989
Cinnamaldehyde	*Cinnamomum zeylanicum*	Antagonist of octopamine	Dayan et al. 2009, Fischer et al. 2013
1,8-Cineol	*Rosmarinus officinalis*	An antagonist of octopamine receptors	Dayan et al. 2009, Fischer et al. 2013
Menthol	*Mentha piperita*	A positive allosteric modulator of the GABAA receptor	Hall et al. 2004
Limonene	Peel of citrus family fruits	Modulate cytochrome p450, epoxide hydrolyase activity, and destabilization of the pathogen's lipid envelope	Zahi et al. 2015, Maltzman et al. 1991
Citronellal	*Cymbopogon nardus*	Antagonist of octopamine	Dayan et al. 2009, Fischer et al. 2013
Thujone	*Artemisia absinthium*	A reversible modulator of GABAA receptors	Höld et al. 2000
Nicotine	*Nicotiana* sp.	Agonist of acetylcholine	Copping and Duke 2007, Isman and Paluch 2011
Geraniol	Various fruits, vegetables, and herbs including rose oil, citronella, lemongrass, lavender, and other aromatic plants	Enhanced penetration of transdermal delivery	Chen et al. 2010

Farmers completely depend on the conventional methods like crop rotation, improved crop varieties, changing crop sowing dates, and integrated pest management (IPM) to minimize crop damages. But such methods are always associated with excessive usage of chemical pesticide/insecticides causing environmental deterioration, and pest-insect resistance development, forcing farmers to look for other replacements (Prasad et al. 2014, 2017a, b). Implementation of botanical encapsulated nanoformulation to achieve target and host-specific delivery could prove to be a more appropriate option for crop protection and lessen pesticide load. Thus, in this chapter, the authors have tried to tabulate the different aspects of agro-nanotechnology to manage the various insect pests (Table 2).

Nanoformulations for insect-pest management

Nanotechnology is providing an opportunity to achieve sustainability in various agricultural practices. Nanotechnology leads to the modernization of agricultural research and developed new tools for the molecular treatment of diseases, early detection of diseases enhancement in the absorption of nutrients by plants. Nanostructured catalysts will lower the doses by increasing the efficiencies of pesticides, insecticides, and herbicides. Pathogens including microorganisms, insects, pests, etc., causing plant diseases and damage to plants are known as phytopathogens. Pesticides and fertilizers are required by the cultivator to control the phytopathogens efficiently achieving the desired crop yield. The effectiveness and solubility of such formulations can be improved using various nanoparticles or adding active ingredients manipulating the size at the nanoscale range (Naderi and Danesh-Shahraki 2013). Nanoparticles encapsulating active ingredients could lead to the protection of plant seeds, foliage, or roots against pathogens like pests, insects, microorganisms, parasites, etc. (Table 3). Various applications of agro-nanotechnology are depicted in (Figure 1).

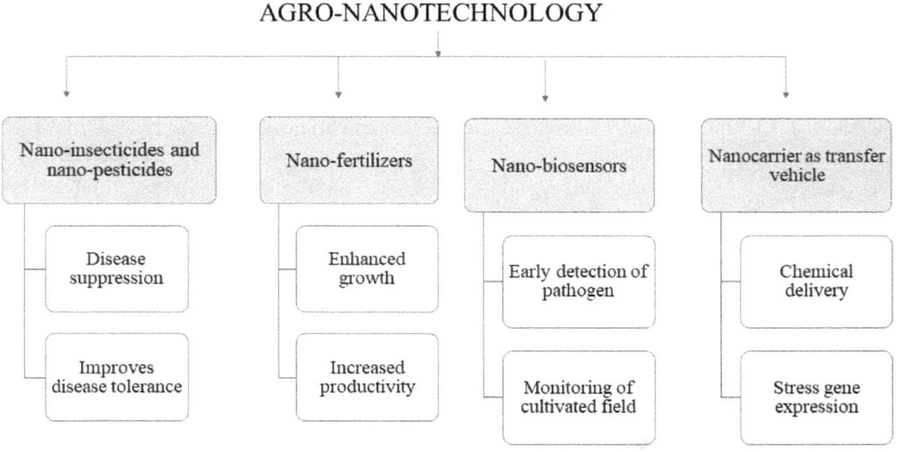

Figure 1: Various aspects of agro-nanotechnology for crop protection.

Table 2: Ideal characteristics, advantages, and constraints of botanicals and perspective of nanotechnology to overcome the constraint.

Ideal characteristics of the botanical-based formulation	Advantage	Constraint	Perspective through nanotechnology
Enhanced efficiency	Content of botanicals (e.g., essential oils) can check insects, pests, diseases through their mode of action	Botanicals may be degraded prematurely by environmental factors like sun rays, temperature, etc., causing shorter span	Nanocarrier system helps in enhancing bioavailability, stability, and protection against premature degradation-controlled release of botanicals due to change in pH, enzyme, temperature, ionic contents, etc.
Fewer effects on non-target species	Natural origin, easy degradation and less toxic to mammals	Some botanicals, e.g., nicotine, may have similar effects to synthetic compounds	Ameliorate efficacy with the least amount of the botanicals through nano-encapsulation and sustained release of botanicals leading to fewer effects to non-target species
No contamination of soil and water resources	Mostly biodegradable, non-persistent in the environment, and low phytotoxicity	–	Biodegradable and biocompatible matrices for better effects
Very few pest resistance	Various AIs are mixed and thus potent action against pests, insects	Amount and effects may vary due to different regions/locations and other abiotic factors; scientific verifications may not be possible to standardize universally the claims and recommendations of the producers	Targeted delivery through functionalization of the carrier system and release of the botanicals when triggered specifically by pH, temperature, moisture, enzymes, etc.
Easy to store	–	Storage of the formulation is hard due to less stability and short shelf life	Improvement in stability during storage and dispersion or solubility of botanicals through nano-encapsulation
Cost-effective	Plenty of plant metabolites. The farmers might have known and used the plant metabolites and extracts. It can be used as a flavouring, for aroma, and medicinal implications	Extraction procedures are generally expensive, and standardization of the metabolites or botanicals may not be possible	Better potency/efficacy, demand, and commercial availability

Table 3: Botanicals encapsulated nanoparticles against insect-pests.

Nanoparticles	Botanicals	Insect-pests	References
PEG	Garlic essential oil	Red flour beetle (*T. castaneum*)	Yang et al. 2009
Chitosan, zinc oxide and chitosan, PCL, PCL, PHB, poly (methyl methacrylate) (PMMA)	Neem extraction and oil (*Azadirachta indica*), Azadirachtin	Tobacco cutworm (*S. litura*) culture ovarian cell lines Sl-1 and Groundnut bruchid (*C. serratus*), *Bemisia tabaci* Biotype B	Lu et al. 2015, Jenne et al. 2018
Silica	α-pinene and Linalool	Tobacco cutworm (*S. litura*) Castor semi-looper (*A. janata*)	Rani et al. 2014
Chitosan	Carvacrol and Linalool	Mite (*T. urticae*), Corn earworm (*H. armigera*) and Mites (*T. urticae*)	Campos et al. 2018
Chitosan/Gum Arabic	Geraniol	Whitefly (*B. tabaei*)	De Oliveira et al. 2018
Chitosan/TPP	*S. hortensis* L. essential oil	Mite (*T. urticae*)	Ahmadi et al. 2018
Zein	Geraniol and R-citronellal essential oils	Mite (*T. urticae*)	Oliveira et al. 2018
PCL	*Rosmarinus officinalis* essential oil	*Tribolium castaneum*	Khoobdel et al. 2017
Chitosan/TPP	Nicotine (4B)	House fly (*M. domestica*)	Yang et al. 2018
PEG and chitosan	*Geranium maculatum* and *Citrus bergamia* essential oils	*Culex pipiens*	Werdin González et al. 2017

Applications of botanicals incorporated nanoformulation in agriculture sector

World population growth and food demand need to stabilize agricultural production to minimize the loss in agriculture fields while transportation and storage. The main loss of cultivation is due to the presence of insects and other pests, which can be controlled by active ingredients derived from natural sources (i.e., botanicals insecticides) and also using nanotechnology to produce new formulations. Therefore, investigation should be targeted towards the development of new formulations (Table 4) which will improve and enhance the effectiveness of botanical active ingredients. In such manner, nanotechnology can make a major contribution to the agriculture sector (Scott and Chen 2012).

Applications of botanicals incorporated nanoformulation in plant disease management

Nanotechnology is a novel agrochemical tool that aids in increasing agricultural yield while reducing the use of hazardous pesticides, controlling plant-pathogen interactions, and adapting plants to provide crop protection (Figure1).

Table 4: Botanical ingredients encapsulated in nanoformulations.

Botanical ingredients	**Nano formulations**	**Applications**	**References**
Garlic essential oil	Polyethylene glycol coated nanoparticles	Control of *Tribolium castaneum* with 80% efficacy	(Yang et al. 2009)
Rotenone	Amphiphilic modified-N-(octadecanol-1-glycidyl ether)-O-sulfate chitosan; sodium caseinate stabilized zein nanoparticles	Enhanced its activity 13,000 times than free rotenone, sodium caseinate stabilized zein NPs loaded with rotenone showed significantly high antimicrobial activity against *P. syringae* and *F. oxysporum*	(Lao et al. 2010) (Bidyarani and Kumar 2019)
β-Cypermethrin	Nano and microemulsion system (oil in water)	Sprayed solution helps to improve its stability	(Wang et al. 2007)
Validamycin	Porous hollow silica nanoparticles	Control release when needed by plant	(Liu et al. 2006)
Azadirachtin	Alginate	Neem oil nanoemulsion	(Jerobin et al. 2012)
Imidacloprid	Alginate (emulsion)	Cytotoxicity in sucking pest	(Kumar et al. 2014)
Extracts of neem	Nano capsules and nanospheres of poly-(ε-caprolactone) polymer	High entrapment efficiency and UV stability	(Rossi et al. 2012)
Azadirachtin	Synthesized 6-*O*-carboxymethylated with anchorage of ricinoleic acid	Improved control of degradation rate and release of insecticide	(Feng and Peng 2012)
Imidacloprid	Poly citric acid and Poly ethylene glycol (PCA-PEG-PCA)	Slow freeing nano capsules	(Memarizadeh et al. 2014)
Carvacrol	Rhamnolipid complexed zein NPs	Antimicrobial activity against *P. syringae* and *F. oxysporum*	Bidyarani et al. 2020

Traditional insecticides are no longer effective against plant pathogenic bacteria and fungus. Due to their high adaptability to environmental changes, this has led to the hunt for more effective active chemicals, particularly in the case of plant pathogenic fungus (Kaur and Garg 2014). The need for more food has led to the use of 2 million tonnes more pesticides around the world each year: 24 percent in the United States, 45 percent in Europe, and 31 percent in the rest of the world (Abhilash and Singh 2009). Only a small percentage of the applied pesticides on field gets utilised for crop protection. The rest (99.9%) is lost through the method of application (air, soil, water) and photodegradation of the active ingredients (Castro et al. 2014). Pesticides and insecticides (Table 5) are nano encapsulated in a favourable form to manage pests and illnesses while reducing the quantity of residues that harm the environment. These technologies, when used in this context, are capable of lowering nitrogen transfer to aquifers while minimising leaching and enhancing plant uptake of nutrients and moderating eutrophication (Liu and Lal 2015). Furthermore, nanoparticles might be used to enhance the stability and function of pesticides by

Table 5: Nanoparticles as carrier of insecticides that target insect pest.

Insecticides	Nanoparticles	Target pest	References
Garlic essential oil	PEG	Red flour beetle	(Yang et al. 2009)
Chlorfenapyr	Silica	Cotton bollworm (*H. armigera*) *P. xylostella*	(Song et al. 2012)
Carvacrol, Linalool	Chitosan	Corn earworm (*H. armigera*) and mites (*T. urticae*)	(Campos et al. 2018)
Geraniol and R-citronellal essential oils	Zein	Mites (*T. urticae*)	(Oliveira et al. 2018)

increasing solubility, resistance towards hydrolysis, improving photodecomposition, and regulated release toward target species (Grillo et al. 2016). Agro-nanotechnology research is progressing because the advantage it might provide could improve crop quality and productivity.

Prospects of nanoformulations

This chapter offers the current possibilities of nanotechnology to improve the efficacy of botanical insecticides, herbicides (Table 6), and fungicides to control the agricultural pests. Use of this strategy is becoming popular because it helps to reduce the effect of agrochemicals on environment as well as on human health. Thus, there is a great demand for its research in this area. The major drawbacks like issue of scalability for nanocarrier production along with its biofunctionality need to be addressed before it gets commercialized. Introducing products that combine the benefits of plants through nanotechnology is less harmful for a wider environment. Still with all these benefits, the use of nanotechnology in food production should be properly studied by doing intensive research to protect any damage to environment as well as human health.

Table 6: Nanoparticles as carrier of herbicides.

Herbicides	Nanoparticles	Target pest	References
Paraquat	Chitosan/tripolyphosphate	Maize Mustard (*Brassica* sp.)	(Grillo et al. 2014)
Imazapic, Imazapyr combined	Alginate/ Chitosan/ tripolyphosphate	Black jack (*B. pilosa*)	(Maruyama et al. 2016)
Atrazine	Poly (lactic-co-glycolic-acid)	Potato plant	(Schnoor et al. 2018)
Atrazine, Simazine	Solid lipid nanoparticles	*R. raphanistrum*	(De Oliveira et al. 2015)

Conclusion

Botanicals are well-known pesticides and insecticides which are eco-friendly and cause either no or minimal harm to humans. Control of insect-pests through chemical pesticides caused extensive harm in many aspects including environment and human. Agro-nanotechnology based nanoformulations could be a promising alternative approach to manage the insect-pests. Botanicals encapsulated nanoformulation could pave out a means to control insect-pests. Moreover, conversion of the pesticides into nanoformulation could definitely lead to further more specificity and targeted delivery. The toxic effects of pesticide can significantly be controlled in a sustainable manner. Thus, harnessing such approach will give a chance to the farmers to sustainably manage insect-pests' population besides providing an opportunity in the agricultural field for extensive research and validation.

References

Abhilash, P.C. and Singh, N. 2009. Pesticide use and application: An Indian scenario. J. Hazard. Mater. 165: 1–12. https://doi.org/10.1016/j.jhazmat.2008.10.061.

Ahmadi, Z., Saber, M., Akbari, A. and Mahdavinia, G.R. 2018. Encapsulation of *Satureja hortensis* L. (Lamiaceae) in chitosan/TPP nanoparticles with enhanced acaricide activity against Tetranychus urticae Koch (Acari: Tetranychidae). Ecotoxicol. Environ. Saf. 161: 111–119.

Bidyarani, N. and Kumar, U. 2019. Synthesis of rotenone loaded zein nano-formulation for plant protection against pathogenic microbes. RSC Adv. 9(70): 40819–40826.

Bidyarani, N., Srivastav, A.K., Gupta, S.K. and Kumar, U. 2020. Synthesis and physicochemical characterization of rhamnolipid-stabilized carvacrol-loaded zein nanoparticles for antimicrobial application supported by molecular docking. J. Nanoparticle Res. 22(10): 1–13.

Campos, E.V., Proença, P.L., Oliveira, J.L., Melville, C.C., Vechia, J.F., Andrade, D.J. and Fraceto, L.F. 2018. Chitosan nanoparticles functionalized with _-cyclodextrin: A promising carrier for botanical pesticides. Sci. Rep. 8: 2067.

Campos, E.V.R., Proença, P.L.F., Oliveira, J.L., Pereira, A.E.S., De Morais Ribeiro, L.N., Fernandes, F.O., Gonçalves, K.C., Polanczyk, R.A., Pasquoto-Stigliani, T., Lima, R., Melville, C.C., Della Vechia, J.F., Andrade, D.J. and Fraceto, L.F. 2018. Carvacrol and linalool co-loaded in β-cyclodextrin-grafted chitosan nanoparticles as sustainable biopesticide aiming pest control. Sci. Rep. 8: 7623. https://doi.org/10.1038/s41598-018-26043-x.

Castro, M.J.L., Ojeda, C. and Cirelli, A.F. 2014. Advances in surfactants for agrochemicals. Environ. Chem. Lett. 12: 85–95. https://doi.org/10.1007/s10311-013-0432-4.

Chen, W. and Viljoen, A.M. 2010. Geraniol—A review of a commercially important fragrance material. South African J. Bot. 76(4): 643–651.

Copping, L.G. and Duke, S.O. 2007. Natural products that have been used commercially as crop protection agents. Pest. Manag. Sci. Former. Pestic. Sci. 63(6): 524–554.

Damalas, C.A. 2009. Understanding benefits and risks of pesticide use. Sci. Res. Essays 4(10): 945–949.

Dayan, F.E., Cantrell, C.L. and Duke, S.O. 2009. Natural products in crop protection. Bioorg. Med. Chem. 17(12): 4022–4034.

De Oliveira, J.L., Campos, E.V.R., Gonçalves Da Silva, C.M., Pasquoto, T., Lima, R. and Fraceto, L.F. 2015. Solid lipid nanoparticles co-loaded with simazine and atrazine: Preparation, characterization, and evaluation of herbicidal activity. J. Agric. Food Chem. 63: 422–432. https://doi.org/10.1021/jf5059045.

De Oliveira, J.L., Campos, E.V.R., Pereira, A.E.S., Nunes, L.E., da Silva, C.C., Pasquoto, T., Lima, R., Smaniotto, G., Polanczyk, R.A. and Fraceto, L.F. 2018. Geraniol encapsulated in chitosan/gum arabic nanoparticles: A promising system for pest management in sustainable agriculture. J. Agric. Food Chem. 66: 5325–5334.

Divya, K. and Sankar, M. 2009. Entomopathogenic nematodes in pest management. Indian J. Sci. Technol. 2(7): 53–60.

Dhaliwal, G.S., Koul, O. and Arora, R. 2004. Integrated pest management: Retrospect and prospect. Integrated Pest Management: Potential, Constraints Challenges, pp. 1–20 (Chapter 1). Edited by Opender Koul, G.S. Dhaliwal, G.W. Cuperus.

Dunn, D.W. and Follett, P.A. 2017. The sterile insect technique (SIT)—An introduction. Entomol. Exp. Appl. 163(3): 151–154.

Enan, E. 2001. Insecticidal activity of Essential Oils: Octopaminergic sites of action. Comp. Biochem. Physiol. Part C Toxicol. Pharmacol. 130(3): 325–337.

Enan, E.E. 2005. Molecular and pharmacological analysis of an octopamine receptor from american cockroach and fruit fly in response to plant essential oils. Arch. Insect. Biochem. Physiol. Publ. Collab. with Entomol. Soc. Am. 59(3): 161–171.

Feng, B.H. and Peng, L.F. 2012. Synthesis and characterization of carboxymethyl chitosan carrying ricinoleic functions as an emulsifier for azadirachtin. Carbohydr. Polym. 88: 576–582. https://doi.org/10.1016/j.carbpol.2012.01.002.

Fischer, A., Mortensen, M.F., Henriksen, P.S., Mathiassen, D.R. and Olsen, J. 2013. Dating the trollesgave site and the bromme culture–chronological fix-points for the lateglacial settlement of southern scandinavia. J. Archaeol. Sci. 40(12): 4663–4674.

Gangemi, S., Miozzi, E., Teodoro, M., Briguglio, G., De Luca, A., Alibrando, C., Polito, I., Libra, M. 2016. Occupational exposure to pesticides as a possible risk factor for the development of chronic diseases in humans (review). Mol. Med. Rep. 14(5): 4475–4488.

García-García, C.R., Parrón, T., Requena, M., Alarcón, R., Tsatsakis, A.M. and Hernández, A.F. 2016. Occupational pesticide exposure and adverse health effects at the clinical, hematological and biochemical level. Life Sci. 145: 274–283.

Grillo, R., Abhilash, P.C. and Fraceto, L.F. 2016. Nanotechnology applied to bio-encapsulation of pesticides. J. Nanosci. Nanotechnol. 16: 1231–1234. https://doi.org/10.1166/jnn.2016.12332.

Grillo, R., Pereira, A.E.S., Nishisaka, C.S., De Lima, R., Oehlke, K., Greiner, R. and Fraceto, L.F. 2014. Chitosan/tripolyphosphate nanoparticles loaded with paraquat herbicide: An environmentally safer alternative for weed control. J. Hazard. Mater. 278: 163–171. https://doi.org/10.1016/j.jhazmat.2014.05.079.

Hall, A.C., Turcotte, C.M., Betts, B.A., Yeung, W.Y., Agyeman, A.S. and Burk, L.A. 2004. Modulation of human GABAA and glycine receptor currents by menthol and related monoterpenoids. Eur. J. Pharmacol. 506(1): 9–16.

Hendrichs, J. and Robinson. A. 2009. Chapter 243 – Sterile Insect Technique A2 - Resh, Vincent H. In Encyclopedia of Insects (2nd Ed).

Höld, K.M., Sirisoma, N.S., Ikeda, T., Narahashi, T. and Casida, J.E. 2000. α-Thujone (the active component of absinthe): γ-aminobutyric acid type A receptor modulation and metabolic detoxification. Proc. Natl. Acad. Sci. 97(8): 3826–3831.

Isman, M.B. and Paluch, G. 2011. Needles in the Haystack: Exploring chemical diversity of botanical insecticides. Green Trends Insect Control. R. Soc. Chem. Cambridge, London, 248–265.

Jenne, M., Kambham, M., Tollamadugu, N.P., Karanam, H.P., Tirupati, M.K., Balam, R.R., Shameer, S. and Yagireddy, M. 2018. The use of slow releasing nanoparticle encapsulated Azadirachtin formulations for the management of Caryedon serratus O. (groundnut bruchid). IET Nanobiotechnol. 12: 963–967.

Jerobin, J., Sureshkumar, R.S., Anjali, C.H., Mukherjee, A. and Chandrasekaran, N. 2012. Biodegradable polymer based encapsulation of neem oil nanoemulsion for controlled release of Aza-A. Carbohydr. Polym. 90: 1750–1756. https://doi.org/10.1016/j.carbpol.2012.07.064.

Kaur, H. and Garg, H. 2014. Pesticides: Environmental impacts and management strategies. Pesticides-Toxic Aspects 8(2014): 187.

Khoobdel, M., Ahsaei, S.M. and Farzaneh, M. 2017. Insecticidal activity of polycaprolactone nanocapsules loaded with *Rosmarinus officinalis* essential oil in *Tribolium castaneum* (Herbst). Entomol. Res. 47: 175–184.

Kumar, S., Bhanjana, G., Sharma, A., Sidhu, M.C. and Dilbaghi, N. 2014. Synthesis, characterization and on field evaluation of pesticide loaded sodium alginate nanoparticles. Carbohydr. Polym. 101: 1061–1067. https://doi.org/10.1016/j.carbpol.2013.10.025.

Lao, S.B., Zhang, Z.X., Xu, H.H. and Jiang, G.B. 2010. Novel amphiphilic chitosan derivatives: Synthesis, characterization and micellar solubilization of rotenone. Carbohydr. Polym. 82: 1136–1142. https://doi.org/10.1016/j.carbpol.2010.06.044.

Le Goff, G. and Giraudo, M. 2019. Effects of pesticides on the environment and insecticide resistance. In Olfactory Concepts of Insect Control – Alternative to Insecticides, pp. 51–78. Springer, Cham.

Liu, F., Wen, L.X., Li, Z.Z., Yu, W., Sun, H.Y. and Chen, J.F. 2006. Porous hollow silica nanoparticles as controlled delivery system for water-soluble pesticide. Mater. Res. Bull. 41: 2268–2275. https://doi.org/10.1016/j.materresbull.2006.04.014.

Liu, R. and Lal, R. 2015. Potentials of engineered nanoparticles as fertilizers for increasing agronomic productions. Sci. Total Environ. 514: 131–139. https://doi.org/10.1016/j.scitotenv.2015.01.104.

Lu, W., Lu, M.L., Zhang, Q.P., Tian, Y.Q., Zhang, Z.X. and Xu, H.H. 2013. Octahydrogenated retinoic acid-conjugated glycol chitosan nanoparticles as a novel carrier of azadirachtin: Synthesis, characterization, and *in vitro* evaluation. J. Polym. Sci. Part A Polym. Chem. 51: 3932–3940.

Maruyama, C.R., Guilger, M., Pascoli, M., Bileshy-José, N., Abhilash, P.C., Fraceto, L.F. and De Lima, R. 2016. Nanoparticles based on chitosan as carriers for the combined herbicides imazapic and imazapyr. Sci. Rep. 6. https://doi.org/10.1038/srep19768.

Maltzman, T.H., Christou, M., Gould, M.N. and Jefcoate, C. 1991. Effects of monoterpenoids on *in vivo* DMBA-DNA adduct formation and on phase I hepatic metabolizing enzymes. Carcinogenesis 12(11): 2081–2087.

Memarizadeh, N., Ghadamyari, M., Adeli, M. and Talebi, K. 2014. Preparation, characterization and efficiency of nanoencapsulated imidacloprid under laboratory conditions. Ecotoxicol. Environ. Saf. 107: 77–83. https://doi.org/10.1016/j.ecoenv.2014.05.009.

Mu, Q., Yu, J., McConnachie, L.A., Kraft, J.C., Gao, Y., Gulati, G.K. and Ho, R.J.Y. 2018. Translation of combination nanodrugs into nanomedicines: Lessons learned and future outlook. J. Drug Target 26(5-6): 435–447.

Naderi, M.R. and Danesh-Shahraki, A. 2013. Nanofertilizers and their roles in sustainable agriculture. Int. J. Agric. Crop Sci. 5(19): 2229–2232.

Nathan, S.S., Choi, M.Y., Seo, H.Y., Paik, C.H., Kalaivani, K. and Kim, J.D. 2008. Effect of Azadirachtin on Acetylcholinesterase (AChE) activity and histology of the brown planthopper nilaparvata Lugens (Stål). Ecotoxicol. Environ. Saf. 70(2): 244–250.

Nikolouli, K., Colinet, H., Renault, D., Enriquez, T., Mouton, L., Gibert, P., Sassu, F., Cáceres, C., Stauffer, C., Pereira, R. et al 2018. Sterile insect technique and Wolbachia symbiosis as potential tools for the control of the invasive species Drosophila suzukii. J. Pest. Sci. 91(2): 489–503.

Oliveira, J.L.D., Campos, E.V.R., Pereira, A.E.S., Pasquoto, T., Lima, R., Grillo, R., Andrade, D.J. De Santos, F.A. Dos Fraceto, L.F. 2018. Zein nanoparticles as eco-friendly carrier systems for botanical repellents aiming sustainable agriculture. J. Agric. Food Chem. 66: 1330–1340. https://doi.org/10.1021/acs.jafc.7b05552.

Pandey, P., Irulappan, V., Bagavathiannan, M.V. and Senthil-Kumar, M. 2017. Impact of combined abiotic and biotic stresses on plant growth and avenues for crop improvement by exploiting physio-morphological traits. Front. Plant Sci. 8: 537.

Patra, J.K., Das, G., Fraceto, L.F., Campos, E.V.R., Rodriguez-Torres, M.d.P., Acosta-Torres, L.S., Diaz-Torres, L.A., Grillo, R., Swamy, M.K., Sharma, S. et al. 2018. Nano based drug delivery systems: Recent developments and future prospects. J. Nanobiotechnol. 16(1): 1–33.

Peralta-Videa, J.R., Huang, Y., Parsons, J.G., Zhao, L., Lopez-Moreno, L., Hernandez-Viezcas, J.A. and Gardea-Torresdey, J.L. 2016. Plant-based green synthesis of metallic nanoparticles: Scientific curiosity or a realistic alternative to chemical synthesis? Nanotechnol. Environ. Eng. 1(1): 4.

Perry, N.S.L., Houghton, P.J., Theobald, A., Jenner, P. and Perry, E.K. 2000. *In-vitro* inhibition of human erythrocyte acetylcholinesterase by salvia lavandulaefolia essential oil and constituent terpenes. J. Pharm. Pharmacol. 52(7): 895–902.

Prasad, R., Kumar, V. and Prasad, K.S. 2014. Nanotechnology in sustainable agriculture: Present concerns and future aspects. Afr. J. Biotechnol. 13(6): 705–713.

Prasad, R., Bhattacharyya, A. and Nguyen, Q.D. 2017a. Nanotechnology in sustainable agriculture: Recent developments, challenges, and perspectives. Front. Microbiol. 8: 1014.

Prasad, R., Gupta, N., Kumar, M., Kumar, V., Wang, S. and Abd-Elsalam, K.A. 2017b. Nanomaterials act as plant defense mechanism. pp. 253–269. *In*: Prasad, R., Kumar, V. and Kumar, M. (eds.). Nanotechnology. Springer, Singapore.

Price, D.N. and Berry, M.S. 2006. Comparison of effects of octopamine and insecticidal essential oils on activity in the nerve cord, foregut, and dorsal unpaired median neurons of cockroaches. J. Insect Physiol. 52(3): 309–319.

Rani, P.U., Madhusudhanamurthy, J. and Sreedhar, B. 2014. Dynamic adsorption of _-pinene and linalool on silica nanoparticles for enhanced antifeedant activity against agricultural pests. J. Pest Sci. 87: 191–200.

Rossi, M., Fernandes da Silva, M.F. das, G. and Batista, J. 2012. Secondary metabolism as a measurement of efficacy of botanical extracts: The use of *Azadirachta indica* (Neem) as a model. Insectic. - Adv. Integr. Pest Manag. https://doi.org/10.5772/27961.

Ryan, M.F. and Byrne, O. 1988. Plant-insect coevolution and inhibition of acetylcholinesterase. J. Chem. Ecol. 14(10): 1965–1975.

Schnoor, B., Elhendawy, A., Joseph, S., Putman, M., Chacón-Cerdas, R., Flores-Mora, D., Bravo-Moraga, F., Gonzalez-Nilo, F. and Salvador-Morales, C. 2018. Engineering atrazine loaded poly (lactic-co-glycolic acid) nanoparticles to ameliorate environmental challenges. J. Agric. Food Chem. 66: 7889–7898. https://doi.org/10.1021/acs.jafc.8b01911.

Scott, M.J., Concha, C., Welch, J.B., Phillips, P.L. and Skoda, S.R. 2017. Review of research advances in the screwworm eradication program over the past 25 years. Entomol. Exp. Appl. 164(3): 226–236.

Scott, N. and Chen, H. 2012. Nanoscale science and engineering for agriculture and food systems. Ind. Biotechnol. 8: 340–343. https://doi.org/10.1089/ind.2012.1549.

Soderlund, D.M. and Bloomquist, J.R. 1989. Neurotoxic actions of pyrethroid insecticides. Annu. Rev. Entomol. 34: 77–96.

Song, M.R., Cui, S.M., Gao, F., Liu, Y.R., Fan, C.L., Lei, T.Q. and Liutt, D.C. 2012. Dispersible silica nanoparticles as carrier for enhanced bioactivity of chlorfenapyr. J. Pestic. Sci. 37: 258–260. https://doi.org/10.1584/jpestics.D12-027.

Shukla, N., Meghvanshi, K. and Shukla, J.N. 2020. Role of nanotechnology in the management of agricultural pests. *In*: Biogenic Nano-Particles and their Use in Agro-ecosystems, pp. 85–98. Springer, Singapore.

Tong, F., Gross, A.D., Dolan, M.C. and Coats, J.R. 2013. The phenolic monoterpenoid carvacrol inhibits the binding of nicotine to the housefly nicotinic acetylcholine receptor. Pest Manag. Sci. 69(7): 775–780.

Vincent, C., Hallman, G., Panneton, B. and Fleurat-Lessard, F. 2003. Management of agricultural insects with physical control methods. Annu. Rev. Entomol. 48: 261–281.

Wang, L., Li, X., Zhang, G., Dong, J. and Eastoe, J. 2007. Oil-in-water nanoemulsions for pesticide formulations. J. Colloid Interface Sci. 314: 230–235. https://doi.org/10.1016/j.jcis.2007.04.079.

Wakeling, E.N., Neal, A.P. and Atchison, W.D. 2012. Pyrethroids and their effects on ion channels. Pestic. Chem. Bot. Pestic. Rijeka, Croat. InTech. 39–66.

Werdin Gonzalez, Jorge Omar, Emiliano Nicolás Jesser, Cristhian Alan Yeguerman, Adriana Alicia Ferrero and Beatriz Fernández Band. 2017. Polymer nanoparticles containing essential oils: New options for mosquito control. Environmental Science and Pollution Research 24(20): 17006–17015.

Yang, F.L., Li, X.G., Zhu, F. and Lei, C.L. 2009. Structural characterization of nanoparticles loaded with garlic essential oil and their insecticidal activity against Tribolium castaneum (Herbst) (Coleoptera: Tenebrionidae). J. Agric. Food Chem. 57: 10156–10162. https://doi.org/10.1021/jf9023118.

Yang, Y., Cheng, J., Garamus, V.M., Li, N. and Zou, A. 2018. Preparation of an environmentally friendly formulation of the insecticide nicotine hydrochloride through encapsulation in chitosan/tripolyphosphate nanoparticles. J. Agric. Food Chem. 66: 1067–1074.

Zahi, M.R., Liang, H. and Yuan, Q. 2015. Improving the antimicrobial activity of d-limonene using a novel organogel-based nanoemulsion. Food Control 50: 554–559. https://doi.org/https://doi.org/10.1016/j.foodcont.2014.10.001.

Zhang, D., Zheng, X., Xi, Z., Bourtzis, K. and Gilles, J.R.L. 2015. Combining the sterile insect technique with the incompatible insect technique: I—Impact of Wolbachia infection on the fitness of triple and double-infected strains of Aedes albopictus. PLoS One 10(4): e0121126.

Section III

Emerging Technologies for the Management of Plant Diseases

Chapter 12

Strategic Applications of CRISPR-Cas Technology for the Management of Plant Diseases

Sarika R. Bhalerao,[1,*] *Namrata D. Patil*,[1] *Pramod G. Kabade*,[2] *Vaishnavi M. Tattapure*[1] and *Mahendra Rai*[3]

Introduction

Agriculture is essential for a society's survival and growth. The objective of agriculture in the twenty-first century is to ensure food security while minimizing environmental impact. Diseases are major and important constraints that threaten agricultural development and global food security. Diseases limit crop yields by impairing plant growth and development, as well as lowering the quality of agricultural goods in the field and during storage. According to statistical data, diseases globally reduce crop yields by as much as 20–40% (FAO, http://www.fao.org/news/story/en/item/280489/icode/). At present, chemical pesticides are still the mainstay of crop disease control. However, these pesticides are frequently detrimental to humans and the environment, either directly or indirectly (Damalas et al. 2011). Because pesticides are not always extremely specific, they can also influence other organisms while killing pathogens, thereby disturbing ecological balances. Furthermore, harmful microbes may acquire pesticide resistance, necessitating the development of new pesticides and/or the administration of additional pesticides. Therefore, reducing the dependence of food production on chemical pesticides is a key goal to avoid their negative environmental impact, especially in developing countries (Tilman et al. 2002).

[1] Department of Plant Biotechnology, Vilasrao Deshmukh College of Agricultural Biotechnology, V.N.M.K.V., Latur, Maharashtra, India.
[2] Centre for Plant Biotechnology and Molecular Biology (CPBMB), Kerala Agricultural University, Vellanikara, Thrissur, Kerala, India.
[3] Department of Biotechnology, Sant Gadge Baba Amravati University, Amravati, Maharashtra, India.
* Corresponding author: sarikasshende@gmail.com

On the other hand, plant breeding is an effective and environmentally beneficial technique for developing disease-resistant crop types for sustainable agriculture. Over the last few decades, conventional disease resistance breeding has been successful, but it has significant drawbacks such as the introgression of resistance (R) genes and labour intensiveness. In addition, crossing/selfing between the two most compatible plants, unavailability of sufficient genetic variation in plant population, and transfer of undesirable genes/traits along with desirable resistance gene/traits are supreme limitations in conventional breeding for resistance (Gao 2018). Therefore, it is a big challenge for conventional breeding in keeping up with pathogen evolution and rising food demand, particularly during an era of global climate change. Although mutation breeding and transgenic technology are also in use, they both have drawbacks that make them less demanding nowadays.

These challenges to our current agricultural practices suggest the need for the introduction of newer and quicker means, i.e., genome editing technologies, referred to as GETs of creating materials for breeding new cultivars. There are three major types of sequence-specific nucleases for genome editing currently available: zinc finger nucleases (ZFNs), transcription activator-like effector nucleases (TALENs), and the clustered regularly interspaced short palindromic repeats/CRISPR-associated protein (CRISPR/Cas) system. Among that, CRISPR is a popular method currently employed to develop desirable plant materials for sustainable food production (Zaidi et al. 2019). Unlike conventional mutagenesis techniques, the CRISPR/CRISPR-associated protein 9 (Cas9) system can achieve efficient and transgene-free gene editing in plants through different ways, i.e., protoplast transformation or direct bombardment of guide RNA (gRNA) and Cas9 to plant cells (Liang et al. 2018), transient expression of CRISPR/Cas9 reagents delivered at the callus stage (Zhang et al. 2016), CRISPR/Cas9-derived cytidine base editors technology (Veillet et al. 2019) and transgene killer CRISPR (TKC) technology (He et al. 2019). Nevertheless, geminiviral DNA replicon can enhance gene targeting frequencies 1-fold to 2-fold higher than conventional *Agrobacterium tumefaciens* T-DNA transformation. Therefore, virus based gRNA delivery system for CRISPR/Cas9 mediated plant genome editing (VIGE) has been used as an effective tool for genome editing (Yin et al. 2015). However, such advancements have substantially increased the CRISPR toolset. The main emphasis of this chapter is the applications, challenges, and promises of genome editing techniques, particularly CRISPR/Cas9, in the enhancement of disease resistance in plants.

Concept

Genome editing is a technique for making precise alterations to the genomic DNA of a cell or organism. The core of genome editing technology is the use of sequence-specific nucleases for recognizing specific DNA sequences and producing double-stranded DNA breaks (DSBs) at targeted sites. DSBs are repaired mainly *via* two pathways: the nonhomologous end-joining (NHEJ) pathway and the homologous recombination (HR) pathway (Voytas and Gao 2014). In most cases, cells use the NHEJ pathway to repair DSBs. However, NHEJ is error-prone and usually results in insertion or deletion mutations. In the presence of a donor DNA template, DSBs are likely to be repaired by the HR pathway, which results in precise base changes or gene replacement (Yin et al. 2019).

The development of efficient genome-editing techniques, particularly CRISPR/SpCas9, has emerged as a broad array of probable uses that could be explored in plant pathogens. The ability to modify plant pathogen genomes offers the possibility to confer desirable phenotypes for numerous purposes (Zhang et al. 2018a). Compared to traditional methods for genetic manipulation of the microbial genome, which are usually associated with inefficient homologous recombination, the CRISPR/SpCas9 tools are highly efficient and much simpler in some cases. Besides, they provide a high-throughput experimental platform to dissect gene function at the whole genome level in plant pathogens.

The CRISPR/Cas system has been applied in a wide range of host plants and plant pathogens for dissecting the molecular mechanisms underlying plant-pathogen interactions and improving host resistance to bacteria, fungi, oomycetes, DNA, and RNA viruses. The CRISPR/Cas system is a useful tool for making gene loss-of-function and gain-of-function in mutants, as well as for better understanding plant-pathogen interactions and decreasing the damage caused by destructive pathogens in agricultural applications (Gosavi et al. 2020).

Genome editing organism vs genetically modified organism

Global demand for food is increasing day by day due to an increase in population and shrinkage of the arable land area. To meet this increasing demand, there is a need to develop high-yielding varieties that are nutritionally enriched and tolerant against environmental stresses. Mutagenesis, intergeneric crosses, and translocation breeding are some of the techniques being developed to improve crop quality. Later, as genetic engineering progressed, genetically modified crops developed the transgene insertion technique, which helps them to resist adverse conditions. For regulating genetically modified crops and their risk analysis on the environment and public health, process or product-focused techniques are applied. Recent advances in gene editing technologies, such as site-directed nucleases, zinc finger nucleases, and the clustered regularly interspaced short palindromic repeats (CRISPR)/CRISPR-associated protein 9 (Cas9) have ushered in a new era of plant breeding by allowing precise gene editing without the transfer of foreign genes. However, there is always a debate about whether or not these tactics should be regulated and whether or not they should be accepted by the general public (Gupta et al. 2021).

Since 1996, when GM crops were first introduced, the area under cultivation has expanded 113 times and is constantly expanding. GM crops are expected to rise in popularity over the next decade, based on current projections. These crops looked to be the most quickly accepted crop in modern agriculture history, with the least amount of pesticide use (Chen and Lin 2013). In the beginning, commercial GM crops such as insect-resistant Bt-cotton and pest-resistant crops were introduced. Growers benefit from these crops because of their higher yield. Some crops, such as golden rice, which has a high β-carotene content, and GM maize, which has had its ascorbate level enhanced to avoid scurvy disorders, have been introduced for large-scale production in the interest of consumer benefit (Naqvi et al. 2009).

In the United States, in terms of GM crop regulation, most governments are dedicated to risk assessment, as proposed by the Food and Agricultural Organization (FAO), the Economic Cooperation and Development (ECD), the World Health Organization (WHO), and other comparable organizations. However, the assessment is largely viewed as inadequate in detecting the unintended environmental consequences of GM crops. In many nations, including India, the health risk is the most significant factor against the commercialization of GM crops. It is claimed that the introduction of high-profile evaluation approaches can reduce unexpected consequences (Gupta et al. 2021).

The rapid advancement of new generation gene editing technologies such as CRISPR-cas9 can make gene editing more accurate and lessen the chance of unintended mutations, lowering the health risk. Gene editing is considered a second-generation technology, with the CRISPR/Cas9 system being the most recent. This method is based on the bacterial immune system's resistance to invading viruses. In traditional GM crops, the first generation has been developed by using different genetic components of plant pests including Cauliflower mosaic virus and *Agrobacterium tumefaciens* for insertion of the gene of interest. In contrast, in the approach of edited plants with the CRISPR/Cas9 and other nucleases, it has been claimed that these are without any pest sequences (free of transgenes) (Doudna and Charpentier 2014). The United States Department of Agriculture (USDA) has exempted many altered plants from GM crop regulation, allowing them to be sold without the pre-market field study required for GM crops. With further exclusions from GMO rules, this revolutionary approach can be applied to a wider range of fields, gaining greater public acceptance.

Table 1: Comparison between genetically modified vs genome edited organisms.

Genetically modified (first generation)	Genome editing (second generation)
The introduction of a gene is unregulated	It is possible to introduce a gene at a specific site
Random integration of the transgene	Editing that is extremely precise
High off effects	Low off effects
Comparatively expensive	Comparatively reliable and cheap

Previous techniques before CRISPR

Mega nucleases (MNs)

Meganucleases are sequence-specific endonucleases with large (> 14 bp) cleavage sites that can deliver DNA double-strand breaks (DSBs) at specific loci in living cells. It is used for genome engineering, the homing endonucleases or meganucleases derived from naturally occurring proteins encoded by mobile introns (Smith et al. 2006, Paques and Duchateau 2007). These endonucleases have large DNA sequence recognition signatures (ranging from 12 to 40 bp) and are natural mediators of gene targeting. One feature that distinguishes mega nucleases from ZFNs and TALENs is that the cleavage and DNA-binding domains overlap (Stoddard 2011). Engineering new DNA sequence specificity is therefore challenging, as the catalytic activity of the enzyme is often compromised when the amino acid sequence is altered to achieve new DNA sequence specificity (Smith et al. 2006). Mega nucleases typically act as dimers of two identical subunits; however, they can be linked to function as a single peptide (Grizot et al. 2009). Despite the need for dimerization, mega nucleases are the smallest genome engineering reagents (~ 165 amino acids for a mega nuclease monomer compared with ~ 310 amino acids for a ZFN monomer and ~ 950 amino acids for a TALEN monomer), which makes them amenable to delivery to cells, particularly if the cargo capacity of the vector is limited. Mega nucleases are difficult to engineer for new target specificities; advances in high-throughput screening and structure-based design will likely make custom mega nucleases easier to create and consequently more accessible.

TALENs

The ability of genome engineering has been further enhanced by transcription activator-like effector nucleases (TALENs) based on bacterial transcription activator-like effectors (TALEs). The TALEN DNA binding domain is derived from proteins generated by the *Xanthomonas* genus of plant bacterial diseases. During infection, *Xanthomonas* delivers transcription activator-like (TAL) effectors to the plant cell that binds to specific plant gene promoters and an expression. This, in turn, leads to increased pathogen virulence (Kay et al. 2007, Romer et al. 2007). This DNA binding region, known as the repeat variable di-residues (RVDs), specifically binds the DNA (Christian et al. 2010, Miller et al. 2011). Shortly after the discovery of TALEs, TALE nucleases (TALENs) were developed that, like ZFNs, are a fusion protein comprised of a TALE and a FokI nuclease (Christian et al. 2010, Miller et al. 2011, Li et al. 2012, Reyon et al. 2012). Unlike ZFNs, the design and engineering of TALENs is much simpler and can be done in a shorter time (Cermak et al. 2011, Reyon et al. 2012). The binding and specificity of TALENs are achieved by the protein's central domain, which comprises 13–28 copies of a tandemly repeated 34 amino acid sequence, with one repeat specifically binding to a single DNA base. With the exception of two sites at amino acids 12 and 13 that define DNA binding. These repeats are nearly identical. The TAL effector's unique property led to the development of TALENs, which combine the TAL effector's modular DNA binding domain with the FokI endonuclease's catalytic domain. Custom TAL arrays are simpler to develop than ZFNs, and they bind to their DNA targets with great efficiency (90%). Due to the 1:1 TALE DNA binding, TALENs are not as confined in target site selection as ZFNs (Zhu et al. 2013).

TALENs have been shown to be more specific and less cytotoxic than ZFNs. TALENs, however, are substantially larger than ZFNs, requiring 3 kb of cDNA encoding for one TALEN versus just 1 kb for a single ZFN. This makes delivery of a pair of TALENs more challenging than a pair of ZF due to delivery vehicle cargo size limitations. Further, packaging and delivery of TALENs in some viral vectors may be problematic due to the high level of repetition in the TALENs sequence. However, the construction of novel TALE arrays can be cumbersome and relatively costly, as a newly designed protein is required for each target sequence. In addition, the large size of TALENs, along with the necessity for a pair of proteins to recognize antiparallel

DNA strands and induce a DSB, makes TALENs less suitable for multiplex gene editing (Mussolino et al. 2011).

ZFNs

The first designed endonucleases that identify and cut chromosomal DNA were zinc-finger proteins. The DNA recognition component of ZFNs is made up of a fusion of modular DNA binding proteins that may be modified to bind specific recognition sequences in double-stranded DNA. They are made up of artificial bipartite enzymes (a ZFN monomer has 310 amino acids) linked together by a linker peptide. Two ZFNs are required to bring FokI monomers into close proximity with the DNA and cleave the target sequence. This further minimizes off-target cleavage and reduces cytotoxic effects that are thought to be the consequence of spurious cleavage by some ZFNs (Voytas 2013). ZFNs that act in pairs, each pair recognizing two sequences flanking the target cleavage site, have been developed to solve off-target problems. One ZFN binds the forward strand, and the second ZFN binds the reverse strand, ZFPs function as dimers, and thus recognize 18–36 bp of target DNA sequence, making ZFNs highly target specific (Urnov et al. 2005). ZF arrays that recognize novel DNA sequences were initially made by stringing together individual ZFs with known sequence specificities (Bae et al. 2003, Segal et al. 2003). This approach, called a modular assembly, met with success for some ZF arrays; in general, however, success rates were low (Ramirez et al. 2008).

In spite of numerous successful demonstrations of ZFN-mediated genome editing in various organisms, the design and construction of large modular proteins that properly match the triplet code is time-consuming and costly. Despite the availability of solutions to address these constraints, the general use of programmable ZFNs is still limited due to the relatively high rate of failure for cleavage of the desired DNA sequence. Zinc finger nucleases (ZFNs) are credited as the first genome editing technology that used programmable nucleases to achieve a breakthrough in genome engineering (Cabaniols et al. 2010).

RNA interference

RNA interference (RNAi) is a plant defense tool or gene silencing technique in which a complementary short RNA attaches to its target gene either before or after transcription (at the promoter), resulting in the downregulation of target gene expression. By expressing an externally introduced RNAi construct containing double-stranded hairpin RNA structures complementary to the targeted gene, RNAi can be used to silence any target gene in plants or their pathogens (Eamens et al. 2010, Rosa et al. 2018). Researchers first demonstrated that RNAi plants with a longer self-complementary hairpin RNA (hpRNA) are more effective in 1998. hpRNA transgenes have been extensively applied to silence plant viral RNAs since then. For the development of fungal and bacterial disease-resistant plants, the host gene silencing-hairpin RNAi (HGS-hpRNAi) approach can be utilised to down-regulate the expression of numerous plant susceptibility genes (Rani et al. 2020).

The superiority of CRISPR over the previous technology

ZFNs and TALENs were the most widely used prior to the introduction of CRISPR/Cas9 as a genome editing technology. ZFNs and TALENs are targeted using DNA-binding proteins that recognize DNA sequences, whereas CRISPR/Cas9 technique requires an RNA-guided targeting approach. CRISPR/Cas9 technology has already been shown to be more versatile, easier to develop and implement, and less expensive than other genome editing technologies because of this unique approach. One of the primary barriers to the scientific community using ZFNs and TALENs as genome editing tools is the complexity and difficulty of protein design and synthesis. In comparison, sgRNA-based cleavage relies on a simple Watson-Crick base pairing with the target DNA sequence, and only 20-nt in the sgRNA needs to be modified to recognize a different target.

Furthermore, despite the numerous advantages of CRISPR/Cas9 as a system for engineering single gene mutations, current research frequently requires generating plants with mutations in multiple genes; the ability to generate plants with multiple mutations is critical for analysing the functions of gene family members for studying epistatic relationships between different genetic pathways. Generating plants carrying multiple mutated genes using ZFNs or TALENs has proven to be

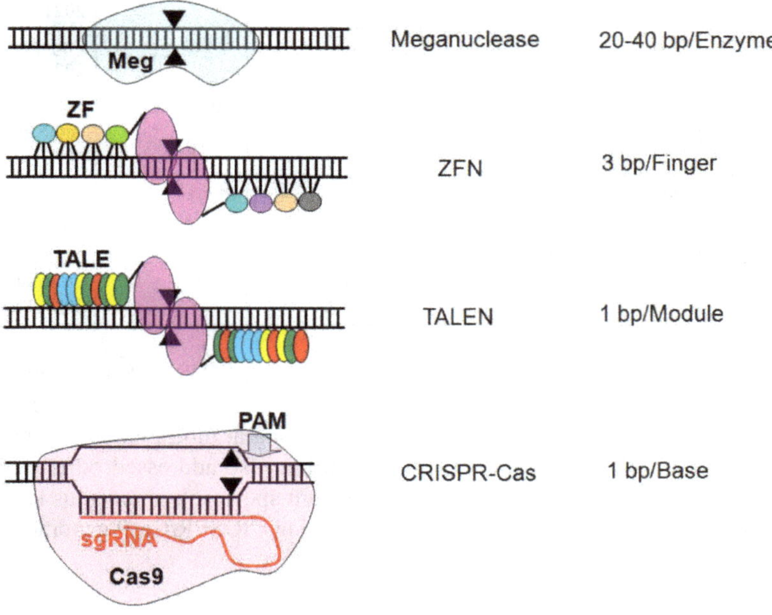

Figure 1: Action of the targeted nucleases and their DNA binding mode. Genomic DNA is shown horizontally in black and double stranded, with the site of DNA cleavage indicated by arrowheads. For meganuclease, the holoenzyme binds and cleaves the target DNA. For ZFN and TALENs, they function as a pair, with one zinc finger DNA binding domain (ZF) binding to the upper strand while the other ZF binding to the lower strand. Once this fused FokI enzyme (purple oval) is oriented to form a homodimer, it is activated to cut DNA. For CRISPR-Cas, the Cas9 holoenzyme (orange oval) is directed to the target site by the guide RNA and cleave the DNA at the position close to the PAM motif (grey arrow). (Adapted from Yu et al. (2016); an open-access book chapter).

Table 2: Comparison of genome editing techniques.

Role	ZFNs	TALENs	MNs	CRISPR/Cas9	References
Efficacy of target recognition	Higher	Higher	Higher	Higher	Rahim et al. 2021
Kind of Action	Double stranded break in target DNA	Double stranded break in target DNA	Direct conversions in targeted regions	Double stranded break in target DNA	Es et al. 2019
Mutagenesis	Higher	Middle	Middle	Lower	Rahim et al. 2021
Multiplexing	Difficult	Difficult	Difficult	Possible	Sauer et al. 2016
Target range	Unlimited	Unlimited	Unlimited	Limited by PAM	Bhardwaj and Nain 2021
Effects	Lower	Lower	Lower	Lower	Khandagale and Nadaf 2016
Cost	Higher	Higher	Higher	Low	Prajapat et al. 2021
Crop improvement	Low	Low	Low	Higher	Prajapat et al. 2021
Range	Narrow	Narrow	Narrow	Broad	Prajapat et al. 2021
Dimerization	Required	Not required	Not required	Not required	Khandagale and Nadaf 2016
Types	One	One	One	Many	Khandagale and Nadaf 2016
Future use	Medium	Medium	Medium	High	Prajapat et al. 2021

inefficient, time-consuming and labour intensive. With CRISPR/Cas9, however, it has been reported that multiple gRNAs bearing different sequences can be used to achieve high-efficiency multiplex genome engineering at multiple loci (Cong et al. 2013, Mali et al. 2013a). CRISPR/Cas9 system has also addressed other issues in plant genome engineering. Critically, in some plant species, homozygous knockout mutants can be generated in a single generation using CRISPR/Cas9 genome editing technology (Zhang et al. 2014).

Timeline of CRISPR

Many scientists from around the world contributed to the discovery of the CRISPR-Cas microbial adaptive immune system and its ongoing development into a genome editing tool. From the initial discovery to the first demonstrations of CRISPR-mediated genome editing, this timeline provides a concise history of seminal contributions and the scientists who pushed this field forward.

Table 3: A timeline of milestones of the CRISPR/Cas9 genome editing.

Sr. No.	Year	Event	References
1.	1987	First record of CRISPR cluster repeats was reported in *Escherichia coli*	Ishino et al. 1987
2.	1995	First insight on CRISPR functionality	Mojica et al. 1995
3.	2000	Recognition and approval that CRISPR families are present throughout prokaryotic organisms	Mojica et al. 2000
4.	2002	Description of term CRISPR along with defined signature Cas genes was coined	Jansen et al. 2002
5.	2005	Identified foreign origin of spacer; Proposed as adaptive immunity function	Mojica et al. 2005, Pourcel et al. 2005
6.	2007	The CRISPR-Cas complex was successfully demonstrated and it was the first experimental evidence for CRISPR adaptive immunity	Barrangou et al. 2007
7.	2008	CRISPR system can act upon specified DNA targets; Spacers are converted into mature crRNAs which act as a small guide RNAs	Brouns et al. 2008
8.	2009	Cmr/Type-IIIB CRISPR complexes cleave RNA	Hale et al. 2009
9.	2010	Cas9 is directed by spacer sequences and cleaves target DNA via double strands breaks	Garneau et al. 2010
10.	2011	Tracer RNA designs a twofold structure with crRNA in association with cas9: Type II CRISPR systems are transposable and can be heterologously expressed in different organisms	Deltcheva et al. 2011
11.	2012	*In vitro* characterization of DNA targeting by cas9	Jinek et al. 2012
12.	2013	First demonstration of cas9 genome engineering in prokaryotic cells	Cong et al. 2013
13.	2014	Description of genome-wide functional screening with cas9: Crystal structure of cas9 in complex with guide RNA can rewrite DNA efficiently	Wang et al. 2014
14.	2015	CRISPR/Cas systems were successfully refined against numerous Gemini viruses	Ali et al. 2015
15.	2016	Evolution of resistance against three ssRNA plant viruses through CRISPR/Cas 9 mediated mutagenesis of host gene elF4F that codes for a protein required for virus replication	Chandrasekaran et al. 2016
16.	2017	Cas 13 used for RNA editing; CRISPR based ultra-sensitive diagnostic platform	Cox et al. 2017, Abudayydeh et al. 2017, Gootenberg et al. 2017

What is meant by CRISPR and Cas proteins?

CRISPR

CRISPR/Cas organization occurs in two forms of classes and is further categorized into six types (I–VI) and 27 subtypes (Makarova et al. 2015, Koonin et al. 2017, Shmakov et al. 2017). A large number of archaea (such as in entire hyperthermophiles) and bacteria are known to consist of the Class 1 CRISPR–Cas system in their genomes,

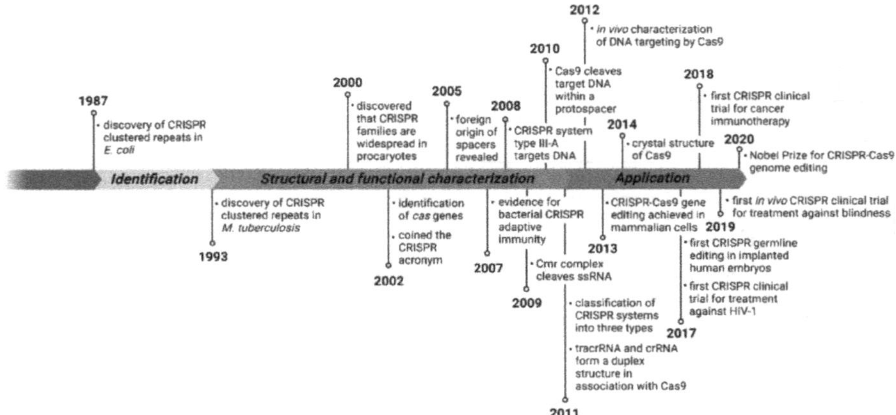

Figure 2: Timeline of the Clustered Regularly Interspaced Palindromic Repeats (CRISPR)–Cas system with important milestones (Adopted from Nidhi et al. 2021; Open access journal).

whereas the Class 2 system is known to exist in bacteria, but not in hyperthermophiles (Chylinski et al. 2014, Makarova et al. 2015). Based on the character of the nuclease effector, Class 1 system comprises types I, III and IV which have multisubunit Cas protein effector complexes whereas Class 2 system includes types II, V and VI with single protein effector modules. The nuclease effector proteins are necessary at interference stage (Makarova et al. 2011, 2013, Shmakov et al. 2015). CRISPR/Cas systems targeting the DNA viruses are type I, II and V, whereas type VI targets the RNA viruses. However, type III is both DNA and RNA targeting CRISPR/Cas system, though the target for type IV system has not yet been identified (Koonin et al. 2017).

Type I CRISPR-Cas system

In addition to the presence of the conserved Cas1 and Cas2 proteins, Type I is defined by the ubiquitous presence of a signature protein, the Cas3 helicase/nuclease. Cas3 is a large multidomain protein with distinct DNA nuclease and helicase activities (Sinkunas et al. 2011). In addition, there are multiple Cas proteins that form CASCADE-like complexes that are involved in the interference step. Many of these proteins are in distinct RAMP superfamilies (Cas5, Cas6, Cas7). Of the three systems, Type I, thus far, is the most diverse with six different subtypes (Type I-A through Type II-F) (Makarova et al. 2011). The Type I CRISPR system is believed to target DNA, and cleavage requires Cas3 [which has a histidine, aspartic acid (HD) nuclease domain] or Cas4, a RecB-family nuclease (Sinkunas et al. 2011). The Type I CRISPR-Cas system in *E. coli* is one of the best characterized and recent experiments using *E. coli* (Jore et al. 2011). Multiple studies in *Pseudomonas aeruginosa* have also shed light on the Type I CRISPR-Cas mechanism of action. For DNA interference, CASCADE associates with processed crRNA to form a ribonucleoprotein complex that drives the formation of R-loops in invasive double-stranded DNA (dsDNA) through seed sequence-driven base pairing.

Type II CRISPR-Cas system

This system is typified by the Cas9 signature protein, a large multifunctional protein with the ability to generate crRNA, as well as target phage and plasmid DNA for degradation (Garneau et al. 2010). Cas9 appears to contain two nuclease domains-one at the N terminus (RuvC-like nuclease) and an HNH (McrA-like) nuclease domain in the middle section (which might be involved in target cleavage based on its endonuclease activity). Type II is the simplest of the three CRISPR/Cas types, with only four genes that comprise the operon which include cas9, cas1, cas2, and either cas4 or csn2. There are two subtypes-Type IIA (or CASS4 that includes csn2) and Type IIB (or CASS4a that includes cas4). The best-studied Type II system is that of *Streptococcus thermophilus*, which has been shown to provide defence against bacteriophage and plasmid DNA (Barrangou et al. 2007). It was also recently established that a trans-encoded small CRISPR RNA (tracrRNA) is involved in the processing of pre-crRNA into crRNA in Type II systems through the formation of a duplex with the CRISPR repeat sequence. Mature crRNA, together with Cas9, interferes with matching invasive dsDNA by homology-driven cleavage within the protospacer sequence, in the direct vicinity of the PAM. Mismatches at the 3' end of the protospacer and/or in the PAM allow phages and plasmids to circumvent CRISPR-encoded immunity (Garneau et al. 2010).

Type III CRISPR-Cas system

This system has several recognizable features, including the signature RAMP protein, Cas10, which is likely involved in the processing of crRNA and possibly also in target DNA cleavage (Anantharaman et al. 2010), and is somewhat functionally analogous to the Type I CASCADE. The Type III system also contains the signature Cas6, involved in crRNA processing and additional RAMP proteins likely to be involved in crRNA trimming. The universal cas1 and cas2 genes are mostly in operon-like structures with the rest of the cas genes but are not always in the same operon as the RAMP proteins in the Type III systems. So far, two type III systems have been distinguished-Type IIIA and IIIB. In *Pyrococcus furiosus*, a Type IIIA system, the target of CRISPR interference is mRNA, whereas in *Staphylococcus epidermidis*, a Type IIIB system, the target is DNA (Marraffini and Sontheimer 2008). This highlights the polymorphic nature of CRISPR/Cas systems, even within the Type III systems. The distribution of the three CRISPR/Cas systems has some notable features, with the Type I system being found in both bacteria and archaea. In contrast, Type II is exclusively present in bacteria, whereas the Type III systems appear more commonly in archaea, although it is also found in bacteria (Makarova et al. 2011).

Although no large-scale detailed distributional or functional analysis is yet available, there are several examples of species that contain more than one CRISPR/Cas type. Horizontal gene transfer *via* plasmids that harbor CRISPR/Cas loci or by other gene transfer mechanisms such as transposon activity has been implicated in the movement of CRISPR/Cas loci across widely diverged lineages (Horvath et al. 2009).

CRISPR RNA (crRNA)

After the recognition of foreign genetic material, especially of bacteriophages, bacterial cell transcribes repetitive DNA to produce CRISPR-RNA (crRNA) along with cas genes, which then produces Cas protein. The crRNA consists of two components, that is, a spacer at the 50 ends, which is a short RNA segment that matches perfectly to a sequence from bacteriophages, and a repetitive DNA sequence at the 30 ends. In addition to these two components, crRNA also contains transactivating RNA (tracrRNA), which provides the binding site for the Cas proteins. This crRNA is then combined with Cas proteins to produce crRNA effector complexes and this assembly will then destroy the foreign genetic material once it matches exactly with the spacer sequence in crRNA. Therefore, the infection is essentially ended, even before it starts to spread in bacteria. Usually, a 25 bp short conserved sequence is situated either immediate upstream or downstream of the target sequence of foreign genetic material, which is also called as a protospacer adjacent motif (PAM sequence; 50-NGG-30 for Cas9). This PAM sequence has an important role to play in target DNA selection.

In the case of novel bacteriophage infection, bacterial cells do not have the complementary spacer DNA sequence with them that will match foreign genetic material. Therefore, under such circumstances, bacterial cell produces class I type of Cas proteins, which degrades the foreign genetic material piece by piece and inserts a copy of it into CRISPR cascade as another spacer DNA sequence. So, the next time whenever this bacteriophage infects again, this spacer DNA will be used along with Cas proteins to degrade the genetic material of foreign invasion (Deveau et al. 2008, 2010, Bhaya et al. 2011, Wagh and Pohare 2019). Doudna and Charpentier (2014) used the mechanism of RNA-led DNA interference to develop a very potent RNA-guided DNA targeting tool for genome editing (Koonin et al. 2017).

Transactivating CRISPR RNA (tracr RNA)

TracrRNA is a small trans-encoded RNA. Emmanuelle Charpentier discovered it while researching the human pathogen *Streptococcus pyogenes*. It is the second component of guideRNA. TracrRNA acts as a scaffold for the endonuclease Cas 9 protein. In other words, tracrRNA serves as a handle to direct Cas9 to the target DNA. TracrRNA is made up of 42 nucleotides. It exists in conjunction with crRNA.

A tracrRNA is necessary for Type II CRISPR/Cas systems because it aids in the maturation of crRNA (Charpentier et al. 2015). In order to create an RNA duplex, the tracrRNA base pairs with a pre-crRNA and is partially complementary to it. RNase III, an RNA-specific ribonuclease, cleaves this to create a hybrid crRNA/tracrRNA. The endonuclease Cas9 uses this hybrid as a guide as it cleaves the invasive nucleic acid (Deltcheva et al. 2011, Brouns 2012, Jinek et al. 2012).

Guide RNA (gRNA)

One of the two main elements of the CRISPR/Cas9 immune system is guide RNA, or gRNA. It is a unique RNA sequence made up of two components, tracrRNA and crRNA. When gRNA detects the target DNA, Cas proteins are urged to break the double-stranded DNA. In order to accomplish this, tracrRNA serves as a handle for Cas proteins whereas crRNA is made up of a complementary sequence of target DNA.

A crucial stage in the CRISPR-Cas9 gene editing technique is designing the ideal gRNA. As a result, the effectiveness of the CRISPR system's editing process depends on the gRNA's proper sequence. From a transfected plasmid, gRNA can be expressed in cells. Host cells produce gRNA when cloned plasmids are introduced into them. The most frequent gRNA is made up of 100 base pairs.

Cas proteins

CRISPR loci often have groups of conserved protein-encoding genes, named cas genes, in their vicinity (Makarova et al. 2002). Based on computational analyses, Cas proteins were predicted to contain identifiable domains characteristic of helicases, nucleases, polymerases, and RNA-binding proteins, which led to the initial speculation that they may be part of a novel DNA repair system (Makarova et al. 2002). The order, orientation, and groupings of cas genes appear to be extremely variable, and this picture grows ever more complex as the number of annotated genomes increases. Attempts to classify Cas proteins have been made, but this has proven difficult because of the diversity of the proteins involved (Haft et al. 2005). Initially, Jansen's group identified four gene families, cas1–4 (Jansen et al. 2002), which were then extended to include cas5 and cas6. Haft and colleagues defined eight subtypes of Cas proteins based on the phylogeny of the highly conserved Cas1 protein and the operonic organization of cas genes, which were named after eight representative organisms that contained a single CRISPR/Cas locus [e.g., *E. coli* Cas proteins were designated cse1 (CRISPR system of *E. coli* gene1); other subtypes included *Aeropyrum* (csa), *Desulfovibrio* (csd), *Haloarcula* (csh), *Mycobacterium* (csm), *Neisseria* (csn), *Thermotoga* (cst), and *Yersinia* (csy)]. These initial categories, although useful, cannot easily handle the relationships between homologous but distantly related Cas proteins, the extensive variability that exists in cas operons, or organisms that contain multiple CRISPR loci (Haft et al. 2005).

In a new and unified classification system based on multiple criteria, including evolutionary relationships of conserved proteins and cas operon organization, several groups (Makarova et al. 2011) working on CRISPR/Cas systems have proposed a consensus view that the CRISPR/Cas system can be divided into two partially independent subsystems. The first consists of an information processing module and requires the universally present core proteins, Cas1 and Cas2, which are involved in new spacer acquisition. The second, or executive, subsystem is required for the processing of primary CRISPR transcripts (crRNA) and recognition and degradation of invading foreign nucleic acid and is quite diverse. For instance, in certain CRISPR sub-types, the multisubunit CASCADE is involved in the processing of the crRNA, whereas in other types a single multifunctional protein may play this role. In addition, there are several repeats associated with mysterious proteins (RAMPs) that constitute a large superfamily of Cas proteins. RAMPs contain at least one RNA recognition motif (RRM; it is also called the ferredoxin-fold domain), and some have been shown to be involved in pre-crRNA processing (Ebihara et al. 2006, Haurwitz et al. 2010).

Types of Cas proteins

Table 4: Selected Cas proteins and their functions (Clark et al. 2019, Makarova et al. 2020).

Protein	Association in		Function
	Type	Subtype	
Cas1	I, II, IV, IV (assumed)	III-A, III-B	DNA nuclease
Cas2	I, II, V	III-A, III-B, VI (some)	RNA nuclease
Cas3	I		DNA nuclease and helicase
Cas4	II, V	I (most)	DNA nuclease
Cas5	IV	I-C, III (some)	pre-crRNA processing
Cas6		I (most), III-A, III-B	pre-crRNA processing
Cas7	I, III, IV		RNA recognition, crRNA binding
Cas8		I (most)	Large subunit of Cascade complex
Cas9	II		DNA nuclease
Cas10		I (some), III (most)	Large subunit of Csm or Cmr complex
Cas11	III	I (some), IV (some)	Small subunit of effector complexes
Cas12	V		crRNA processing, DNA nuclease
Cas13	VI		crRNA processing, RNA nuclease

Cas9 protein

The Cas9 endonuclease from *Streptococcus pyogenes* (Class II category) has emerged as a very powerful platform for genome engineering because it requires only a single effector Cas protein, which is much easier to handle than class I CRISPR systems, which require a multiprotein effector complex for genome modifications (GMs). Cas9 can be easily directed to new targets by changing the sequence of its single guide RNA (sgRNA), making it a plug-and-play tool for GMs. Notably, sgRNA consists of two components, that is, tracrRNA, which is a Cas9 binding domain, and crRNA, which binds to the target sequence specifically (Jinek et al. 2012, Doudna and Charpentier 2014). Cas9-mediated GMs involve two steps, namely, DNA double-strand breaks (DSB), which is followed by DNA repair. Multiplexing stage to generate multiple mutations can be achieved by using co-expression of Cas9 and several RNA guides (Cong et al. 2013).

Cas9 activity

Several Cas9 proteins have been found and exploited for gene editing in various species, like *Staphylococcus aureus* (SaCas9) (Ran et al. 2013), *Neisseria meningitidis* (NmCas9) (Hou et al. 2013), and *Staphylococcus thermophiles* (St1Cas9) (Kleinstiver et al. 2015). Each has its own PAM sequences and activity levels. As a result, selecting a certain Cas9 ortholog for a given target sequence may improve gene editing efficiency and should be considered as part of the gene editing system design. Other factors, in addition to the intrinsic activity of a Cas9 protein, have been proven to influence activity. For gene editing in eukaryotic cells, Cas9 must translocate into the nucleus. In these systems, the nuclear location signal (NLS) is connected to the Cas9 protein. Increasing access to the NLS by adding a 32 amino acid spacer between the NLS and Cas9 was shown to increase DNA cleavage activity (Shen et al. 2013). Increased on-target cutting activity was also shown when the relative concentration of sgRNA to Cas9 protein was increased, likely because all Cas9 proteins formed the active ribonucleoprotein complex (Kim et al. 2014). However, excessive sgRNA was also shown to increase off-target effects (Fu et al. 2013). Finally, in comparison with other enzymes, the activity of Cas9 is quite low, with a single turnover rate of 0.3–1.0 min^{-1} (Jinek et al. 2012). Once bound to the target DNA sequence, displacement of Cas9 from the DNA strand, even after DSB formation, is challenging –1 nM Cas9 cleaved 2.5 nM plasmid DNA after 120 min (Jinek et al. 2012). Cas9 is thus more like a single-shot actuator than a catalytic enzyme. While this property may be advantageous in some situations, such as gene activation/inhibition or short-lived activity for gene editing with fewer off-target effects, it may be inconvenient in other situations when catalytic activity is required.

Novel and enhanced CRISPR/Cas systems

Researchers have worked to enhance CRISPR systems in order to increase their specificity, efficacy, and consistency despite the aforementioned limitations.

Cas12a (Cpf1)

Type V Cas12a and Cas9 are similar in that they both rely only on RNA molecules to create double-strand breaks (DSBs), and as a result, they are both categorized as Class 2 CRISPR systems (Shmakov et al. 2015, Fonfara et al. 2016). However, unlike Cas9, it just needs a crRNA molecule to direct it towards its target, as opposed to Cas9's dual guidance of a crRNA and tracrRNA. Additionally, the resulting DSBs are staggered cuts with 5-nt 5′-overhangs as opposed to the blunt cuts produced by Cas9 (Zetsche et al. 2015). Additionally, while Cas9 enzymes recognize PAMs with a high G content, Cas12a prefers to attach to targets with a high T-rich PAM site (Gao et al. 2017). The fact that Cas12a outperforms Cas9 has additional benefits. One pre-crRNA template can be delivered to the cell, where it is then cleaved by Cas12a into various crRNA molecules targeting different genes, facilitating multiplex gene editing (Zetsche et al. 2016). Additionally, Cas12a has advantages over Cas9 in that it has lower mismatch tolerance, which reduces off-target effects. As staggering breaks are preferentially repaired using this method rather than NHEJ, the overhangs left after Cas12a cleaves the target DNA also enable HDR (Bothmer et al. 2017).

Cas13a (C2c2)

Although type VI Cas13a is likewise a class 2 CRISPR system, it can only cleave RNA by the action of two HEPN domains, in contrast to Cas9 and Cas12a's capacity to cleave DNA (Abudayyeh et al. 2016). It has a similar capability as Cas12a to produce its own crRNA, allowing it to target numerous loci with a single pre-crRNA template (Abudayyeh et al. 2017). For post-transcriptional suppression, Cas13a's RNA-cleaving abilities can be used. All mRNA isoforms are impacted when DNA is targeted using CRISPR systems because alternative splicing causes the transcription of a single DNA sequence to result in several splicing isoforms. It is feasible to specifically target one isoform utilizing Cas13a in order to investigate its function or block its impact without impairing the activity of the other isoforms (Mahas et al. 2018).

Cas9n

A Cas9 nickase variant (Cas9n) is produced by inserting a particular mutation into the RuvC domain of Cas9. Instead of creating double-stranded breaks (DSB), Cas9n nicks the target DNA (Jinek et al. 2012, Cong et al. 2013). Cas9n can be utilized to increase the effectiveness of the procedure by lowering the frequency of indel mutations brought on by unintended NHEJ repairs since single-nicked DNA is best repaired through base excision repair (Dianov and Hübscher 2013). With neighboring gRNA targets offset by a specific number of base pairs, researchers created a double-nicking technique that uses two Cas9ns to target opposite strands (Mali et al. 2013a). When Cas9n systems are paired, DSBs with gRNA-defined overhangs are created, which can then be combined with HDR to produce highly specific gene edits or with NHEJ to create exact deletions in essential alleles (Ren et al. 2014, Shen et al. 2014).

dCas9

The Cas9 system loses its ability to cleave DNA but retains the capacity to attach to specific regions when both the RuvC and HNH catalytic domains are altered through two silencing mutations (Bikard et al. 2013, Qi et al. 2013). According to research, this catalytically inactive Cas9 variant (dCas9) can obstruct transcription on its own. This issue is likely accomplished by either preventing the pairing of the RNA-polymerase with the promoter sequences that dCas9 targets, or by stopping the elongation process if the target sequence is a part of an open reading frame region.

The dCas9 system can be further altered in a number of ways, such as by fusing it with transcription activators or repressors to boost the effectiveness of dCas9-mediated transcription inhibition or with direct or indirect transcription activators, like VP64, to increase the expression of a particular DNA sequence. Since the genomic DNA is not permanently altered, the alteration of gene expression by dCas9 is a transient process. However, the combination of epigenetic modifiers with dCas9 allows for precise and durable changes to genomic expression (Lo and Qi 2017).

eSpCas9, SpCas9-HF1, and HypaCas9

Modifying the contacts between the Cas9 system and the attached DNA strands is one method for increasing CRISPR targeting specificity. According to Slaymaker et al.

(2016) hypothesis, Cas9 cleavage is more effective when the separation of the target and non-target strands is stable; hence, undermining this separation in undesirable targets would minimize off-target consequences. Re-hybridization between the target and non-target strand is facilitated by weakening contacts on the non-target strand by lowering positive charges. Since careful base pairing between gRNA and the target DNA is necessary to establish a stable separation of the target and non-target strands, off-target effects are therefore diminished. The two "enhanced specificity" SpCas9 variants (eSpCas9(1.0) and eSpCas9(1.1)) showed comparable on-target efficiency to WT SpCas9 while significantly lower levels of off-target cleavage were created through the engineering of SpCas9 mutants with a single positively charged amino acid residue substitution to weaken groove interactions.

Kleinstiver et al. (2016) created the high fidelity SpCas9-HF1 variant, which resulted in undetectable genome-wide off-target cleavage, by concentrating also on the binding between Cas9 and the target locus. However, Kleinstiver and his associates altered four SpCas9 residues that created hydrogen bonds with the phosphate backbone of the target strand instead of disturbing the non-target strand contacts. This impaired gRNA binds to DNA targets in the presence of any mismatches. SpCas9-HF1, which was created *via* alanine replacements in all four residues, along with eSpCas9, also demonstrated comparable on-target activity to WT SpCas9 without significantly negative off-target effects.

The most recent study by Chen et al. investigated how SpCas9-HF1 and eSpCas9(1.1) distinguish between targets using single-molecule Förster resonance energy transfer (smFRET) (Chen et al. 2017). Researchers have discovered that SpCas9-HF1 and eSpCas9(1.1) attach to mismatched sequences and then stall in an inactive state. They also defined the roles of Cas9's non-catalytic REC3 domain, which controls target complementarity and HNH catalytic activity. They created a hyper-accurate Cas9 variation (HypaCas9) using this newly learned information by inducing mutations in the REC3 domain. This variant had the same on-target efficacy as WT Cas9 and similar to or enhanced specificity when compared to SpCas9-HF1 or eSpCas9 (1.1).

CRISPR Cas9 mechanism for plant disease management

In prokaryotes, such as archaea and bacteria, a variety of mechanisms exist to give resistance against invading foreign agents, primarily viruses and plasmids. CRISPR/Cas system is found in prokaryotic genomes, it is found in roughly 83 percent of archaea and 45 percent of bacteria (Barrangou and Marrafni 2014). Innate immunity and adaptive (acquired) immunity are the two types of immunity used by prokaryotic organisms. In innate immunity, prokaryotes sense foreign invaders on first contact and contribute to the first line of defence, whereas adaptive immunity provides immunological responses and stores immunogenic memory for the second encounter. The presence of the CRISPR/Cas structure in archaeal and bacterial genomes is owing to the fact that CRISPR and Cas proteins target foreign mobile genetic elements (MGEs), which are eventually eradicated (Barrangou et al. 2007, van der Oost et al. 2014). The CRISPR/Cas system equips bacteria with immunogenic memory to protect them from foreign invaders on a second exposure (Marrafni and Sontheimer 2010).

Table 5: Cas9 orthologs and engineered Cas9 variants (Adopted from Wada et al. 2020; open access journal).

Cas9 Nuclease	Origin	PAM	Notes	References	Application to plant genome editing
SpCas9	*S. pyogenes*	NGG		Osakabe and Osakabe 2015, Jaganathan et al. 2018, Najera et al. 2019	+
NmCas9	*N. meningitidis*	NNNNGMTT		Hou et al. 2013	–
StCas9	*S. thermophilus*	NNAGAAW		Müller et al. 2016, Steinert et al. 2015	+
SaCas9	*S. aureus*	NNGRRT, NNNRRT		Steinert et al. 2015, Hua et al. 2019a, Ran et al. 2015	+
CjCas9	*C. rjejuni*	NNNNRYAC		Kim et al. 2017	–
FnCas9	*F. novicida*	No		Hirano et al. 2016, Zhang et al. 2018b	+
SpCas9-VQR	*S. pyogenes*	NGA	Altered PAM	Kleinstiver et al. 2015, Hu et al. 2016, Wu et al. 2019	+
SpCas9-EQR	*S. pyogenes*	NGAG	Altered PAM	Kleinstiver et al. 2015, Yamamoto et al. 2019	+
SpCas9-VRER	*S. pyogenes*	NGCG	Altered PAM	Kleinstiver et al. 2015, Hu et al. 2016	+
SpCas9-NG	*S. pyogenes*	NG	Altered PAM	Nishimasu et al. 2018, Endo et al. 2019, Zhong et al. 2019, Ge et al. 2019, Hua et al. 2019b, Niu et al. 2020, Negishi et al. 2019	+
SpCas9-HF1	*S. pyogenes*	NGG	High fidelity	Kleinstiver et al. 2016, Zhang et al. 2017a	+
eSpCas9	*S. pyogenes*	NGG	High fidelity	Zhang et al. 2017a, Slaymaker et al. 2016	+
HypaCas9	*S. pyogenes*	NGG	High fidelity	Chen et al. 2017	–
evoCas9	*S. pyogenes*	NGG	High fidelity	Casini et al. 2018	–
Sniper-Cas9	*S. pyogenes*	NGG	High fidelity	Lee et al. 2018	–
xCas9	*S. pyogenes*	NG, GAA, GAT	Altered PAM, high fidelity	Zhong et al. 2019, Ge et al. 2019, Hua et al. 2019b, Niu et al. 2020, Hu et al. 2018, Li et al. 2019a, Wang et al. 2019a	+

The CRISPR/Cas organization mechanism is divided into three distinct phases: (1) adaptation, (2) expression and maturation, and (3) interference (Amitai and Sorek 2016, Puschnik et al. 2017).

Adaptation of CRISPR–Cas spacer sequences

The adaptation phase is divided into two parts: first, the bacterium's Cas proteins identify the invader and acquire specific sequences from foreign nucleic acids, which are referred to as 'protospacers,' and second, the protospacer is incorporated in the extremity of the leader sequence in the CRISPR array as 'spacer,' causing the array's first repeat to be extended (Pourcel et al. 2005, Mojica et al. 2009, Yosef et al. 2012). These spacers are in charge of instilling immunological memory in archaea and bacteria so that they can defend themselves if they come into contact with MGEs again (Bolotin et al. 2005, Mojica et al. 2005, Pourcel et al. 2005).

The process of spacer acquisition can be further divided into three steps: (a) recognizing the invasive nucleic acid and scanning foreign DNA for potential PAMs that identify protospacers, (b) processing the nucleic acid to generate a new repeat spacer, and (c) integrating the new CRISPR repeat spacer unit at the leader end of the CRISPR locus. Only the first phase of these processes has been identified, and the method by which a new spacer is integrated into the host genome remains unknown. PAMs, also known as CRISPR motifs, have been discovered in close proximity to several protospacers (Mojica et al. 2009). Depending on the system, these conserved sections are brief (usually 2 to 5 nt long) and occur within 1 to 4 bp of the protospacer sequence on either side.

Expression and maturation of CRISPR/Cas system

During the expression and maturation phase, the leader sequence, which is located upstream of the CRISPR loci, functions as a promoter and stimulates transcription of the loci, producing lengthy precursor CRISPR RNA or pre-crRNA, which is then processed into small and mature units known as crRNA (Pougach et al. 2010, Yosef et al. 2012, Wei et al. 2015). A spacer region at the 5′ end is joined to a repeat sequence at the 3′ end to represent crRNA.

CRISPR locus transcription/regulation and crRNA processing, both of which are essential for interference to occur, have been categorized into two - transcription and regulation of the CRISPR locus. In a few organisms, the transcription of a CRISPR locus into the main transcript, or pre-crRNA, has been studied. The Gram-negative bacterium *E. Coli*, the plant pathogen *Xanthomonas oryzae*, the thermophilic bacterium *T. thermophilus*, and two archaeal species, *P. furiosus* and *Sulfolobus*, are among them (Garneau et al. 2010, Barrangou 2015).

Interference of CRISPR/Cas system

During the interference phase, Cas–crRNA complex, formed as a result of recruitment of Cas proteins to crRNA, detects the foreign MGEs via Watson–Crick base pairing of sequences that is complementary to the crRNA and hence, the targeted element is subjected to cleavage (Amitai and Sorek 2016). The existence of a small conserved sequence (2–5 bp) called protospacer adjacent motif (PAM) juxtaposed to the target site in the invading nucleic acid is essential for identification between self and non-self nucleic acids by the Cas–crRNA complex. The processed crRNA, together with specific Cas proteins, form a CRISPR ribonucleoprotein (crRNP) complex that facilitates spacer base pairing to the target or matching invasive nucleic acid. The crRNA serves as a guide (hence the term guide RNA has also been used) to allow for specific base pairing between the exposed crRNA within the ribonucleoprotein interference complex and the corresponding protospacer on the foreign DNA (Mojica et al. 2009). It is likely that crRNA interacts directly with complementary sequences in the target. The unique occurrence of the PAM sequence on the invading foreign DNA (and conversely, its absence in the host spacer sequence) is likely to play a dual role: first, in spacer selection and acquisition and second, in the interference process for discrimination of self versus non-self, which highlights its importance. Indeed, it has been demonstrated that despite perfect matches between the spacer and protospacer sequences, mutations in the PAM can circumvent CRISPR encoded immunity (Deveau et al. 2008, Mojica et al. 2009, Westra et al. 2013). The CRISPR/Cas system is versatile and has the ability to interfere with foreign dsDNA and single-stranded (ss) mRNA. This is reflected by the diversity of Cas proteins and their enzymatic activity in various species.

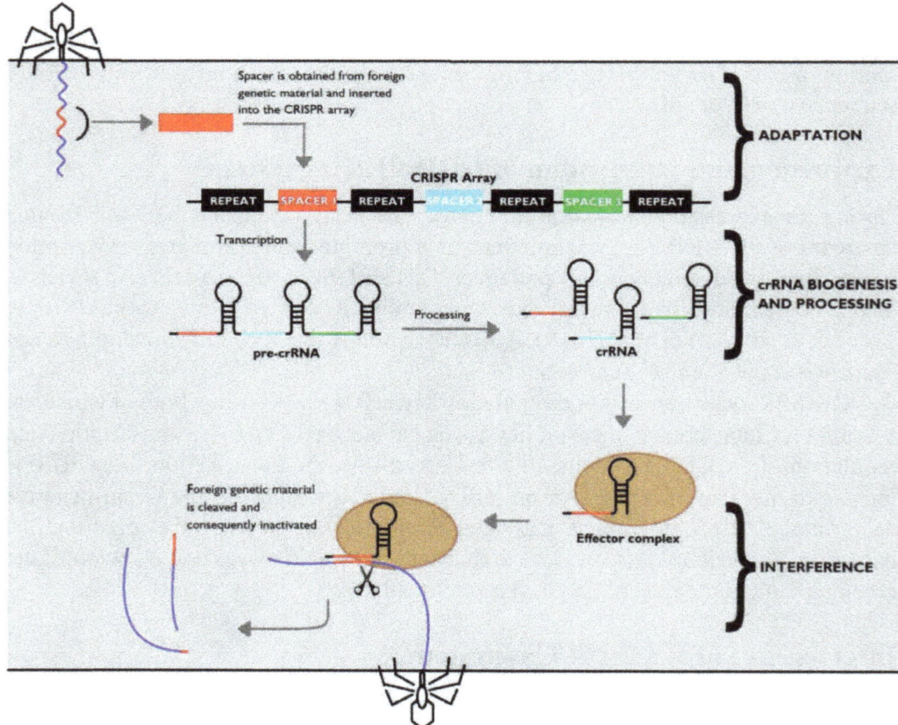

Figure 3: CRISPR-Cas adaptive immunity. Upon injection of genetic material from a virus or a plasmid into the bacteria, part of the invading sequence is cleaved and incorporated into the CRISPR locus, forming a new spacer within the locus. The CRISPR array is transcribed into a precursor to crRNA molecules (pre-crRNA), which is then cleaved into mature crRNA, which form effector complexes with type-specific Cas proteins (brown). When a foreign sequence matches a CRISPR spacer, the matching crRNA binds to the invading strand, activating Cas proteins with nuclease activity which silences the invader (Adapted from Loureiro and daSilva (2019); an open-access article).

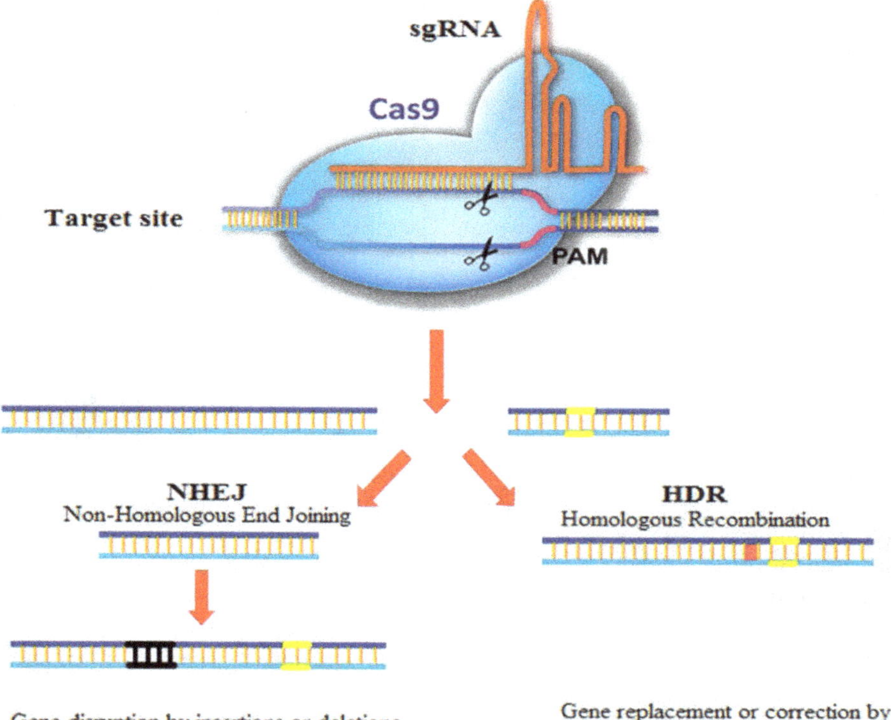

Figure 4: Schematic of CRISPR-Cas9 genome editing system (Adapted from Tavakoli et al. (2021); an open-access article).

Genome editing for resistance against fungal pathogen

Fungal pathogens are one of the most harmful causal agents of plant diseases, affecting both the quantity and quality of agricultural goods. Fungi are considered to be responsible for around 30% of new diseases (Giraud et al. 2010). Fungal infections cause significant reductions in crop production and quality, resulting in significant global economic losses, and are the primary causes of the most serious crop diseases. Mildews, smuts, blights, rusts, and rots are among the diseases they cause. In addition, some fungal diseases create mycotoxins, which can cause serious health problems in humans and animals (Borrelli et al. 2018, Yin and Qiu 2019, Zaynab et al. 2020). Diverse fungal lifestyles and high genetic flexibility allow fungi to adapt to new hosts, break resistance (R) gene-mediated resistance, and evolve resistance to fungicides, making disease control challenging (Yin and Qiu 2019).

Powdery mildew is a global fungal disease that infects a wide range of plants and persists in all mono and dicot plants. Breeding-resistant cultivars are the most effective, economical, and eco-friendly approach to control powdery mildew. The traditional method for producing resistant cultivars relies on introducing resistance (via R genes) from alien species into elite varieties by hybridization. Since most of these resistance genes are race-specific, their resistance gradually diminishes as new races of wheat powdery mildew evolve in the field. Therefore, breeding wheat varieties with a broad spectrum and durable resistance is highly desired.

The Mildew resistant LOCUS (MLO) is one of the key genes and it was edited through CRISPR/Cas9 and triggered the resistance against powdery mildew disease caused by the pathogen *Blumeria graminis* f. sp. *hordei* (Bgh) in wheat (Wang et al. 2014). Both TALENs and CRISPR tools have been exploited to induce mutations in three homoeoalleles encoding for MLO loci (TaMLO-A1, TaMLO-B1 and TaMLO-D1) in hexaploid bread wheat. These results show that these copies of TaMlo function redundantly and demonstrate that genome editing is a superior tool for modifying targets within polyploidy genomes. The mutated plants conferred a broad spectrum of resistance to *B. graminis* f. sp. *tritici* (Wang et al. 2014).

In another study, the same CRISPR-mediated MLO mutation in barley exhibited resistance to powdery mildew (*Blumeria graminis* f. sp. *hordei*) but it enhanced the susceptibility to rice blast fungus *M. grisea* (Jarosch et al. 1999). Similarly, downy mildew and powdery mildew are major fungal diseases in grapes. Two 'S' genes, MLO-6 and DMR were edited using CRISPR/Cas9 and exhibited enhanced resistance to powdery mildew (*Erysiphe necator*) and downy mildew in grapevine (Giacomelli et al. 2018).

Nekrasov et al. (2017) described that there are 16 MLO genes in tomato plants and SlMlo1 is a major gene responsible for powdery mildew disease susceptibility. The CRISPR/Cas9 technology has been employed to knockout SlMlo1 in tomato plants. CRISPR mediated transgene-free 'Tomelo' generated by deleting 48 bp region from SlMLO1 locus and the resulted plants acquired resistance to powdery mildew pathogen *Oidium neolycopersici* without affecting phenotypic features and yield parameters (Nekrasov et al. 2017). Second-generation progeny (F_1) were cultivated by selfing the first-generation (F_0) resistant mutants, which resulted in the CRISPR/Cas9 transfer DNA being removed *via* segregation. The F_1 progeny also exhibited resistance to *O. neolycopersici*. Likewise, the susceptibility gene Powdery Mildew Resistance 4 (PMR4), which functions as a callose synthase, has also been mutated by the CRISPR/Cas9 system, which resulted in resistance to *O. neolycopersici* in tomato (Santillán Martínez et al. 2020). However, CRISPR/Cas9 mediated mutation in the native tomato PMR4 gene led to a reduction in hyphal growth but not complete loss of susceptibility to *O. neolycopersici* in tomato (Martinez et al. 2020).

Similarly, CRISPR/Cas9 mediated gene knockout of Solyc08g075770 reduced mycorrhizal colonization (rmc) in tomatoes and exhibited immunity against *Fusarium* (Prihatna et al. 2018). *Agrobacterium*-mediated transient expression of cacao, nonexpression of pathogenesis-related 3 (TcNPR3) gene through CRISPR/Cas9 showed enhanced immunity against Cacao pathogen *Phytophthora tropicalis* (Fister et al. 2018). The targeted mutation in the putative extracellular cystatin like cysteine protease inhibitor (PpalEPIC8) gene *via* CRISPR/Cas9 was generated through *Agrobacterium* mediated transformation and witnessed enhanced resistance against *P. palmivora* (Gumtow et al. 2018).

The knockdown of Gh14-3-3d gene *via* CRISPR/Cas9 exhibited the resistance to *Verticillium dahliae* in cotton (Zhang et al. 2018b). Similarly, CRISPR/Cas9 mediated editing in sdhB (succinate dehydrogenase B inhibitor) gene of plant fungal pathogen *B. cinerea* was introduced by Cas-gRNA-RNP complex with telomere vector. The editing rate in *B. cinerea* was greatly superior to the existing CRISPR/Cas mediated approach in *M. oryzae*, which facilitates new molecular research against fungal pathogens (Leisen et al. 2020).

Li et al. (2020) demonstrated knockout mutation in VvPR4b, which led to activation of defense mechanism to downy mildew caused by *Plasmopara viticola*. Mutation in Clpsk1 gene enhanced watermelon resistance to *Fusarium oxysporum* (FON) and loss-of-function rendered watermelon seedlings more resistant to FON (Zhang et al. 2020). Gao et al. (2021) reported CRISPR/Cas9 mediated knockouts mutants lacking the SlymiR482e-3p in tomato, and exhibited an increased resistance against *F. oxysporum* f. sp *lycopersici*, the causal agent of tomato wilt disease; thus, the targeted protein of this miRNA enhanced the expression of nucleotide binding site leucine-rich repeat (NBS-LRR) protein. Thus, editing of host susceptible genes through CRISPR/Cas tool may pave the way for achieving robust and durable plant resistance against different kinds of fungal pathogens.

Rice blast caused by *Magnaporthe oryzae* is one of the most devastating diseases that affect rice production worldwide (Dean et al. 2012). Ethylene responsive factors (ERFs) of the APETELA2/ERF (AP2/ERF) superfamily play pivotal roles in rice adaptation to multiple biotic and abiotic stresses (Mizoi et al. 2012). The expression of OsERF922 is induced not only by abscisic acid (ABA) and salt but also by *M. oryzae*. Knockdown of OsERF922 by RNAi leads to increased resistance to *M. oryzae*, indicating that OsERF922 is a negative regulator of rice blast resistance (Liu et al. 2012). Targeted modification of OsERF922 using CRISPR/Cas9 generated rice Oserf922 knockout mutants. These null mutants showed enhanced resistance to rice blast without affecting other major agronomic traits. Therefore, the targeted knockout of negative regulators or/and susceptibility genes *via* genome editing represents a powerful approach for plant disease resistance breeding. One example of positive CRISPR/Cas9 grapevine defense is the WRKY52 transcription factor targeted by four gRNAs (Wang et al. 2018). About 21% had biallelic mutations and were more resistant to *Botrytis cinerea* than monoallelic mutants. The CRISPR/Cas9 strategy for woody plantations showed no major changes to phenotypes between the wild type and the biallelic mutant crops.

Another method to speed up genome editing on tree crops of slow production is the use of transient leaf and disease monitoring as demonstrated in *Theobroma cacao*. The mechanisms by which these fungi may antagonize pathophytic funguses are another strategy, for example, the release of a broad range of degrading enzymes from the cell wall and secondary metabolites like antibiotics (Khalid 2017). Genetic metabolic pathways, which are responsible for the biosynthetic way of the secreted proteins and secondary compounds, can lead to increased biocontrol activities of new fungal strains. The silence of the Ace1 gene contributes to an increase in the production of antibiotics and other secondary metabolites, leading to a substantial increase in the ability to act as a biocontrol agent for *Fusarium oxysporum*, as opposed to *Rhizoctonia solani* (Fang and Chen 2018). Under this strategy, unknown fungal cluster activation can be inducted by the use of CRISPR/Cas9 and new secondary metabolites that may interfere with plants or phytopathogens can be identified. It may mean the release of new interesting biocontrol strains on the planet, stopping transgenes from being introduced into the ecosystem. The results show the clear and advantaged use of the CRISPR/Cas9 method to boost crop resistance to fungal diseases.

Furthermore, the coding region of mitogen-activated protein kinase 5, mutated by CRISPR/Cas9 in rice improves resistance against *M. grisea* (Xie and Yang 2013). The editing of multiple gRNAs in another transcription factor, VvWRKY52, led to enhanced resistance in grapes against *Botrytis cinerea* (Leisen et al. 2020).

In *Arabidopsis*, *EDR1* (enhanced disease resistance 1) negatively regulates resistance against the powdery mildew fungus *Erysiphe cichoracearum*, but only slightly affects plant growth (Frye et al. 2001), suggesting that *EDR1* is an ideal target for improving resistance to powdery mildew. Moreover, similar to *Mlo*, *EDR1* is highly conserved across plant species (Frye et al. 2001). The CRISPR/Cas9 system was used to generate *Taedr1* wheat plants by targeting all three homoeologs of wheat *EDR1* simultaneously. The resultant *Taedr1* mutant plants were resistant to *Bgt* but without mildew-induced cell death (Zhang et al. 2017b). As *Arabidopsis edr1* mutant plants are resistant to bacteria and oomycetes, it is reasonable to speculate that *Taedr1* plants might also be resistant to other wheat pathogens.

In grape, the CRISPR/Cas9 mediated targeted mutagenesis of the transcription factor VvWRKY52 generated biallelic mutation mutant lines and knock-out of VvWRKY52 enhanced resistance to gray mold disease caused by *Botrytis cinerea* (Wang et al. 2018). Also, introduced mutations into the coding regions of the thermosensitive male sterile gene *TMS5*, proline-rich protein *Pi21* gene, and bacterial blight resistance a recessive gene *Xa13* in rice *via* CRISPR/Cas9 improved resistance against rice blast and bacterial blight. This is because the knock-out of recessive genes, *Pi21* and *Xa13*, positively mediated resistance to rice blast and bacterial blight, respectively (Chu et al. 2006, Fukuoka et al. 2009, Li et al. 2019b).

Genome editing for resistance against viruses

Controlling viral infections is difficult due to the dynamic and rapid evolution of viral pathogens. Several researchers have contributed significantly to the development of resistant plants by expressing viral and non-viral proteins, host resistant (R) genes, and gene silencing by RNA interference (Ahmad et al. 2020, Zhao et al. 2020). CRISPR's advantage has had a significant impact on the creation of plants that are resistant to DNA and RNA viruses. To engineer virus resistance, primarily two ways are used: directly addressing viral genomes and targeting plant recessive 'S' genes that are useful for virus replication in plants (Zaidi et al. 2020).

The CRISPR/Cas9 technique was used to modify viral genomes to improve plant tolerance to viral infections. Viruses of the Geminiviridae family pose a threat to almost all crops' productivity. These viruses, which have single-strand DNA-A and B genomes, can replicate to double-strands in the nucleus of host plants through the rolling circle mechanism of amplification (RCA). In the first experiment, researchers used CRISPR to build viral tolerance in *N. benthamiana* and *A. thaliana* against beet severe curly top virus (BSCTV) and yellow dwarf virus (BeYDV) (Ji et al. 2015, Liang et al. 2016). Yellow dwarf virus (YDV) overexpression of gRNAs coding for

Table 6: Applications of CRISPR/Cas9 system for plant resistance to fungal pathogens.

Sr. no.	Gene modified	Promoter used	Target plant	Effector protein	Strategy of transformation	Trait improvement	Reference
1.	TaMLO-A1	ZmUbi, TaU6	*Triticum aestivum*	SpCas9	Biolistic bombardment	Heritable broad-spectrum resistance to *Blumeriagraminis* f. sp. *tritici* (powdery mildew)	Wang et al. 2014
2.	MLO-7	No promoter	*Vitis vinifera*	SpCas9	CRISPR-RNPs/ PEG-protoplast transformation	Resistance against grape wine powdery mildew (*Erysiphe necator*)	Malnoy et al. 2016
3.	OsERF922	Ubi, OsU6a	*Oryza sativa* L. *Japonica*	SpCas9	Agrobacterium EHA105	Enhanced resistance against *Magnaporthe oryzae*-blast resistance	Wang ct al. 2016
4.	SlMlo1	pAGM4723	*Solanum lycopersicum*	SpCas9	Agrobacterium	Resistance against *Oidium neolycopersici* (tomato powdery mildew)	Nekrasov et al. 2017
5.	TaEDR1	pJIT163-UbiCas9, TaU6	*Triticum aestivum*	SpCas9	Biolistic bombardment	Resistance against *Blumeriagraminis* f. sp. *tritici* (powdery mildew)	Zhang et al. 2017
6.	Gh14-3-3d	AtU3	Cotton	SpCas9	Agrobacterium VIGS-mediated GV3101	Resistance to *Verticillium dahliae* infestation	Zhang et al. 2018a
7.	TcNPR3, a suppressor of the defense response	AtU6-26	*Theobroma cacao*	SpCas9	Agrobacterium infiltration	Increased resistance to infection with the cacao pathogen *Phytophthora tropicalis*	Fister et al. 2018
8.	Solyc08g075770/ genomic region	CaMV35S	Tomato	SpCas9	Agrobacterium GV3101	Enhanced susceptibility to disease fusarium wilt (*Fusarium oxysporum* f. sp. *lycopersici*)	Prihatna et al. 2018
9.	alEPIC8/cds region	Hsp70, HAM34 RPL41	Papaya	SpCas9	Agrobacterium EHA105	Increased resistance against *P. palmivora*	Gumtow et al. 2018
10.	MLO6, DMR	–	*Vitis vinifera*	SpCas9	Agrobacterium	Resistance against grape wine powdery mildew (*Erysiphe necator*) and downy mildew	Giacomelli et al. 2018

Table 6 contd. ...

...Table 6 contd.

Sr. no.	Gene modified	Promoter used	Target plant	Effector protein	Strategy of transformation	Trait improvement	Reference
11.	VvWRKY52	AtU3d, AtU3b, AtU6-1, AtU6-29	*Vitis vinifera*	SpCas9	Agrobacterium EHA105	Resistance to Graymold (*Botrytis cinerea*)	Leisen et al. 2020
12.	Clpsk1	AtU6	Watermelon (*Citrullus lanatus*)	SpCas9	Agrobacterium EHA105	Loss of function rendered watermelon seedlings more resistant to infection by *Fusarium oxysporum* f.sp. *niveum*	Zhang et al. 2020
13.	PMR4	AtU6	*Solanum lycopersicum* L.	SpCas9	Agrobacterium AGL1	Knock-down of SlPMR4 enhances resistance against powdery mildew	Martínez et al. 2020
14.	VvPR4b	AtU6	*Vitis vinifera*	SpCas9	Agrobacterium GV3101	Editing decreases downy mildew (*Plasmoparaviticola*) resistance in grapevine	Li et al. 2020
15.	StDMR6-1 and StDMR6-2	–	*Solanum tuberosum* L. *Vitis vinifera*	SpCas9	Agrobacterium C58	Editing confer increased late blight resistance	Kieu et al. 2021

replication and cell mobility showed enhanced resistance to various Geminiviruses in tobacco (Baltes et al. 2015). Through CRISPR/Cas9, transgenic tobacco plants expressing dual gRNAs targeting the C1 (replication associated protein) and IR regions of the Cotton Leaf Curl Multan virus (CLCuMuV) showed total resistance to leaf curl disease and prevented viral infections (Yin et al. 2019). Targeting multiple locations like intergenic (TAATATTAC), coding and non-coding regions could effectively curb multiple viral pathogens including TYLCV, beet curly top virus (BCTV) and Merremia mosaic virus (MeMV) in *N. benthamiana* (Ali et al. 2015). The editing of coat protein (CP)/replicase (Rep) in TYLCV genome induced interference of viral genome and subsequently reduced viral load in transgenic tomato and tobacco. Moreover, durable resistance was exhibited against different viruses even during the next generations (Tashkandi et al. 2018). Using the same gRNAs from conserved domains or intergenic sections of distinct viral genomes, numerous DNA viruses were inhibited at the same time. When compared to non-coding targets, gRNAs targeting coding regions were ineffective in inducing viral interference (Ali et al. 2016).

The multiplexed gRNA strategy was used to target various portions of the Chilli leaf curl virus genome, and the resulting tobacco plants showed high resistance to ChiLCV (Roy et al. 2019). Tobacco and *Arabidopsis* plants were given engineered gRNAs that targeted the tobacco rattle virus (TRV) and the pea early browning virus (PEBV) (Ali et al. 2018). The gRNAs homologue to the wheat dwarf virus (WDV) was delivered into rice cells and achieved up to 19.4% knock in the frequency of transgenic plants (Wang et al. 2017). Expression of multiple gRNAs targeting the coat protein of CaMV exhibited resistance against Caulimovirus without any symptoms in tobacco (Liu et al. 2018). Endogenous banana streak virus (eBSV), a double-stranded DNA Badnavirus belonging to the Caulimoviridae family, lives inside *Musa* spp., and the virus was inactivated by using the CRISPR/Cas9 technique to express sgRNA targeting the coding region of eBSV. In compared to non-edited control plants, the transgenic banana plants displayed resistance to eBSV and showed mild symptoms (Tripathi et al. 2019). In Cassava, however, CRISPR mutants were unable to gain immunity to the African Cassava mosaic virus (Mehta et al. 2019).

Cucumber mosaic virus (CMV) and Tobacco mosaic virus (TMV) were induced in *N. benthamiana* and *A. thaliana* by expressing gRNAs and FnCas9, and these plants showed a significant reduction in virus accumulation with minimal symptoms. The genome editing of potyvirus, and Turnip mosaic virus (TuMV) in tobacco using CRISPR/Cas13a led to the development of immunity (Aman et al. 2018). In tobacco and *Arabidopsis*, CRISPR-mediated editing of single or multiple gRNA of Tobacco rattle virus (TRV) and Pea early browning virus (PEBV) resulted in resistance (Ali et al. 2018). Different strains of potato virus Y (PVY) were suppressed in transgenic potato plants expressing Cas13a/sgRNA, and disease symptoms in potato plants were minimized (Zhan et al. 2019). LshCas13a system was used for developing resistance to Southern rice black-streaked dwarf virus (SRBSDV) and Rice stripe mosaic virus (RSMV) in rice (Zhang et al. 2019a). The advantage of Cas13 for targeting RNA viral genomes in plants is yet to be studied completely (Zaidi et al. 2020).

Table 7: Applications of CRISPR/Cas9 system for plant resistance against viruses.

Sr. no.	Gene modified	Promoter used	Virus type	Target plant	Effector protein	Strategy of Transformation	Trait improvement	Reference
1.	TYLCV-IR, RCA region/IR	PEBV pea early browning virus promoter	ssDNA virus	*Nicotiana benthamiana*	SpCas9	Agrobacterium	Resistance to leaf curl disease	Ali et al. 2015
2.	BeYDV (short intergenic region, transacting replication initiation protein)	CaMV35S, AtU6	ssDNA virus	*Nicotiana benthamiana*	SpCas9	Agrobacterium GV3101	Resistance against Bean yellow dwarf virus and observed mild symptoms due to minimal expression of Cas9	Baltes et al. 2015
3.	SgRNAs targeting Beet severe curly top virus (BSCTV)	AtU6		*Nicotiana* and *Arabidopsis*	SpCas9	Agrobacterium EHA105	Reduction in viral load of 97% and severe relief from leaf curly symptoms	Ji et al. 2015
4.	gRNAs-pCVA and pCVB of Cabbage leaf curl virus (CaLCuV) CaMV35S	AtU6	ssDNA virus	*Nicotiana benthamiana/ Arabidopsis*	SpCas9	Agrobacterium -infiltration GV3101	Demonstrated gemini viral mediated VIGE	Yin et al. 2015
5.	IR, CP, Rep-CLCuKoV, TYLCV, TYLCSV, MeMV, BCTV-Logan, BCTV-Worland	PEBV promoter		*Nicotiana benthamiana*	SpCas9	Agrobacterium GV3103	Disease resistance not confirmed but resistance to Gemini viruses CLCuKoV	Ali et al. 2016
6.	sgRNA-Rep of Wheat dwarf virus	ZmUbi	ssDNA virus	Rice	SpCas9	Agrobacterium and Biolistic	To increase gene targeting	Wang et al. 2017
7.	P3-membrane protein, CI-cytoplasmic inclusion body, Nib-RNA dependent RNA polymerase (RdRp) and CP-Coat Protein	U10, AtU6		*Solanum tuberosum* (potato)	LshCas13a	Agrobacterium GV3101	Resistance to Potato virus Y disease	Zhang et al. 2018a
8.	TuMV/CP	CaMV35S	ssRNA virus	*Nicotiana benthamiana*	LshCas13a	Agrobacterium GV3101	Resistance to Leaf curl disease	Aman et al. 2018

9.	gRNA1 and gRNA2 of Tobacco rattle virus (TRV) and Pea early browning virus (PEBV)	pPEBV	ssRNA virus	Tobacco and *Arabidopsis*	SpCas9	Agrobacterium GV3101	Increased the targeted genome editing	Ali et al. 2018
10.	Coat protein (CP) or Replicase (Rep) TYLCV	U6	ssDNA virus	Tomato and Tobacco	SpCas9	Agrobacterium GV3101	Low accumulation of TYLCV in tomato and tobacco transgenic plants	Tashkandi et al. 2018
11.	CP region of Caulimovirus	AtU6		*Arabidopsis*	SpCas9	Agrobacterium Floral dip recA strain	Resistance to CaMV	Liu et al. 2018
12.	gRNAs-pCVA and pCVB of Cabbage leaf curl virus (CaLCuV) CLCuMuV C1 (Rep) and IR	CaMV35S, AtU6 35S, U6	ssDNA virus	*Arabidopsis* *Nicotiana benthamiana*	SpCas9 SpCas9	Agrobacterium GV3101 Agrobacterium GV2260	Demonstrated gemini viral mediated VIGE for genome editing Resistance to Cotton leaf curl multan virus	Yin et al. 2019
13.	TMV (Tobacco mosaic virus) SRBDSV (Southern rice black streaked dwarf virus) RSMV (Rice stripe mosaic virus)	CaMV35S, OsU6, AtU6	ssRNA virus	Tobacco and Oryza sativa	LshCas13a		Resistance to viruses, i.e., TMV SRBDSV, RSMV	Zhang et al. 2019a
14.	ChiLCV	U3, U6 and CaMV35S	ssDNA virus	*Nicotiana benthamiana*	SpCas9	Agrobacterium transient assay GV3103	Resistant to ChiLCV reduced viral accumulation	Roy et al. 2019
15.	ORF1, 2, 3 and IR of BSV	OsU6	Open circular dsDNA virus	*Musa* spp. (banana)	SpCas9	Agrobacterium EHA105	Generated resistant banana plants against endogenous Banana Streak Virus	Tripathi et al. 2019
16.	Four WDV-specific sgRNAs-MP, CP, Rep/Rep, IR/WDV	ZmUbi1, CaMV35S	ssDNA virus	Barley	SpCas9	Agrobacterium C58C1	Resistant to Wheat dwarf virus	Kis et al. 2019
17.	AC2 and AC3 of African Cassava Mosaic Virus	CaMV35S, T7, U6	ssDNA virus	Cassava	SpCas9	Agrobacterium LBA4404	Failed to show clear disease resistance against ACMV in cassava	Mehta et al. 2019

Genome editing for resistance against bacterial pathogens

Bacterial pathogens are extremely varied, multiply quickly, and can spread in a variety of ways; as a result, diseases caused by bacterial pathogens are extremely difficult to control, especially after epidemics have emerged. Many plant 'S' genes are involved in host-bacterial pathogen interactions, making them interesting candidates for bacterial disease resistance gene editing. 'S' gene(s) are becoming prominent targets for developing crops that are resistant to bacterial illnesses *via* genome editing because they may be more durable in the field.

Type III effectors are secreted into the plant cell during the bacterial infection process (Buttner and He 2009). For disease development, these effectors primarily disrupt the host's defence pathways and/or activate the 'S' genes (Zaidi et al. 2018). As a result, CRISPR/Cas9-mediated gene editing of 'S' genes and negative regulators of plant innate immune response is a good target site for improving plant resistance.

TALENs have already demonstrated genome editing of SWEET genes. Rice bacterial blight (BB) is a serious disease caused by the bacterium *Xanthomonas oryzae* pv. *oryzae* (Xoo). It results in a 10–20 percent yield loss, which can reach 50 percent under pathogen-friendly conditions (i.e., high humidity), and sometimes an entire loss. TALE-mediated activation of at least one member of the SWEET family of sugar-transporter genes is required for BB. TALE proteins are secreted into host cells by Xoo *via* the type III secretion system (Makino et al. 2006), and SWEET genes are activated to establish a favourable environment for disease infection (Doyle et al. 2013). CRISPR/Cas9 was used to modify vulnerable sucrose transporter genes like *OsSWEET13* and *OsSWEET14* in rice to achieve bacterial blight resistance (Zhou et al. 2015, Zeng et al. 2020).

Due to their substantial virulence effect, SWEET genes (*SWEET11/xa13*, *SWEET1/Xa25*, and *SWEET14/Os11N3*) are known to be important TALEs. Strains encoding the TALEs PthXo1, PthXo2, and any one of many TALEs, namely AvrXa7, PthXo3, TalC, and TalF (formerly Tal5), respectively, promote the expression of *SWEET11*, *SWEET13*, and *SWEET14* (Streubel et al. 2013, Tran et al. 2018). SWEET genes encode sucrose-efflux transporters, which are hijacked by bacteria to transfer sugars from rice cells to promote pathogen development and virulence, as demonstrated by the interaction between *PXO86* and *SWEET14* (Chen et al. 2012). Scientists used the CRISPR/Cas9 system to target effector-binding regions of TALE proteins to disrupt their binding to SWEET genes (EBEs).

The broad spectrum of resistance to bacterial blight has been revealed by recent work on promoter editing. Mutations in the EBE regions of the promoters of the *SWEET11*, *SWEET13*, and *SWEET14* genes, for example, conferred long-term resistance to bacterial blight in rice (Oliva et al. 2019). The rice line Kitaake, as well as the elite mega varieties IR64 and Ciherang-Sub1, were all given five promoter mutations at the same time. Two SWEET promoters (*OsSWEET11* and *OsSWEET14*) were altered using CRISPR/Cas9 in rice and gave broad and long-lasting resistance to BB (Xu et al. 2019). CRISPR/Cas9 genome editing in the promoter region EBEs of the *OsSWEET14* gene has recently been shown to give resistance in Super Basmati rice lines to the Xoo strain expressing the AvrXa7 TALE (Zafar et al. 2020). With a few exceptions, the modified MS14K line showed broad-spectrum resistance to most Xoo strains. These findings also reveal that multiplex targeting with CRISPR/Cas9 can modify many EBEs at the same time.

CRISPR/Cas9 mutations in the mitogen-activated protein kinase 5 (OsMPK5) gene increased rice resistance to *Bukholderia glumae* (Xie and Yang, 2013). Using CRISPR/Cas9, researchers knocked out the *TMS5* (thermosensitive male sterility), *Pi21* (*Pyricularia oryzae* resistance protein 21), and *Xa13* (*Xanthomonas oryzae* 13) genes, resulting in thermosensitive male sterility and increased resistance to rice blast and bacterial blight (Li et al. 2019b).

Another significant invasive disease produced by the phytopathogen *Erwinia amylovora* in apple and other commercial and ornamental plants is fire blight. However, in Golden Delicious fruit crop plants, efficient targeted mutagenesis was investigated, in which direct pure CRISPR/Cas9 ribonucleoproteins were given to the apple protoplast and mutations were successfully detected. Disease-specific interacting proteins of Malus domestica (DIPM) genes may operate as *E. amylovora* susceptibility factors. The resistance to fire blight disease in apples was accomplished by direct CRISPR/Cas9 - RNPs delivery of mutations in the *DIPM-1*, *DIPM-2*, and *DIPM-4* genes (Malnoy et al. 2016).

Despite being one of the most commercially significant crops worldwide, numerous main diseases, such as *Pseudomonas syringae*, *Phytophthora* spp., and *Xanthomonas* spp., continue to limit tomato output and quality (Wang et al. 2014). *DMR6* (downy mildew resistance 6) mutations in *Arabidopsis* resulted in elevated salicylic acid levels and resistance to a variety of plant diseases, including bacteria and oomycetes. According to a recent study (Zhang et al. 2017b), SlDMR6-1, a tomato orthologue, is also elevated in response to *P. syringae* pv. tomato and *Phytophthora capsici* infection. The CRISPR/Cas9 method was used to create null mutants of SlDMR6-1 that displayed resistance to *P. syringae*, *P. capsici*, and *Xanthomonas* spp. without affecting tomato growth and development (Nekrasov et al. 2017).

The transcription factor, Lateral Organ Boundaries 1 (CsLOB1), facilitates the proliferation of bacterium *Xanthomonas citri* spp. *citri* (Xcc), which causes citrus canker in citrus species. CRISPR/Cas9 mediated disruption of EBEs in *CsLOB1* promoter and coding region of *CsLOB1* gene conferred immunity to citrus canker disease in orange (*Citrus sinensis*) and grapefruit (*C. paradisi*). The phenotypes in both plants showed no canker symptoms upon infestation with Xcc under controlled conditions (Jia et al. 2017, Peng et al. 2017). Simultaneously, EBE-CsLOB promoter region was also edited by employing LbCas13a (Cpf1) and CRISPR/SpCas9P tools to enhance resistance to citrus canker (Jia et al. 2019, Jia and Wang 2020). Another transcription factor, *CsWRKY22* gene, was mutated through CRISPR/Cas9 in Wanjincheng orange (*C. sinensis* (L.) Osbeck) and exhibited resistance to citrus canker (Wang et al. 2019b). Multiple gRNA targets in *Os8N3* through CRISRP/Cas9 can simultaneously knockout multiple EBEs in rice, exhibiting broad-spectrum resistance to most Xoo strains and BB (Kim et al. 2019). In another study, Tripathi et al. (2021) generated resistant banana against Xanthomonas Wilt (BXW) by editing *Musa dmr6* gene, an orthologue in *Musa* species. Altogether, CRISPR/Cas9 system is

Table 8: Applications of CRISPR/Cas9 system for plant resistance to bacterial pathogens.

Sr. no.	Gene modified	Promoter used	Target plant	Effector protein	Strategy of transformation	Trait improvement	Reference
1.	OsMPK5/CDS region	U3 or U6	*O. sativa*	SpCas9	Protoplast transformation	Disease resistance to *Burkholderia glumae* pathogen not confirmed	Xie and Yang 2013
2.	effector/TALE PthXo2 of OsSWEET13	CaMV35S	*O. sativa*	SpCas9	Agrobacterium infiltration	Resistance to bacterial blight	Zhou et al. 2015
3.	DIPM-1, 2 and 4	No promoter	Apple	SpCas9	CRISPR-RNPs/ PEG protoplast transformation	Resistance to fire blast disease (*Erwinia amylovora*)	Malnoy et al. 2016
4.	SlDMR6	pPZP200	*Solanum lycopersicum*	SpCas9	Agrobacterium GV3101	Enhanced resistance against *Xanthomonas*, *Pseudomonas*, and *Phytophthora capsici*	Thomazella et al. 2016
5.	CsLOB1	CaMV35S	*Citrus paradisi*	SpCas9	Agrobacterium infiltration	Resistance against *Xanthomonas citri* subsp. to citrus canker	Jia et al. 2017
6.	EBE of three promoters, i.e., SWEET11, SWEET13 and SWEET14	U6	*O. sativa*	SpCas9	Agrobacterium	Enhanced broad-spectrum disease resistance to *Xanthomonas oryzae* pv. *oryzae*	Oliva et al. 2019
7.	S1JAZ2	Ubiquitin	*Solanum lycopersicum*	SpCas9	Agrobacterium	Resistance against bacterial speck caused by *Pseudomonas syringae*, DC3000	Ortigosa et al. 2019
8.	EBE of OsSWEET11 and OsSWEET14	SWEET11, SWEET14	*O. sativa*	SpCas9	Agrobacterium	Enhanced broad-spectrum disease resistance (*Xanthomonas oryzae* pv. *oryzae*)	Xu et al. 2019

9.	DMR6	2XCaMV35S OsU6 CaMV35S	Banana (*Musa* spp.)	SpCas9	Agrobacterium EHA105	Musa dmr6 mutants of banana showed enhanced resistance to BXW (Banana Xanthomonas Wilt) and did not show any detrimental effect on plant growth	Tripathi et al. 2021
10.	EBEs of OsSWEET14 gene	OsU3,U6	Super Basmati (Pakistan's indigenous rice variety	SpCas9	Biolistic	Resistance in Super Basmati rice lines against Xoo strain carrying AvrXa7 TALE	Zafar et al. 2020
11.	OsSWEET14	AtU6	*O. sativa* Zhonghua 11	SpCas9	Agrobacterium EHA105	Disruption of OsSWEET14 in the Zhonghua 11 background confer strong resistance to African Xoo strain AXO1947 and Asian Xoo strain PXO86	Zeng et al. 2020

an effective tool, which offers resistance against several devastating bacterial disease in plants. Identification and research on new 'S' genes will give new dimension towards sustainable disease resistance through genome editing technologies such as CRISPR/Cas9 in crop plants.

Designing of CRISPR experiments

The effectiveness of the CRISPR/Cas9 system depends critically on the construction of the plasmid-based sgRNAs/Cas9 cassette since it dictates the appropriate expression of the Cas9 nuclease and the specificity of the targeting sgRNAs. To effectively use the genome editing system in plants and prevent any off-target effects, a number of considerations must be made, including the selection of the target sequence, PAM compatibility, the promoter activity, as well as the overall architecture of the expression cassettes.

Selection of target-specific site (*In silico* analysis)

↓

Designing and synthesising sgRNA

↓

Construction of a binary expression vector containing cas9 and sgRNA with appropriate promoters

↓

Introduction of CRISPR cassette into the plants either by *Agrobacterium*-mediated transformation or by particle bombardment

↓

Screening by either PCR and restriction digestion or by DNA sequencing

↓

Removal of CRISPR cassette by segregation

Multiplexing strategies

The CRISPR-Cas9 technology has a number of key advantages, one of which is its programmable capacity to simultaneously target several loci, or multiplexed genome editing. Multiple sgRNAs must be expressed concurrently with the Cas9 protein in order to enable multiplexed genome editing with this approach. Multiple strategies have been created and are currently in use for the simultaneous and effective generation of many sgRNAs, making multiplexed genome editing simple.

Limitations

The 3′ end of the target sequence must have a brief PAM sequence next to it in CRISPR/Cas systems (Mojica et al. 2009, Deltcheva et al. 2011). For instance,

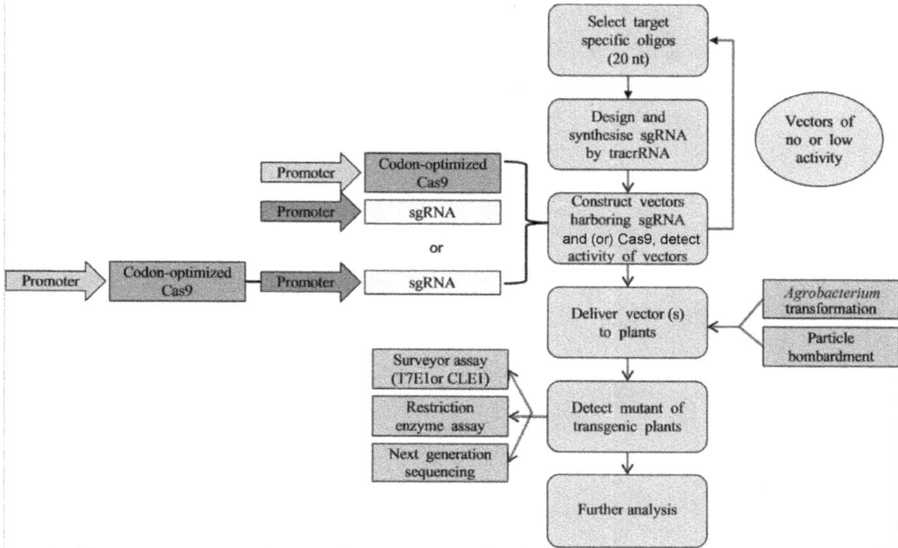

Figure 5: The basic flow of CRISPR/Cas9 editing of target genes (Adopted from Liu et al. 2017b; Open access journal).

because the most popular Cas9 system, SpCas9, exclusively recognizes NGG motifs, only sequences close to those motifs can be targeted (Jiang et al. 2013). Because of this limitation, CRISPR cannot be used to target loci when there are no PAMs in the area. A SpCas9 variation called xCas9 was recently created by Hu and colleagues (Fu et al. 2013). It recognizes new PAM sites like NG, GAA, and GAT while also showing noticeably fewer off-target effects.

The tendency of CRISPR to produce off-target consequences is one of the biggest barriers to widespread use. Cas9 is guided by a 20 nt fraction of the gRNA, and it maintains cleaving activity even with 3–5 mismatches at the PAM-distal end of the gRNA molecule (Cong et al. 2013, Hsu et al. 2013, Liu et al. 2017a). As a result of the error-prone repair of DSB by NHEJ, defective off-target binding and cleaving can cause collateral mutagenesis (Fu et al. 2013, Cho et al. 2014). The efficacy of CRISPR can also be hampered by several features of HDR or DSBs.

The increasing availability of crop and fungal/oomycete genome sequences, as well as user-friendly CRISPR/Cas9 technologies with open resources (e.g., procedures and plasmids), has hastened disease resistance research in a variety of crops. There are, however, a few constraints that must be addressed. Because crop plants have a longer life cycle than model plants, precision and frequency of mutant generation are important considerations. The CRISPR/Cas9 system was recently applied to cocoa by agroinfiltration, and it was found to produce a disease resistance phenotype (Fister et al. 2018). As a result, more study is needed into convenient and simple approaches to confirm the resistance phenotype by transient production of CRISPR/Cas9 components. More basic challenges, such as a lack of selectable markers, limit genome editing in fungi/oomycetes. The selectable marker

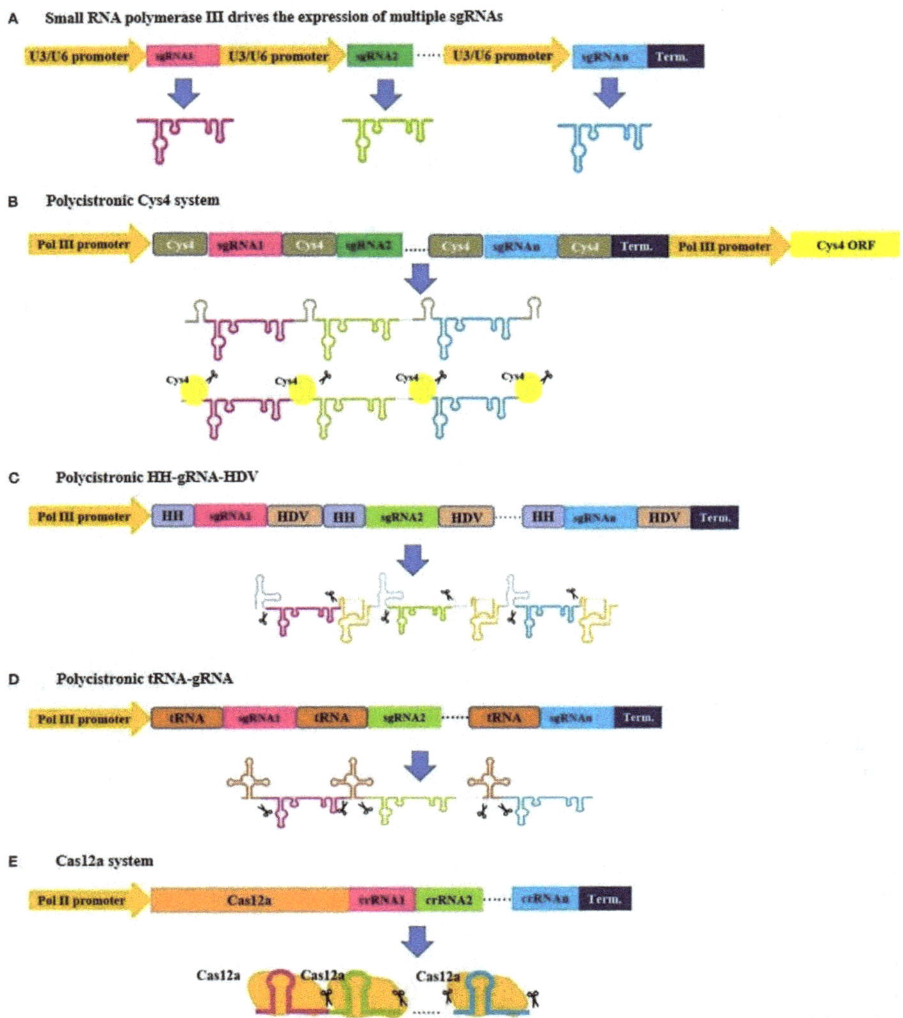

Figure 6: An illustration demonstrating various strategies for expressing multiplex gRNA cassettes in plants. (A) Small gRNAs are cloned after U3 or U6 promoters and derived from small RNA polymerase III to generate individual gRNAs. (B–D) Small gRNAs are cloned to be transcribed into a single transcript, and subsequent post-transcriptional processing is needed for gRNA separation, where Csy4, tRNAs, and hammerhead ribozyme regulate this separation. Similarly, a single transcript is generated in the (E) Cas12a system, but this system has a gRNA self-cleaving feature and does not require additional elements for post-transcriptional processing. (Adopted from Abdelrahman et al. 2021; Open access journal).

(oxathiapiprolin-resistance gene) for *Phytophthora* sp. was recently produced and found to have a significantly lower minimal inhibitory concentration than geneticin (G418) or hygromycin B (Wang et al. 2019c); however, other oomycetes are yet to be investigated.

Conclusions and future prospects

CRISPR/Cas9 is an essential revolutionary tool for gene editing. Therefore, in the future, the use of this technique to enhance yield, quality, and disease resistance in crops may be a significant field of research. During the last 5 years, it has been applied dynamically in myriad plant systems for conducting practical studies, combating stress-induced responses, and increasing significant agronomic characteristics. However, multiple modifications to this tool should lead to an increase in target effectiveness, and most studies are introductory and warrant development. Nonetheless, CRISPR/Cas9-based genome manipulation will attain a reputation and will be a critical method to attain the generation of "suitably manipulated" plants to help achieve the zero starvation objective and to realize sustainable food production for the increasing human population. The CRISPR/Cas9 tool and speed breeding programs can be used to ensure global food security. CRISPR/Cas9-based genome editing system has the advantage of mixing it with next-generation sequencing. Now researchers can conduct comprehensive mutational screening. Optimization and proper designing of gRNAs are very important at each phase to avoid or reduce the deleterious effects while doing off-target gene editing. Therefore, the use of the CRISPR/CAS9 library has several advantages like high multiplexing, specificity as well as high throughout targeting of a gene. To reduce or nullify negative results, it is important to do a quality check of the CRISPR library at each point during the screening procedure. Analysis of gene function by the above method is critical to recognize the function of genes. Newly exposed CRISPR/Cas9 methods and the development of novel tools are being uninterruptedly described, signifying that CRISPR/Cas9 toolbox for plant engineering will increase further in the future. This set of tools will deliver new methods to achieve defined genome editing without any bits of transgenes residual in genome-edited plants.

References

Abdelrahman, M., Wei, Z., Rohila, J.S. and Zhao, K. 2021. Multiplex genome-editing technologies for revolutionizing plant biology and crop improvement. Front. Plant Sci. 12: 721203. Doi: 10.3389/fpls.2021.721203. PMID: 34691102; PMCID: PMC8526792.

Abudayyeh, O.O., Gootenberg, J.S., Essletzbichler, P., Han, S., Joung, J., Belanto, J.J., Verdine, V., Cox, D.B.T., Kellner, M.J. and Regev, A. 2017. RNA targeting with CRISPR-Cas13. Nature 550: 280–284. [Google Scholar] [CrossRef] [PubMed].

Abudayyeh, O.O., Gootenberg, J.S., Konermann, S., Joung, J., Slaymaker, I.M., Cox, D.B.T., Shmakov, S., Makarova, K.S., Semenova, E. and Minakhin, L. 2016. C2c2 is a single-component programmable RNA-guided RNA-targeting CRISPR effector. Science 353, aaf5573. [Google Scholar] [CrossRef] [PubMed][Green Version].

Ahmad, S., Wei, X., Sheng, Z., Hu, P. and Tang, S. 2020. CRISPR/Cas9 for development of disease resistance in plants: Recent progress, limitations and future prospects. Brief. Funct. Genom. 19: 26–39.

Ali, Z., Abulfaraj, A., Idris, A., Ali, S., Tashkandi, M. and Mahfouz, M.M. 2015. CRISPR/Cas9-mediated viral interference in plants. Genome Biol. 16: 238. Doi.org/10.1186/s13059-015-0799-6.

Ali, Z., Ali, S., Tashkandi, M., Zaidi, S.S.E.A. and Mahfouz, M.M. 2016. CRISPR/Cas9-mediated immunity to geminiviruses: Differential interference and evasion. Sci. Rep. 6: 1–13.

Ali, Z., Eid, A., Ali, S. and Mahfouz, M.M. 2018. Pea early-browning virus-mediated genome editing via the CRISPR/Cas9 system in *Nicotiana benthamiana* and *Arabidopsis*. Virus Res. 244: 333–337.

Aman, R., Ali, Z., Butt, H., Mahas, A., Aljedaani, F., Khan, M.Z., Ding, S. and Mahfouz, M. 2018. RNA virus interference via CRISPR/Cas13a system in plants. Genome Biol. 19: 1–9.

Amitai, G. and Sorek, R. 2016. CRISPR-Cas adaptation: Insights into the mechanism of action. Nat. Rev. Microbiol. 14(2): 67–76. Doi: 10.1038/nrmicro.2015.14. Epub 2016 Jan 11. PMID: 26751509.

Anantharaman, V., Iyer, L.M. and Aravind, L. 2010. Presence of a classical RRM-fold palm domain in Thg1- type 3–5 nucleic acid polymerases and the origin of the GGDEF and CRISPR polymerase domains. Biol. Direct 5: 43.

Bae, K.H., Kwon, Y.D., Shin, H.C., Hwang, M.S. and Ryu, E.H. 2003. Human zinc fingers as building blocks in the construction of artificial transcription factors. Nat. Biotechnol. 21: 275–80.

Baltes, N.J., Hummel, A.W., Konecna, E., Cegan, R., Bruns, A.N., Bisaro, D.M. and Voytas, D.F. 2015. Conferring resistance to geminiviruses with the CRISPR–Cas prokaryotic immune system. Nat. Plants 1: 1–4.

Barrangou, R. 2015. The roles of CRISPR-Cas systems in adaptive immunity and beyond. Curr. Opin. Immunol. 32: 36–41.

Barrangou, R. and Marraffini, L.A. 2014. CRISPR-Cas systems: Prokaryotes upgrade to adaptive immunity. Mol. Cell. 54(2): 234–44. Doi: 10.1016/j.molcel.2014.03.011. PMID: 24766887; PMCID: PMC4025954.

Barrangou, R., Fremaux, C., Deveau, H., Richards, M., Boyaval, P., Moineau, S., Romero, D.A. and Horvath, P. 2007. CRISPR provides acquired resistance against viruses in prokaryotes. Science 315: 1709–12. Doi: 10.1126/science.1138140.

Bhardwaj, A. and Nain, V. 2021. TALENs-An indispensable tool in the era of CRISPR: A mini review. J. Genet. Eng. Biotechnol. 19: 1–10.

Bhaya, D., Davison, M. and Barrangou, R. 2011. CRISPR-Cas systems in bacteria and archaea: Versatile small RNAs for adaptive defense and regulation. Annu. Rev. Genet. 45: 273–97. Doi: 10.1146/annurev-genet-110410-132430. PMID: 22060043.

Bikard, D., Jiang, W., Samai, P., Hochschild, A., Zhang, F. and Marraffini, L.A. 2013. Programmable repression and activation of bacterial gene expression using an engineered CRISPR-Cas system. Nucleic Acids Res. 41: 7429–7437. [Google Scholar] [CrossRef] [PubMed][Green Version].

Bolotin, A., Ouinquis, B., Sorokin, A. and Ehrlich, S.D. 2005. Clustered regularly interspaced short palindrome repeats (CRISPRs) have spacers of extrachromosomal origin. Microbiol. 151: 2551–2561.

Borrelli, V.M.G., Brambilla, V., Rogowsky, P., Marocco, A. and Lanubile, A. 2018. The enhancement of plant disease resistance using CRISPR/Cas9 technology. Front. Plant Sci. 9: 1245. Doi: 10.3389/fpls.2018.01245.

Bothmer, A., Phadke, T., Barrera, L.A., Margulies, C.M., Lee, C.S., Buquicchio, F., Moss, S., Abdulkerim, H.S., Selleck, W. and Jayaram, H. 2017. Characterization of the interplay between DNA repair and CRISPR/Cas9-induced DNA lesions at an endogenous locus. Nat. Commun. 8: 1–12. [Google Scholar] [CrossRef] [PubMed].

Brouns, S.J. 2012. A swiss army knife of immunity. Science 337(6096): 808–809. Bibcode:2012Sci.337.808B. Doi:10.1126/science.1227253.

Brouns, S.J., Jore, M.M., Lundgren, M., Westra, E.R., Slijkhuis, R.J. and Dickman, M.J. 2008. Small CRISPR RNAs guide antiviral defense in prokaryotes. Science 321: 960–964.

Büttner, D. and He, S.Y. 2009. Type III protein secretion in plant pathogenic bacteria. Plant Physiol. 150(4): 1656–64. Doi: 10.1104/pp.109.139089. Epub 2009 May 20. PMID: 19458111; PMCID: PMC2719110.

Cabaniols, J.P., Ouvry, C., Lamamy, V., Fery, I. and Craplet, M.L. 2010. Meganuclease-driven targeted integration in CHO-K1 cells for the fast generation of HTS-compatible cell-based assays. J. Biomol. Screen. 15: 956–67.

Casini, A., Olivieri, M., Petris, G., Montagna, C., Reginato, G., Maule, G. et al. 2018. A highly specific SpCas9 variant is identified by *in vivo* screening in yeast. Nat. Biotechnology 36(3): 265–71.

Cermak, T., Doyle, E.L., Christian, M., Wang, L. and Zhang, Y. 2011. Efficient design and assembly of custom TALEN and other TAL effector-based constructs for DNA targeting. Nucleic Acids Res. 39: e82. Doi: 10.1093/nar/gkr218.

Chandrasekaran, J., Brumin, M., Wolf, D., Leibman, D., Klap, C. and Pearlsman, M. 2016. Development of broad virus resistance in non-transgenic cucumber using CRISPR/Cas9 technology. Mol. Plant Pathol. 17(7): 1140–53. Doi: 10.1111/mpp.12375.

Charpentier, E., Richter, H., van der Oost, J. and White, M.F. 2015. Biogenesis pathways of RNA guides in archaeal and bacterial CRISPR-Cas adaptive immunity. FEMS Microbiol. Reviews 39(3): 428–441. Doi: 10.1093/femsre/fuv023. PMC 5965381. PMID 25994611.

Chen, H. and Lin, Y. 2013. Promise and issues of genetically modified crops. Curr. Opin. Plant Biol. 16(2): 255–60. Doi: 10.1016/j.pbi.2013.03.007.

Chen, J.S., Dagdas, Y.S., Kleinstiver, B.P., Welch, M.M., Sousa, A.A., Harrington, L.B., Sternberg, S.H., Joung, J.K., Yildiz, A. and Doudna, J.A. 2017. Enhanced proofreading governs CRISPR-Cas9 targeting accuracy. Nature 550: 407–410.

Chen, L.Q., Qu, X.Q., Hou, B.H., Sosso, D., Osorio, S., Fernie, A.R. and Frommer, W.B. 2012. Sucrose efflux mediated by SWEET proteins as a key step for phloem transport. Science 335(6065): 207–11. Doi: 10.1126/science.1213351.

Cho, S.W., Kim, S., Kim, Y., Kweon, J., Kim, H.S., Bae, S. and Kim, J. 2014. Analysis of off-target effects of CRISPR/Cas-derived RNA-guided endonucleases and nickases. Genome Res. 24: 132–141. [Google Scholar] [CrossRef] [PubMed].

Christian, M.L., Cermak, T., Doyle, E.L., Schmidt, C. and Zhang, F. 2010. Targeting DNA double-strand breaks with TAL effector nucleases. Genetics 186: 757–61.

Chu, Z., Fu, B., Yang, H., Xu, C., Li, Z., Sanchez, A., Park, Y.J., Bennetzen, J.L., Zhang, Q. and Wang, S. 2006. Targeting xa13, a recessive gene for bacterial blight resistance in rice. Theor. Appl. Genet. 112: 455–461. Doi: 10.1007/s00122-005-0145-6.

Chylinski, K., Makarova, K.S., Charpentier, E. and Koonin, E.V. 2014. Classification and evolution of type II CRISPR-Cas systems. Nucleic Acids Res. 42(10): 6091–105. Doi: 10.1093/nar/gku241.

Clark, D.P., Pazdernik, N.J. and McGehee, M.R. 2019. Chapter 20-Genome Defense. pp. 622–653. *In*: Clark, D.P., Pazdernik, N.J. and McGehee, M.R. (eds.). Molecular Biology, 3rd ed. Elsevier: Amsterdam, The Netherlands. ISBN 978-0-12-813288-3.

Cong, L., Ran, F.A., Cox, D., Lin, S., Barretto, R., Habib, N., Hsu, P.D., Wu, X., Jiang, W., Marraffini, L.A. and Zhang, F. 2013. Multiplex genome engineering using CRISPR/Cas systems. Science 339(6121): 819–23. Doi: 10.1126/science.1231143.

Cox, D.B.T., Gootenberg, J.S., Abudayyeh, O.O., Franklin, B., Kellner, M.J., Joung, J. and Zhang, F. 2017. RNA editing with CRISPR-Cas13. Science 358(6366): 1019–1027. Doi: 10.1126/science.aaq0180. Epub 2017 Oct 25. PMID: 29070703; PMCID: PMC5793859.

Damalas, C.A. and Eleftherohorinos, I.G. 2011. Pesticide exposure, safety issues, and risk assessment indicators. Int. J. Environ. Res. Public Health 8(5): 1402–1419. Doi: 10.3390/ijerph8051402.

De Toledo Thomazella, D.P., Brail, Q., Dahlbeck, D. and Staskawicz, B.J. 2016. CRISPR-Cas9 mediated mutagenesis of a DMR6 ortholog in tomato confers broad-spectrum disease resistance. doi: 10.1101/064824.

Dean, R., Van Kan, J.A., Pretorius, Z.A., Hammond-Kosack, K.E., Di Pietro, A., Spanu, P.D., Rudd, J.J., Dickman, M., Kahmann, R., Ellis, J. and Foster, G.D. 2012. The top 10 fungal pathogens in molecular plant pathology. Mol. Plant Pathol. 13: 414–430. Doi: 10.1111/j.1364-3703.2011. 00783.x.

Deltcheva, E., Chylinski, K., Sharma, C.M., Gonzales, K., Chao, Y., Pirzada, Z.A., Eckert, M.R., Vogel, J. and Charpentier, E. 2011. CRISPR RNA maturation by trans-encoded small RNA and host factor RNase III. Nature 471(7340): 602–607. Doi: 10.1038/nature09886.

Deveau, H., Barrangou, R., Garneau, J.E., Labonté, J., Fremaux, C., Boyaval, P., Romero, D.A., Horvath, P. and Moineau, S. 2008. Phage response to CRISPR-encoded resistance in Streptococcus thermophilus. J. Bacteriol. 190(4): 1390–400. Doi: 10.1128/JB.01412-07.

Deveau, H., Garneau, J.E. and Moineau, S. 2010. CRISPR/Cas system and its role in phage-bacteria interactions. Annu. Rev. Microbiol. 64: 475–93. doi: 10.1146/annurev.micro.112408.134123. PMID: 20528693.

Dianov, G.L. and Hübscher, U. 2013. Mammalian base excision repair: The forgotten archangel. Nucleic Acids Res. 41: 3483–3490. [Google Scholar] [CrossRef] [PubMed].

Doudna, J.A. and Charpentier, E. 2014. The new frontier of genome engineering with CRISPR-Cas9. Science 346(6213): 1258096. Doi: 10.1126/science.1258096.

Doyle, E.L., Stoddard, B.L., Voytas, D.F. and Bogdanove, A.J. 2013. TAL effectors: Highly adaptable phytobacterial virulence factors and readily engineered DNA-targeting proteins. Trends Cell Biol. 23(8): 390–8. Doi: 10.1016/j.tcb.2013.04.003.

Eamens, A., Curtin, S.J. and Waterhouse, P.M. 2010. RNA silencing in plants. *In*: Plant Developmental Biology. pp. 277–294. https://doi.org/10.1007/978-3- 642-04670-4_15.

Ebihara, A., Yao, M., Masui, R., Tanaka, I., Yokoyama, S. and Kuramitsu, S. 2006. Crystal structure of hypothetical protein TTHB192 from *Thermus thermophilus* HB8 reveals a new protein family with an RNA recognition motif-like domain. Protein Sci. 15: 1494–99.

Endo, M., Mikami, M., Endo, A., Kaya, H., Itoh, T., Nishimasu, H. et al. 2019. Genome editing in plants by engineered CRISPR-Cas9 recognizing NG PAM. Nat. Plants 5(1): 14–7.

Eş, I., Gavahian, M., Marti-Quijal, F.J., Lorenzo, J.M., Khaneghah, A.M., Tsatsanis, C., Kampranis, S.C. and Barba, F.J. 2019. The application of the CRISPR-Cas9 genome editing machinery in food and agricultural science: Current status, future perspectives, and associated challenges. Biotech. Adv. 37: 410–421.

Fang, C. and Chen, X. 2018. Potential biocontrol efficacy of *Trichoderma atroviride* with cellulase expression regulator ace1 gene knockout. Biotechnol. 8: 302.

FAO. http://www.fao.org/news/story/en/item/280489/icode/.

Fister, A.S., Landherr, L., Maximova, S.N. and Guiltinan, M.J. 2018. Transient expression of CRISPR/Cas9 machinery targeting TcNPR3 enhances defense response in *Theobroma cacao*. Front. Plant Sci. 9: 268. doi.org/10.3389/fpls.2018.00268.

Fonfara, I., Richter, H., BratoviÄ, M., Le Rhun, A. and Charpentier, E. 2016. The CRISPR-associated DNA-cleaving enzyme Cpf1 also processes precursor CRISPR RNA. Nature 532: 517–521. [Google Scholar] [CrossRef] [PubMed].

Frye, C.A., Tang, D. and Innes, R.W. 2001. Negative regulation of defense responses in plants by a conserved MAPKK kinase. Proc. Natl. Acad. Sci. USA 98: 373–378. Doi: 10.1073/pnas.98.1.373.

Fu, Y., Foden, J.A., Khayter, C., Maeder, M.L., Reyon, D., Joung, J.K. and Sander, J.D. 2013. High-frequency off-target mutagenesis induced by CRISPR-Cas nucleases in human cells. Nat. Biotechnol. 31: 822–826. [Google Scholar] [CrossRef] [PubMed][Green Version].

Fukuoka, S., Saka, N., Koga, H., Ono, K., Shimizu, T., Ebana, K., Hayashi, N., Takahashi, A., Hirochika, H., Okuno, K. and Yano, M. 2009. Loss of function of a proline-containing protein confers durable disease resistance in rice. Science 325: 998–1001. Doi: 10.1126/science.1175550.

Gao, C. 2018. The future of CRISPR technologies in agriculture. Nat. Rev. Mol. Cell Biol. 19: 275–6.

Gao, L., Cox, D.B.T., Yan, W.X., Manteiga, J.C., Schneider, M.W., Yamano, T., Nishimasu, H., Nureki, O., Crosetto, N. and Zhang, F. 2017. Engineered Cpf1 variants with altered PAM specificities. Nat. Biotechnol. 35: 789–792. [Google Scholar] [CrossRef] [PubMed].

Gao, Y., Li, S.J., Zhang, S.W., Feng, T., Zhang, Z.Y., Luo, S.J., Mao, H.Y., Borkovich, K.A. and Ouyang, S.Q. 2021. SlymiR482e-3p mediates tomato wilt disease by modulating ethylene response pathway. Plant Biotechnol. J. 1: 17–19. Doi: 10.1111/pbi.13439.

Garneau, J.E., Dupuis, M.E., Villion, M., Romero, D.A., Barrangou, R., Boyaval, P., Fremaux, C., Horvath, P., Magadán, A.H. and Moineau, S. 2010. The CRISPR/Cas bacterial immune system cleaves bacteriophage and plasmid DNA. Nature 468(7320): 67–71. Doi: 10.1038/nature09523.

Ge, Z., Zheng, L., Zhao, Y., Jiang, J., Zhang, E.J., Liu, T. et al. 2019. Engineered xCas9 and SpCas9-NG variants broaden PAM recognition sites to generate mutations in Arabidopsis plants. Plant Biotechnol. J. 17(10): 1865.

Giacomelli, L., Zeilmaker, T., Malnoy, M., Rouppe van der Voort, J. and Moser, C. 2018. Generation of mildew-resistant grapevine clones via genome editing. Acta Hortic. 1248: 195–200.

Giraud, T., Gladieux, P. and Gavrilets, S. 2010. Linking the emergence of fungal plant diseases with ecological speciation. Trends Ecol. Evol. 25(7): 387–95.

Gootenberg, J.S., Abudayyeh, O.O., Lee, J.W., Essletzbichler, P., Dy, A.J., Joung, J., Verdine, V., Donghia, N., Daringer, N.M., Freije, C.A., Myhrvold, C., Bhattacharyya, R.P., Livny, J., Regev, A., Koonin, E.V., Hung, D.T., Sabeti, P.C., Collins, J.J. and Zhang, F. 2017. Nucleic acid detection with CRISPR-Cas13a/C2c2. Science 356(6336): 438–442. Doi: 10.1126/science.aam9321. Epub 2017 Apr 13. PMID: 28408723; PMCID: PMC5526198.

Gosavi, G., Yan, F., Ren, B., Kuang, Y., Yan, D., Zhou, X. and Zhou, H. 2020. Applications of CRISPR technology in studying plant-pathogen interactions: Overview and perspective. Phytopathol. Res. 2(1): 1–9. doi.org/10.1186/s42483-020-00060-z.

Grizot, S., Smith, J., Daboussi, F., Prieto, J., Redondo, P., Merino, N., Villate, M., Thomas, S., Lemaire, L., Montoya, G., Blanco, F.J., Pâques, F. and Duchateau, P. 2009. Efficient targeting of a SCID gene by an engineered single-chain homing endonuclease. Nucleic Acids Res. 37(16): 5405–19. Doi: 10.1093/nar/gkp548.

Gumtow, R., Wu, D., Uchida, J. and Tian, M. 2018. A *Phytophthora palmivora* extracellular cystatin-like protease inhibitor targets papain to contribute to virulence on papaya. Mol. Plant-Microbe Interact. 31: 363–373.

Gupta, S., Kumar, A., Patel, R. and Kumar, V. 2021. Genetically modified crop regulations: Scope and opportunity using the CRISPR-Cas9 genome editing approach. Mol. Biol. Rep. 48(5): 4851–4863.

Haft, D.H., Selengut, J., Mongodin, E.F. and Nelson, K.E. 2005. A guild of 45 CRISPR-associated (Cas) protein families and multiple CRISPR/Cas subtypes exist in prokaryotic genomes. PLoS Comput. Biol. 1(6): e60. Doi: 10.1371/journal.pcbi.0010060.

Hale, C.R., Zhao, P., Olson, S., Duff, M.O., Graveley, B.R., Wells, L., Terns, R.M. and Terns, M.P. 2009. RNA-guided RNA cleavage by a CRISPR RNA-Cas protein complex. Cell 139(5): 945–56. doi: 10.1016/j.cell.2009.07.040. PMID: 19945378; PMCID: PMC2951265.

Haurwitz, R.E., Jinek, M., Wiedenheft, B., Zhou, K. and Doudna, J.A. 2010. Sequence- and structure-specific RNA processing by a CRISPR endonuclease. Science 329: 1355–58.

He, Y., Zhu, M., Wang, L., Wu, J., Wang, Q., Wang, R. and Zhao, Y. 2019. Improvements of TKC technology accelerate isolation of transgene-free CRISPR/Cas9-edited rice plants. Rice Sci. 26(2): 109–117. doi.org/10.1016/j.rsci.2018.11.001.

Horvath, P., Coute-Monvoisin, A.C., Romero, D.A., Boyaval, P., Fremaux, C. and Barrangou, R. 2009. Comparative analysis of CRISPR loci in lactic acid bacteria genomes. Int. J. Food Microbiol. 131: 62–70.

Hou, Z., Zhang, Y., Propson, N.E., Howden, S.E., Chu, L.F., Sontheimer, E.J. and Thomson, J.A. 2013. Efficient genome engineering in human pluripotent stem cells using Cas9 from *Neisseria meningitidis*. Proc. Natl. Acad. Sci. 110(39): 15644–9.

Hsu, P.D., Scott, D.A., Weinstein, J.A., Ran, F.A., Konermann, S., Agarwala, V., Li, Y., Fine, E.J., Wu, X., Shalem, O., Cradick, T.J., Marraffini, L.A., Bao, G. and Zhang, F. 2013. DNA targeting specificity of RNA-guided Cas9 nucleases. Nature Biotechnol. 31(9): 827–832. Doi: 10.1038/nbt.2647.

Hu, J.H., Miller, S.M., Geurts, M.H., Tang, W., Chen, L., Sun, N. et al. 2018. Evolved Cas9 variants with broad PAM compatibility and high DNA specificity. Nature 556(7699): 57–63.

Hu, X., Wang, C., Fu, Y., Liu, Q., Jiao, X. and Wang, K. 2016. Expanding the range of CRISPR/Cas9 genome editing in rice. Mol. Plant 9(6): 943–5.

Hua, K., Tao, X. and Zhu, J.K. 2019a. Expanding the base editing scope in rice by using Cas9 variants. Plant Biotechnol. J. 17(2): 499–504.

Hua, K., Tao, X., Han, P., Wang, R. and Zhu, J.K. 2019b. Genome engineering in rice using Cas9 variants that recognize NG PAM sequences. Mol. Plant 12(7): 1003–14.

Ishino, Y., Shinagawa, H., Makino, K., Amemura, M. and Nakata, A. 1987. Nucleotide sequence of the iap gene, responsible for alkaline phosphatase isozyme conversion in *Escherichia coli*, and identification of the gene product. J. Bacteriol. 169: 5429–5433.

Jaganathan, D., Ramasamy, K., Sellamuthu, G., Jayabalan, S. and Venkataraman, G. 2018. CRISPR for crop improvement: An update review. Front. Plant Sci. 9: 985.

Jansen, R., Embden, J.D., Gaastra, W. and Schouls, L.M. 2002. Identification of genes that are associated with DNA repeats in prokaryotes. Mol. Microbiol. 43(6): 1565–75. Doi: 10.1046/j.1365-2958.2002.02839.x.

Jarosch, B., Kogel, K.H. and Schaffrath, U. 1999. The ambivalence of the barley Mlo locus: Mutations conferring resistance against powdery mildew (*Blumeriagraminis* f. sp. *hordei*) enhance susceptibility to the rice blast fungus *Magnaporthe grisea*. Mol. Plant Microbe Interact. 12: 508–514.

Ji, X., Zhang, H., Zhang, Y., Wang, Y. and Gao, C. 2015. Establishing a CRISPR–Cas-like immune system conferring DNA virus resistance in plants. Nat. Plants 1: 15144. doi.org/10.1038/nplants.2015.144.

Jia, H. and Wang, N. 2020. Generation of homozygous canker resistant citrus in the T0 generation using CRISPRSpCas9p. Plant Biotechnol. J. 18: 1990–1992.

Jia, H., Orbović, V. and Wang, N. 2019. CRISPR-LbCas12a-mediated modification of citrus. Plant Biotechnol. J. 17: 1928–1937.

Jia, H., Zhang, Y., Orbović, V., Xu, J., White, F.F., Jones, J.B. and Wang, N. 2017. Genome editing of the disease susceptibility gene Cs LOB 1 in citrus confers resistance to citrus canker. Plant Biotechnol. J. 15: 817–823.

Jiang, W., Bikard, D., Cox, D., Zhang, F. and Marraffini, L.A. 2013. RNA-guided editing of bacterial genomes using CRISPR-Cas systems. Nat. Biotechnol. 31: 233–239.

Jinek, M., Chylinski, K., Fonfara, I., Hauer, M., Doudna, J.A. and Charpentier, E.A. 2012. Programmable dual- RNA-guided DNA endonuclease in adaptive bacterial immunity. Science 337: 816–821.

Jore, M.M., Lundgren, M., van Duijn, E., Bultema, J.B., Westra, E.R., Waghmare, S.P., Wiedenheft, B., Pul, U., Wurm, R., Wagner, R., Beijer, M.R., Barendregt, A., Zhou, K., Snijders, A.P., Dickman, M.J., Doudna, J.A., Boekema, E.J., Heck, A.J., van der Oost, J. and Brouns, S.J. 2011. Structural basis for CRISPR RNA-guided DNA recognition by CASCADE. Nat. Struct. Mol. Biol. 18(5): 529–36. Doi: 10.1038/nsmb.2019.

Kay, S., Hahn, S., Marois, E., Hause, G. and Bonas, U. 2007. A bacterial effector acts as a plant transcription factor and induces a cell size regulator. Science 318: 648–51.

Khalid, S.A. 2017. *Trichoderma* as biological control weapon against soil borne plant pathogens. Afr. J. Biotech. 16(50): 2299–2306.

Khandagale, K. and Nadaf, A. 2016. Genome editing for targeted improvement of plants. Plant Biotechnol. Rep. 10: 327–343.

Kieu, N.P., Lenman, M., Wang, E.S., Petersen, B.L. and Andreasson, E. 2021. Mutations introduced in susceptibility genes through CRISPR/Cas9 genome editing confer increased late blight resistance in potatoes. Sci. Rep. 11: 1–12. Doi: 10.1038/s41598-021-83972-w.

Kim, E., Koo, T., Park, S.W., Kim, D., Kim, K., Cho, H.Y. et al. 2017. *In vivo* genome editing with a small Cas9 orthologue derived from Campylobacter jejuni. Nat. Commun. 8: 14500.

Kim, S., Kim, D., Cho, S.W., Kim, J. and Kim, J.S. 2014. Highly efficient RNA-guided genome editing in human cells via delivery of purified Cas9 ribonucleoproteins. Genome Res. 24(6): 1012–9. Doi: 10.1101/gr.171322.113.

Kim, Y.A., Moon, H. and Park, C.J. 2019. CRISPR/Cas9-targeted mutagenesis of Os8N3 in rice to confer resistance to *Xanthomonas oryzae* pv. *oryzae*. Rice 12(67): doi.org/10.1186/s12284-019-0325-7.

Kis, A., Hamar, É., Tholt, G., Bán, R. and Havelda, Z. 2019. Creating highly efficient resistance against wheat dwarf virus in barley by employing CRISPR/Cas9 system. Plant Biotechnol. J. Jun;17(6): 1004–1006. Doi: 10.1111/pbi.13077. Epub 2019 Feb 5. PMID: 30633425; PMCID: PMC6523583.

Kleinstiver, B.P., Pattanayak, V., Prew, M.S., Tsai, S.Q., Nguyen, N., Zheng, Z. and Joung, J.K. 2016. High-fidelity CRISPR-Cas9 variants with undetectable genome-wide off-targets. Nature 529: 490–495. Doi: 10.1038/nature16526.

Kleinstiver, B.P., Prew, M.S., Tsai, S.Q., Topkar, V.V., Nguyen, N.T., Zheng, Z., Gonzales, A.P., Li, Z., Peterson, R.T., Yeh, J.R., Aryee, M.J. and Joung, J.K. 2015. Engineered CRISPR-Cas9 nucleases with altered PAM specificities. Nature 523: 481–485. Doi: 10.1038/nature14592.

Koonin, E.V., Makarova, K.S. and Zhang, F. 2017. Diversity, classification and evolution of CRISPR-Cas systems. Curr. Opin. Microbiol. 37: 67–78. Doi: 10.1016/j.mib.2017.05.008.

Lee, J.K., Jeong, E., Lee, J., Jung, M., Shin, E., Kim, Y.H. et al. 2018. Directed evolution of CRISPR-Cas9 to increase its specificity. Nat. Commun. 9(1): 3048.

Leisen, T., Bietz, F., Werner, J., Wegner, A., Schaffrath, U., Scheuring, D., Willmund, F., Mosbach, A., Scalliet, G. and Hahn, M. 2020. CRISPR/Cas with ribonucleoprotein complexes and transiently selected telomere vectors allows highly efficient marker free and multiple genome editing in *Botrytis cinerea*. PLoS Pathog. 16(8): e1008326. Doi: 10.1371/journal.ppat.1008326.

Li, J., Luo, J., Xu, M., Li, S., Zhang, J., Li, H. et al. 2019a. Plant genome editing using xCas9 with expanded PAM compatibility. J. Genet. Genomics 46(5): 277–80.

Li, M.Y., Jiao, Y.T., Wang, Y.T., Zhang, N., Wang, B.B., Liu, R.Q., Yin, X., Xu, Y. and Liu, G.T. 2020. CRISPR/Cas9-mediated VvPR4b editing decreases downy mildew resistance in grapevine (*Vitis vinifera* L.). Hortic Res. 7: 149. Doi: 10.1038/s41438-020-00371-4.

Li, S., Shen, L., Hu, P., Liu, Q., Zhu, X., Qian, Q., Wang, K. and Wang, Y. 2019b. Developing disease resistant thermosensitive male sterile rice by multiplex gene editing. J. Integr. Plant Biol. 61(12): 1201–1205. Doi: 10.1111/jipb.12774.

Li, T., Liu, B., Spalding, M.H., Weeks, D.P. and Yang, B. 2012. High-efficiency TALEN-based gene editing produces disease-resistant rice. Nat. Biotechnol. 30: 390–92.

Liang, G., Zhang, H., Lou, D. and Yu, D. 2016. Selection of highly efficient sgRNAs for CRISPR/Cas9-based plant genome editing. Sci. Rep. 6: 1–8.

Liang, Z., Chen, K., Zhang, Y., Liu, J., Yin, K., Qiu, J.L. and Gao, C. 2018. Genome editing of bread wheat using biolistic delivery of CRISPR/Cas9 *in vitro* transcripts or ribonucleoproteins. Nat. Protoc. 13(3): 413–430. doi: 10.1038/nprot.2017.145.

Liu, D., Chen, X., Liu, J., Ye, J. and Guo, Z. 2012. The rice ERF transcription factor OsERF922 negatively regulates resistance to *Magnaporthe oryzae* and salt tolerance. J. Exp. Bot. 63: 3899–3911. Doi:10. 1093/jxb/ers079.

Liu, H., Soyars, C.L., Li, J., Fei, Q., He, G., Peterson, B.A., Meyers, B.C., Nimchuk, Z.L. and Wang, X. 2018. CRISPR/Cas9-mediated resistance to cauliflower mosaic virus. Plant Direct 2(3): e00047. Doi: 10.1002/pld3.47.

Liu, L., Li, X., Ma, J., Li, Z., You, L., Wang, J., Wang, M., Zhang, X. and Wang, Y. 2017a. The molecular architecture for RNA-guided RNA cleavage by Cas13a. Cell 170(4): 714–726.e10. Doi: 10.1016/j. cell.2017.06.050. [Google Scholar] [CrossRef] [PubMed].

Liu, X., Wu, S., Xu, J., Sui, C. and Wei, J. 2017b. Application of CRISPR/Cas9 in plant biology. Acta Pharm. Sin. B 7(3): 292–302. Doi: 10.1016/j.apsb.2017.01.002. Epub 2017 Mar 11. PMID: 28589077; PMCID: PMC5443236.

Lo, A. and Qi, L. 2017. Genetic and epigenetic control of gene expression by CRISPR–Cas systems. F1000Res. 6: F1000 Faculty Rev-747. Doi: 10.12688/f1000research.11113.1.

Loureiro, A. and daSilva, G.J. 2019. CRISPR-Cas: Converting a bacterial defense mechanism into a state-of- art-genetic manipulation tool. Antibiotics 8(1): 18.

Mahas, A., Neal Stewart, C. and Mahfouz, M.M. 2018. Harnessing CRISPR/Cas systems for programmable transcriptional and post-transcriptional regulation. Biotechnol. Adv. 36: 295–310. [Google Scholar] [CrossRef] [PubMed].

Makarova, K.S., Aravind, L., Grishin, N.V., Rogozin, I.B. and Koonin, E.V. 2002. A DNA repair system specific for thermophilic Archaea and bacteria predicted by genomic context analysis. Nucleic Acids Res. 30: 482–96.

Makarova, K.S., Haft, D.H., Barrangou, R., Brouns, S.J., Charpentier, E., Horvath, P, Moineau, S., Mojica, F.J., Wolf, Y.I., Yakunin, A.F., van der Oost, J. and Koonin, E.V. 2011. Evolution and classification of the CRISPR/Cas systems. Nat. Rev. Microbiol. 9(6): 467–77. Doi: 10.1038/nrmicro2577.

Makarova, K.S., Wolf, Y.I. and Koonin, E.V. 2013. Comparative genomics of defense systems in archaea and bacteria. Nucleic Acids Res. 41(8): 4360–77. Doi: 10.1093/nar/gkt157.

Makarova, K.S., Wolf, Y.I., Alkhnbashi, O.S., Costa, F., Shah, S.A., Saunders, S.J., Barrangou, R., Brouns, S.J., Charpentier, E., Haft, D.H., Horvath, P., Moineau, S., Mojica, F.J., Terns, R.M., Terns, M.P., White, M.F., Yakunin, A.F., Garrett, R.A., van der Oost, J., Backofen, R. and Koonin, E.V. 2015. An updated evolutionary classification of CRISPR-Cas systems. Nat. Rev. Microbiol. 13(11): 722–36. doi: 10.1038/nrmicro3569.

Makarova, K.S., Wolf, Y.I., Iranzo, J., Shmakov, S.A., Alkhnbashi, O.S., Brouns, S.J.J., Charpentier, E., Cheng, D., Haft, D.H., Horvath, P. et al. 2020. Evolutionary classification of CRISPR–cas systems: A burst of class 2 and derived variants. Nat. Rev. Microbiol. 18: 67–83. [Google Scholar] [CrossRef].

Makino, S., Sugio, A., White, F. and Bogdanove, A.J. 2006. Inhibition of resistance gene-mediated defense in rice by *Xanthomonas oryzae* pv. *oryzicola*. Mol. Plant-Microbe Interact. 19(3): 240–9. Doi: 10.1094/MPMI-19-0240.

Mali, P., Aach, J., Stranges, P.B., Esvelt, K.M., Moosburner, M., Kosuri, S., Yang, L. and Church, G.M. 2013b. CAS9 transcriptional activators for target specificity screening and paired nickases for cooperative genome engineering. Nat. Biotechnol. 31: 833–838. [Google Scholar] [CrossRef] [PubMed][Green Version].

Mali, P., Yang, L., Esvelt, K.M., Aach, J., Guell, M., DiCarlo, J.E., Norville, J.E. and Church, G.M. 2013a. RNA-guided human genome engineering via Cas9. Science 339: 823–826.

Malnoy, M., Viola, R., Jung, M.H., Koo, O.J., Kim, S., Kim, J.S., Velasco, R. and Nagamangala Kanchiswamy, C. 2016. DNA-free genetically edited grapevine and apple protoplast using CRISPR/Cas9 ribonucleoproteins. Front. Plant Sci. 7: 1904. Doi: 10.3389/fpls.2016.01904.

Marraffini, L.A. and Sontheimer, E.J. 2008. CRISPR interference limits horizontal gene transfer in staphylococci by targeting DNA. Science 322: 1843–45.

Marraffini, L.A. and Sontheimer, E.J. 2010. CRISPR interference: RNA-directed adaptive immunity in bacteria and archaea. Nat. Rev. Genet. 11(3): 181–90. Doi: 10.1038/nrg2749.

Martinez, M.I.S., Bracuto, V., Koseoglou, E., Appiano, M., Jacobsen, E., Visser, R.G., Wolters, A.M.A. and Bai, Y. 2020. CRISPR/Cas9-targeted mutagenesis of the tomato susceptibility gene PMR4 for resistance against powdery mildew. BMC Plant Biol. 20: 284. doi.org/10.1186/s12870-020-02497-y.

Mehta, D., Sturchler, A., Anjanappa, R.B., Zaidi, S.S.E.A., Hirsch-Hoffmann, M., Gruissem, W. and Vanderschuren, H. 2019. Linking CRISPR-Cas9 interference in cassava to the evolution of editing-resistant geminiviruses. Genome Biol. 20: 80. doi.org/10.1186/s13059-019-1678-3.

Miller, J.C., Tan, S., Qiao, G., Barlow, K.A., Wang, J., Xia, D.F., Meng, X., Paschon, D.E., Leung, E., Hinkley, S.J., Dulay, G.P., Hua, K.L., Ankoudinova, I., Cost, G.J., Urnov, F.D., Zhang, H.S., Holmes, M.C., Zhang, L., Gregory, P.D. and Rebar, E.J. 2011. A TALE nuclease architecture for efficient genome editing. Nat. Biotechnol. 29: 143–48. Doi: 10.1038/nbt.1755.

Mizoi, J., Shinozaki, K. and Yamaguchi-Shinozaki, K. 2012 AP2/ERF family transcription factors in plant abiotic stress responses. Biochim. Biophys Acta 1819(2): 86–96. Doi: 10.1016/j.bbagrm.2011.08.004.

Mojica, F.J., Diez-Villasenor, C., Garcia-Martinez, J. and Almendros, C. 2009. Short motif sequences determine the targets of the prokaryotic CRISPR defence system. Microbiol. (Reading) 155(Pt 3): 733–740. Doi: 10.1099/mic.0.023960-0.

Mojica, F.J., Diez-Villasenor, C., Garcia-Martinez, J. and Soria, E. 2005. Intervening sequences of regularly spaced prokaryotic repeats derive from foreign genetic elements. J. Mol. Evol. 60: 174–182.

Mojica, F.J., Díez-Villaseñor, C., Soria, E. and Juez, G. 2000. Biological significance of a family of regularly spaced repeats in the genomes of Archaea, Bacteria and mitochondria. Mol. Microbiol. 1: 244–6. Doi: 10.1046/j.1365-2958.2000.01838.x. PMID: 10760181.

Mojica, F.J., Ferrer, C., Juez, G. and Rodríguez-Valera, F. 1995. Long stretches of short tandem repeats are present in the largest replicons of the Archaea Haloferax mediterranei and Haloferax volcanii and could be involved in replicon partitioning. Mol. Microbiol. 1: 85–93. Doi: 10.1111/j.1365-2958.1995.mmi_17010085.x. PMID: 7476211.

Mojica, F.J.M., Díez-Villaseñor, C., García-Martínez, J. and Almendros, C. 2009. Short motif sequences determine the targets of the prokaryotic CRISPR defence system. Microbiol. 155: 733–740. [Google Scholar] [PubMed].

Müller, M., Lee, C.M., Gasiunas, G., Davis, T.H., Cradick, T.J., Siksnys, V. et al. 2016. Streptococcus thermophilus CRISPR-Cas9 systems enable specific editing of the human genome. Mol. Ther. 24(3): 636–44.

Mussolino, C., Morbitzer, R., Lutge, F., Dannemann, N., Lahaye, T. and Cathomen, T. 2011. A novel TALE nuclease scaffold enables high genome editing activity in combination with low toxicity. Nucleic Acids Res. 39: 9283–93.

Najera, V.A., Twyman, R.M., Christou, P. and Zhu, C. 2019. Applications of multiplex genome editing in higher plants. Curr. Opin. Biotechnol. 59: 93–102.

Naqvi, S., Zhu, C., Farre, G., Ramessar, K., Bassie, L., Breitenbach, J., Conesa, D.P., Ros, G., Sandmann, G., Capell, T. and Christou, P. 2009. Transgenic multivitamin corn through biofortifcation of endosperm with three vitamins representing three distinct metabolic pathways. Proc. Natl. Acad. Sci. USA 106: 7762–7767.

Negishi, K., Kaya, H., Abe, K., Hara, N., Saika, H. and Toki, S. 2019. An adenine base editor with expanded targeting scope using SpCas9-NGv1 in rice. Plant Biotechnology J. 17(8): 1476.

Nekrasov, V., Wang, C., Win, J., Lanz, C., Weigel, D. and Kamoun, S. 2017. Rapid generation of a transgene free powdery mildew resistant tomato by genome deletion. Sci. Rep. 7: 482. Doi: 10.1038/s41598-017- 00578-x.

Nidhi, S., Anand, U., Oleksak, P., Tripathi, P., Lal, J.A., Thomas, G., Kuca, K. and Tripathi, V. 2021. Novel CRISPR–Cas systems: An updated review of the current achievements, applications, and future research perspectives. Int. J. Mol. Sci. 22(7): 3327. https://doi.org/10.3390/ijms22073327.

Nishimasu, H., Shi, X., Ishiguro, S., Gao, L., Hirano, S., Okazaki, S. et al. 2018. Engineered CRISPR-Cas9 nuclease with expanded targeting space. Science 361(6408): 1259–62.

Niu, Q., Wu, S., Li, Y., Yang, X., Liu, P., Xu, Y. et al. 2020. Expanding the scope of CRISPR/Cas9-mediated genome editing in plants using an xCas9 and Cas9-NG hybrid. J. Integr. Plant Biol. 62(4): 398–402.

Oliva, R., Ji, C., Atienza-Grande, G., Huguet-Tapia, J.C., Perez-Quintero, A., Li, T., Eom, J.S., Li, C., Nguyen, H., Liu, B. and Auguy, F. 2019. Broad-spectrum resistance to bacterial blight in rice using genome editing. Nat. Biotechnol. 37: 1344–1350.

Ortigosa, A., Gimenez-Ibanez, S., Leonhardt, N. and Solano, R. 2019. Design of a bacterial speck resistant tomato by CRISPR/Cas9-mediated editing of SlJAZ2. Plant Biotechnol. J. Mar;17(3): 665–673. Doi: 10.1111/pbi.13006. Epub 2018 Oct 5. PMID: 30183125; PMCID: PMC6381780.

Osakabe, Y. and Osakabe, K. 2015. Genome editing with engineered nucleases in plants. Plant Cell Physiol. 56(3): 389–400.

Paques, F. and Duchateau, P. 2007. Meganucleases and DNA double-strand break-induced recombination: Perspectives for gene therapy. Curr. Gene Therapy 7: 49–66.

Peng, A., Chen, S., Lei, T., Xu, L., He, Y., Wu, L., Yao, L. and Zou, X. 2017. Engineering canker-resistant plants through CRISPR/Cas9-targeted editing of the susceptibility gene Cs LOB 1 promoter in citrus. Plant Biotechnol. J. 15: 1509–1519.

Pougach, K., Semenova, E., Bogdanova, E., Datsenko, K.A., Djordjevic, M., Wanner, B.L. and Severinov, K. 2010. Transcription, processing and function of CRISPR cassettes in Escherichia coli. Mol. Microbiol. 77(6): 1367–79. Doi: 10.1111/j.1365-2958.2010.07265.x. PMID: 20624226; PMCID: PMC2939963.

Pourcel, C., Salvignol, G. and Vergnaud, G. 2005. CRISPR elements in Yersinia pestis acquire new repeats by preferential uptake of bacteriophage DNA, and provide additional tools for evolutionary studies. Microbiol. 151: 653–663.

Prajapat, R.K., Mathur, M., Upadhyay, T.K., Lal, D., Maloo, S. and Sharma, D. 2021. Genome editing for crop improvement. In Crop Improvement; CRC Press: Boca Raton, FL, USA. pp. 111–123.

Prihatna, C., Larkan, N.J., Barbetti, M.J. and Barker, S.J. 2018. Tomato CYCLOPS/IPD3 is required for mycorrhizal symbiosis but not tolerance to Fusarium wilt in mycorrhiza deficient tomato mutant RMC. Mycorrhiza 28: 495–507.

Puschnik, A.S., Majzoub, K., Ooi, Y.S. and Carette, J.E. 2017. A CRISPR toolbox to study virus-host interactions. Nat. Rev. Microbiol. 15(6): 351–364. Doi: 10.1038/nrmicro.2017.29. Epub 2017 Apr 19. PMID: 28420884; PMCID: PMC5800792.

Qi, L.S., Larson, M.H., Gilbert, L.A., Doudna, J.A., Weissman, J.S., Arkin, A.P. and Lim, W.A. 2013. Repurposing CRISPR as an RNA-Guided platform for sequence-specific control of gene expression. Cell 152: 1173–1183.

Rahim, J., Gulzar, S., Zahid, R. and Rahim, K.A. 2021. Systematic review on the comparison of molecular gene editing tools. Int. J. Innov. Sci. Res. Tech. 6(8): 1–8.

Ramirez, C.L., Foley, J.E., Wright, D.A., Muller-Lerch, F., Rahman, S.H., Cornu, T.I., Winfrey, R.J., Sander, J.D., Fu, F., Townsend, J.A. et al. 2008. Unexpected failure rates for modular assembly of engineered zinc fingers. Nat. Methods 5: 374–375.

Ran, F.A., Cong, L., Yan, W.X., Scott, D.A., Gootenberg, J.S., Kriz, A.J. et al. 2015. *In vivo* genome editing using Staphylococcus aureus Cas9. Nature 520(7546): 186–91.

Ran, F.A., Hsu, P.D., Lin, C.Y., Gootenberg, J.S., Konermann, S., Trevino, A.E., Scott, D.A., Inoue, A., Matoba, S., Zhang, Y. and Zhang, F. 2013. Double nicking by RNA-guided CRISPR Cas9 for enhanced genome editing specificity. Cell 154: 1380–9.

Rani, M., Tyagi, K. and Jha, G. 2020. Advancements in plant disease control strategies. *In*: Advancement in Crop Improvement Techniques. Woodhead Publishing, pp. 141–157.

Ren, X., Yang, Z., Mao, D., Chang, Z., Qiao, H.H., Wang, X., Sun, J., Hu, Q., Cui, Y., Liu, L.P. Ji, J.Y., Xu, J. and Ni, J.Q. 2014. Performance of the Cas9 Nickase system in *Drosophila melanogaster*. G3 Genes|Genomes|Genetics. 4(10): 1955–1962. doi.org/10.1534/g3.114.013821 [Google Scholar] [CrossRef] [PubMed].

Reyon, D., Tsai, S.Q., Khayter, C., Foden, J.A., Sander, J.D. and Joung, J.K. 2012. FLASH assembly of TALENs for high-throughput genome editing. Nat. Biotechnol. 30: 460–65.

Romer, P., Hahn, S., Jordan, T., Strauss, T., Bonas, U. and Lahaye, T. 2007. Plant pathogen recognition mediated by promoter activation of the pepper Bs3 resistance gene. Science 318: 645–48.

Rosa, C., Kuo, Y.W., Wuriyanghan, H. and Falk, B.W. 2018. RNA interference mechanisms and applications in plant pathology. Annu. Rev. Phytopathol. 56: 581–610. doi.org/10.1146/annurev-phyto-080417-050044.

Roy, A., Zhai, Y., Ortiz, J., Neff, M., Mandal, B., Mukherjee, S.K. and Pappu, H.R. 2019. Multiplexed editing of a begomovirus genome restricts escape mutant formation and disease development. PLoS One 14: e0223765.

Santillán Martínez, M.I., Bracuto, V., Koseoglou, E., Appiano, M., Jacobsen, E., Visser, R.G.F. et al. 2020. CRISPR/Cas9-targeted mutagenesis of the tomato susceptibility gene PMR4 for resistance against powdery mildew. BMC Plant Biol. 20: 284. Doi: 10.1186/s12870-020-02497-y.

Sauer, N.J., Mozoruk, J., Miller, R.B., Warburg, Z.J., Walker, K.A., Beetham, P.R., Schöpke, C.R. and Gocal, G.F. 2016. Oligonucleotide directed mutagenesis for precision gene editing. Plant Biotech. J. 14: 496–502.

Segal, D.J., Beerli, R.R., Blancafort, P., Dreier, B., Effertz, K. et al. 2003. Evaluation of a modular strategy for the construction of novel polydactyl zinc finger DNA-binding proteins. Biochemistry 42: 2137–48.

Shen, B., Zhang, J., Wu, H. et al. 2013. Generation of gene-modified mice via Cas9/RNA-mediated gene targeting. Cell Res. 23: 720–3.

Shen, B., Zhang, W., Zhang, J., Zhou, J., Wang, J., Chen, L., Wang, L., Hodgkins, A., Iyer, V., Huang, X. and Skarnes, W.C. 2014. Efficient genome modification by CRISPR-Cas9 nickase with minimal off-target effects. Nat. Methods 11: 399–402.

Shmakov, S., Abudayyeh, O.O., Makarova, K.S., Wolf, Y.I., Gootenberg, J.S., Semenova, E., Minakhin, L., Joung, J., Konermann, S., Severinov, K., Zhang, F. and Koonin, E.V. 2015. Discovery and functional characterization of diverse Class 2 CRISPR-Cas systems. Mol. Cell 60: 385–397. Doi: 10.1016/j.molcel.2015.10.008.

Shmakov, S., Smargon, A., Scott, D., Cox, D., Pyzocha, N., Yan, W., Abudayyeh, O.O., Gootenberg, J.S., Makarova, K.S., Wolf, Y.I., Severinov, K., Zhang, F. and Koonin, E.V. 2017. Diversity and evolution of class 2 CRISPR-Cas systems. Nat. Rev. Microbiol. 15(3): 169–182. Doi: 10.1038/nrmicro.2016.184.

Sinkunas, T., Gasiunas, G., Fremaux, C., Barrangou, R., Horvath, P. and Siksnys, V. 2011. Cas3 is a single-stranded DNA nuclease and ATP-dependent helicase in the CRISPR/Cas immune system. EMBO J. 30: 1335–42.

Slaymaker, I.M., Gao, L., Zetsche, B., Scott, D.A., Yan, W.X. and Zhang, F. 2016. Rationally engineered Cas9 nucleases with improved specificity. Science 351(6268): 84–88. Doi: 10.1126/science.aad5227.[Google Scholar] [CrossRef] [PubMed].

Smith, J., Grizot, S., Arnould, S., Duclert, A., Epinat, J.C., Chames, P., Prieto, J., Redondo, P., Blanco, F.J., Bravo, J., Montoya, G., Pâques, F. and Duchateau, P.A. 2006. A combinatorial approach to create artificial homing endonucleases cleaving chosen sequences. Nucleic Acids Res. 34(22): e149. Doi: 10.1093/nar/gkl720.

Steinert, J., Schiml. S., Fauser, F. and Puchta, H. 2015. Highly efficient heritable plant genome engineering using Cas9 orthologues from Streptococcus thermophilus and Staphylococcus aureus. Plant J. 84(6): 1295–305.

Stoddard, B.L. 2011. Homing endonucleases: From microbial genetic invaders to reagents for targeted DNA modification. Structure 19: 7–15.

Streubel, J., Pesce, C., Hutin, M., Koebnik, R., Boch, J. and Szurek, B. 2013. Five phylogenetically close rice SWEET genes confer TAL effector-mediated susceptibility to *Xanthomonas oryzae* pv. *oryzae*. New Phytol. 200(3): 808–19. Doi: 10.1111/nph.12411.

Tashkandi, M., Ali, Z., Aljedaani, F., Shami, A. and Mahfouz, M.M. 2018. Engineering resistance against Tomato yellow leaf curl virus via the CRISPR/Cas9 system in tomato. Plant Signal Behav. 13: e1525996.

Tavakoli, K., Pour-Aboughadareh, A., Kianersi, F., Poczai, P., Etminan, A. and Shooshtari, L. 2021. Applications of CRISPR-Cas9 as an advanced genome editing system in life sciences. BioTech. 10: 14. https://doi.org/10.3390/biotech10030014.

Tilman, D., Cassman, K.G., Matson, P.A., Naylor, R. and Polasky, S. 2002 Agricultural sustainability and intensive production practices. Nature 418: 671–677. Doi: 10.1038/nature01014.

Tran, T.T., Perez-Quintero, A.L., Wonni, I., Carpenter, S.C.D., Yu, Y., Wang, L., Leach, J.E., Verdier, V., Cunnac, S., Bogdanove, A.J., Koebnik, R., Hutin, M. and Szurek, B. 2018. Functional analysis of African *Xanthomonas oryzae* pv. *oryzae* TALomes reveals a new susceptibility gene in bacterial leaf blight of rice. PLoS Path. 14(6): 1–25. e1007092. Doi: 10.1371/journal.ppat.1007092.

Tripathi, J.N., Ntui, V.O., Ron, M., Muiruri, S.K., Britt, A. and Tripathi, L. 2019. CRISPR/Cas9 editing of endogenous *banana streak virus* in the B genome of *Musa* spp. overcomes a major challenge in banana breeding. Commun. Biol. 2: 46. Doi: 10.1038/s42003-019-0288-7.

Tripathi, J.N., Ntui, V.O., Shah, T. and Tripathi, L. 2021. CRISPR/Cas9-mediated editing of DMR6 orthologue in banana (*Musa* spp.) confers enhanced resistance to bacterial disease. Plant Biotechnol. J. 19: 1291–1293.

Urnov, F.D., Miller, J.C., Lee, Y.L., Beausejour, C.M., Rock, J.M., Augustus, S. and Holmes, M.C. 2005. Highly efficient endogenous human gene correction using designed zinc-finger nucleases. Nature 435(7042): 646–51. Doi: 10.1038/nature03556.

van der Oost, J., Westra, E.R., Jackson, R.N. and Wiedenheft, B. 2014. Unravelling the structural and mechanistic basis of CRISPR-Cas systems. Nat. Rev. Microbiol. 12(7): 479–92. Doi: 10.1038/nrmicro3279.

Veillet, F., Perrot, L., Chauvin, L., Kermarrec, M.-P., Guyon-Debast, A., Chauvin, J.-E., Nogué, F. and Mazier, M. 2019. Transgene-free genome editing in tomato and potato plants using Agrobacterium mediated delivery of a CRISPR/Cas9 cytidine base editor. Int. J. Mol. Sci. 20(2): 402. Doi.org/10.3390/ijms20020402.

Voytas, D.F. 2013. Plant genome engineering with sequence-specific nucleases. Annual Rev. of Plant Biol. 64: 327–350.

Voytas, D.F. and Gao, C. 2014. Precision genome engineering and agriculture: Opportunities and regulatory challenges. PLoS Biol. 12: e1001877. Doi: 10.1371/journal.pbio.1001877.

Wada, N., Ueta, R., Osakabe, Y. and Osakabe, K. 2020. Precision genome editing in plants: State-of-the-art in CRISPR/Cas9-based genome engineering. BMC Plant Biol. 20: 234. Doi.org/10.1186/s12870-020-02385-5.

Wagh, S.G. and Pohare, M.B. 2019. Current and future prospects of plant breeding with CRISPR/Cas. Current Journal of Applied Science and Technology 38(3): 1–17. https://doi.org/10.9734/cjast/2019/v38i330360.

Wang, F., Wang, C., Liu, P., Lei, C., Hao, W., Gao, Y., Liu, YG. and Zhao, K, 2016. Enhanced rice blast resistance by CRISPR/Cas9-targeted mutagenesis of the ERF transcription factor gene OsERF922. PLoS ONE 11(4): e0154027.

Wang, J., Meng, X., Hu, X., Sun, T., Li, J., Wang, K. et al. 2019a. xCas9 expands the scope of genome editing with reduced efficiency in rice. Plant Biotechnology J. 17(4): 709–11.

Wang, L., Chen, S., Peng, A., Xie, Z., He, Y. and Zou, X. 2019b. CRISPR/Cas9-mediated editing of CsWRKY22 reduces susceptibility to *Xanthomonas citri* subsp. *citri* in Wanjincheng orange (*Citrus sinensis* (L.) Osbeck). Plant Biotechnol. Rep. 13: 501–510.

Wang, M., Lu, Y., Botella, J.R., Mao, Y., Hua, K. and Zhu, J.K. 2017. Gene targeting by homology-directed repair in rice using a geminivirus-based CRISPR/Cas9 system. Mol. Plant 10: 1007–1010.

Wang, W., Pan, Q., He, F., Akhunova, A., Chao, S., Trick, H. et al., 2018. Transgenerational CRISPR-Cas9 activity facilitates multiplex gene editing in allopolyploid wheat. CRISPR J. 1: 6574.

Wang, W., Xue, Z., Miao, J., Cai, M., Zhang, C., Li, T., Zhang, B., Tyler, B.M. and Liu, X. 2019c. *PcMuORP1*, an oxathiapiprolin-resistance gene, functions as a novel selection marker for *Phytophthora* transformation and CRISPR/Cas9 mediated genome editing. Front. Microbiol. 10: 2402. Doi: 10.3389/fmicb.2019.02402. PMID: 31708886; PMCID: PMC6821980.

Wang, Y., Cheng, X., Shan, Q. et al. 2014. Simultaneous editing of three homoeoalleles in hexaploid bread wheat confers heritable resistance to powdery mildew. Nat. Biotechnol. 32(9): 947–951.

Wei, Y., Terns, R.M. and Terns, M.P. 2015. Cas9 function and host genome sampling in Type II-A CRISPR-Cas adaptation. Genes Dev. 29(4): 356–61. Doi: 10.1101/gad.257550.114. PMID: 25691466; PMCID: PMC4335292.

Westra, E.R., Semenova, E., Datsenko, K.A., Jackson, R.N., Wiedenheft, B., Severinov, K. and Brouns, S.J. 2013. Type I-E CRISPR-cas systems discriminate target from non-target DNA through base pairing-independent PAM recognition. PLoS Genet. 9(9): e1003742. Doi: 10.1371/journal.pgen.1003742.

Wu, Y., Xu, W., Wang, F., Zhao, S., Feng, F., Song, J. et al. 2019. Increasing cytosine base editing scope and efficiency with engineered Cas9-PmCDA1 fusions and the modified sgRNA in rice. Front. Genet. 10: 379.

Xie, K. and Yang, Y. 2013. RNA-guided genome editing in plants using a CRISPR–Cas system. Mol. Plant 6: 1975–1983.

Xu, Z., Xu, X., Gong, Q., Li, Z., Li, Y., Wang, S. and Chen, G. 2019. Engineering broad-spectrum bacterial blight resistance by simultaneously disrupting variable TALE-binding elements of multiple susceptibility genes in rice. Molecular Plant. Doi: 10.1016/j.molp.2019.08.006.

Yamamoto, A., Ishida, T., Yoshimura, M., Kimura, Y. and Sawa, S. 2019. Developing heritable mutations in *Arabidopsis thaliana* using a modified CRISPR/Cas9 toolkit comprising PAM-Altered Cas9 variants and gRNAs. Plant Cell Physiol. 60(10): 2255–62.

Yin, K. and Qiu, J.L. 2019. Genome editing for plant disease resistance: Applications and perspectives. Philosophical Transactions of the Royal Society B 374(1767): 20180322.

Yin, K. and Qiu, J.L. 2019. Genome editing for plant disease resistance: Applications and perspectives. Philos. Trans. R. Soc. Lond. B Biol. Sci. 374: 20180322. Doi: 10.1098/rstb.2018.0322

Yin, K., Han, T., Liu, G. et al. 2015. A geminivirus-based guide RNA delivery system for CRISPR/Cas9 mediated plant genome editing. Sci. Rep. 5: 14926.

Yin, K., Han, T., Xie, K., Zhao, J., Song, J. and Liu, Y. 2019. Engineer complete resistance to *Cotton Leaf Curl Multan virus* by the CRISPR/Cas9 system in *Nicotiana benthamiana*. Phytopathol. Res. 1: 9. doi.org/10.1186/s42483-019-0017-7.

Yosef, I., Goren, M.G. and Qimron, U. 2012. Proteins and DNA elements essential for the CRISPR adaptation process in Escherichia coli. Nucleic Acids Res. 40: 5569–5576.

Yu, L., Batara, J. and Lu, B. 2016. Application of genome editing technology to MicroRNA research in mammalians. *In*: Michael Kormann (ed.). Modern Tools for Genetic Engineering. IntechOpen. https://doi.org/10.5772/64330.

Zafar, K., Khan, M.Z., Amin, I., Mukhtar, Z., Yasmin, S., Arif, M., Ejaz, K. and Mansoor, S. 2020. Precise CRISPR-Cas9 mediated genome editing in super basmati rice for resistance against bacterial blight by targeting the major susceptibility gene. Front. Plant Sci. 11: 575.

Zaidi, S.S.E.A., Mahas, A., Vanderschuren, H. and Mahfouz, M.M. 2020. Engineering crops of the future: CRISPR approaches to develop climate-resilient and disease-resistant plants. Genome Biol. 21: 1–19.

Zaidi, S.S.E.A., Mukhtar, M.S. and Mansoor, S. 2018. Genome editing: Targeting susceptibility genes for plant disease resistance. Trends Biotechnol. 36: 898–906.

Zaidi, S.S.E.A., Vanderschuren, H., Qaim, M. et al. 2019. New plant breeding technologies for food security. Science 363: 1390–1.

Zaynab, M., Sharif, Y., Fatima, M., Afzal, M.Z., Aslam, M.M., Raza, M.F. et al. 2020. CRISPR/Cas9 to generate plant immunity against pathogen. Microb. Pathog. 141: 103996. Doi: 10.1016/j.micpath.2020.103996.

Zeng, X., Luo, Y., Vu, N.T.Q., Shen, S., Xia, K. and Zhang, M. 2020. CRISPR/Cas9-mediated mutation of OsSWEET14 in rice cv. Zhonghua11 confers resistance to *Xanthomonas oryzae* pv. *oryzae* without yield penalty. BMC Plant Biol. 20: 1–11.

Zetsche, B., Gootenberg, J.S., Abudayyeh, O.O., Slaymaker, I.M., Makarova, K.S., Essletzbichler, P., Volz, S.E., Joung, J., van der Oost, J., Regev, A., Koonin, E.V. and Zhang, F. 2015. Cpf1 is a single RNA-guided endonuclease of a class 2 CRISPR-Cas system. Cell 163: 759–771. [Google Scholar] [CrossRef] [PubMed].

Zetsche, B., Heidenreich, M., Mohanraju, P., Fedorova, I., Kneppers, J., DeGennaro, E.M., Winblad, N., Choudhury, S.R., Abudayyeh, O.O., Gootenberg, J.S. et al. 2016. Multiplex gene editing by CRISPR-Cpf1 using a single crRNA array. Nat. Biotechnol. 35: 31–34.

Zhan, X., Zhang, F., Zhong, Z., Chen, R., Wang, Y., Chang, L., Bock, R., Nie, B. and Zhang, J. 2019. Generation of virus-resistant potato plants by RNA genome targeting. Plant Biotechnol J. 17: 1814–1822.

Zhang, D., Zhang, H., Li, T., Chen, K., Qiu, J.L. and Gao, C. 2017a. Perfectly matched 20-nucleotide guide RNA sequences enable robust genome editing using high-fidelity SpCas9 nucleases. Genome Biol. 18(1): 191.

Zhang, Y., Bai, Y., Wu, G., Zou, S., Chen, Y., Gao, C. and Tang, D. 2017b. Simultaneous modification of three homoeologs of TaEDR1 by genome editing enhances powdery mildew resistance in wheat. Plant J. 91: 714–724. Doi: 10.1111/tpj.13599.

Zhang, Y., Liang, Z., Zong, Y. et al. 2016. Efficient and transgene-free genome editing in wheat through transient expression of CRISPR/Cas9 DNA or RNA. Nat. Commun. 7: 12617.

Zhang, H., Zhang, J., Wei, P., Zhang, B., Gou, F., Feng, Z., Mao, Y., Yang, L., Zhang, H. and Xu, N. 2014. The CRISPR/Cas9 system produces specific and homozygous targeted gene editing in rice in one generation. Plant Biotech. J. 12: 797–807.

Zhang, M., Liu, Q., Yang, X., Xu, J., Liu, G., Yao, X., Ren, R., Xu, J. and Lou, L. 2020. CRISPR/Cas9-mediated mutagenesis of Clpsk1 in watermelon to confer resistance to *Fusarium oxysporum* f. sp. *niveum*. Plant Cell Rep. 39(5): 589–595. Doi: 10.1007/s00299-020-02516-0. Epub 2020 Mar 9. PMID: 32152696.

Zhang, R., Meng, Z., Abid, M.A. and Zhao, X. 2019a. Novel pollen magnetofection system for transformation of cotton plant with magnetic nanoparticles as gene carriers transgenic cotton. Humana Press, New York, pp. 47–54.

Zhang, T., Zheng, Q., Yi, X., An, H., Zhao, Y., Ma, S. et al. 2018b. Establishing RNA virus resistance in plants by harnessing CRISPR immune system. Plant Biotechnology J. 16(8): 1415–23.

Zhang, Z.T., Jiménez-Bonilla, P., Seo, S.O., Lu, T., Jin, Y.S., Blaschek, H.P. et al. 2018a. Bacterial genome editing with CRISPR-Cas9: Taking Clostridium beijerinckii as an example. pp. 297–325. *In*: Braman, J.C. (ed.). Synthetic Biology. Methods in Molecular Biology, Vol. 1772. New York: Humana Press.

Zhong, Z., Sretenovic, S., Ren, Q., Yang, L., Bao, Y., Qi, C. et al. 2019. Improving plant genome editing with high-fidelity xCas9 and non-canonical PAM-targeting Cas9-NG. Mol Plant. 12(7): 1027–36.

Zhao, Y., Yang, X., Zhou, G. and Zhang, T. 2020. Engineering plant virus resistance: from RNA silencing to genome editing strategies. Plant Biotechnol. J. 18: 328–336.

Zhou, J., Peng, Z., Long, J., Sosso, D., Liu, B., Eom, J.S., Huang, S., Liu, S., Vera Cruz, C., Frommer, W.B. and White, F.F. 2015. Gene targeting by the TAL effector PthXo2 reveals cryptic resistance gene for bacterial blight of rice. Plant J. 82: 632–643.

Zhu, H., Lau, C.H., Goh, S.L., Liang, Q., Chen, C., Du, S. and Wang, S. 2013. Baculoviral transduction facilitates TALEN-mediated targeted transgene integration and Cre/LoxP cassette exchange in human-induced pluripotent stem cells. Nucleic Acids Res. 41(19): e180–e180. Doi: 10.1093/nar/gkt721.

Chapter 13
Can Nitric Oxide Overcome Biotic Stress in Plants?

Amedea Barozzi Seabra

Introduction

As primary producers of the ecosystem, plants are exposed to abiotic and biotic stress conditions. The latter is the attack of phytopathogens that impair plant growth and development, reducing crop productivity and yield affecting human development and global food production (Palmgren et al. 2015, Falak et al. 2021). Biotic stress is caused by macro/microorganisms, such as pathogens and/or parasites (fungi, insects, viruses, nematodes, and bacteria) (Gimenez et al. 2018). Plants face several adverse environmental and biotic stress conditions. These stress conditions cannot be avoided by plants, changing plant metabolism, and damaging physiological processes, hence impairing crop productivity. In this scenario, food production is under continuous challenge, mainly with the increasing global population, while the natural resources are the same (González de Molina et al. 2017). Although several signs of progress have been achieved in the combat of phytopathogens, plant diseases still cause a 30% loss in annual production (Oerke 2006). Thus, there is a need for the technological development of new strategies to efficiently combat biotic stress in agriculture.

To cope with biotic stresses, plants have developed their immune system. They have developed physical barriers to protect themselves from pathogens, such as thick cuticles and waxes. Moreover, plants synthesize chemicals to avoid pathogen infection (Taiz et al. 2006). The immunological system of plants has developed pathogen recognition system that activates plant defense response. Consequently,

Center for Natural and Human Sciences (CCNH), Federal University of ABC (UFABC), Santo André, SP, Brazil; Center of Natural and Human Sciences, Universidade Federal do ABC, Av. dos, Estados 5001, CEP 09210-580, Santo André, SP, Brazil.
Email: amedea.seabra@ufabc.edu.br

plant cells enhance the generation of reactive oxygen/nitrogen species and the activation of Mitogen-Activated Protein Kinases (MAPKs) (Muthamilarasan and Prasad 2013). This cascade defense pathway involves the increase of phytohormones, such as jasmonic acid (JA), salicylic acid (SA), and ethylene (ET).

Although plants have developed defense mechanisms to mitigate pathogen attack (biotic stress), there is a strong need to design new methodologies and active agents to significantly improve plant defense against biotic stress. Losses of crop production are still a great issue in agriculture, imposing severe economic losses. The increase of the global population, and therefore the necessity to create new versatile approaches to enhance food production have triggered the intense study of strategies to combat biotic stress. The use of traditional methods to combat biotic stress, the use of agrochemicals, has been facing several restrictions in recent years, in the light of sustainability and green chemistry. New green approaches are welcome to increase food production, from an environmentally friendly perspective.

Nitric oxide (NO): NO news is good news for plant defense

The free radical nitric oxide (NO) is a signaling endogenously found molecule in plant cells, which plays a key role in plant growth and development, in addition to the alleviation of plant abiotic and biotic stresses (Corpas et al. 2020). In mammals, NO is synthesized from L-arginine by the action of nitric oxide synthase (NOS), which have 3 isoforms: endothelial (eNOS), constitutive (cNOS), and inducible (iNOS). In plants, the sources of NO have been hampered by drawbacks. Currently, it is assumed that NO is endogenously synthesized in plants by the non-enzymatic reduction of nitrite (NO_2^-); many enzymes in plant act as NOS-like activity, generating NO, such as nitrate reductase (NR), and nitrite reductase (Kolbert et al. 2019). In plants, an analogous NOS-like enzyme with proprieties similar to animal NOS has been suggested in involving NO production in various stages of pathogen-plant interactions (Rasul et al. 2012). Despite several progress made in the elucidation of NO signaling in plant-pathogen interactions, its origin and exact mechanisms of production and actions in plant defense in biotic stress conditions still need to be further investigated (Nabi et al. 2020).

The administration of exogenous NO donors, and more recently NO-releasing nanomaterials, in plants, have been used to combat abiotic stress (salinity, drought, metal toxicity, extreme temperature, UV light) and biotic stress (pathogen attack) (Corpas et al. 2020). The most common NO donor applied in agriculture is sodium nitroprusside (SNP), due to its low cost, in addition to S-nitrosothiols (such as S-nitrosoglutathione, and S-nitrosomercaptossunic acid), and NONOates. S-nitrosothiol-containing chitosan nanoparticles were recently applied in plants under abiotic stress conditions with promising results (Oliveira et al. 2016, Lopes-Oliveira et al. 2019, Silveira et al. 2021).

Recent evidence suggests that endogenous NO production and the generation of reactive oxygen (ROS) and nitrogen (RNS) species are related to the pathogen – plant interactions (Sedlářová et al. 2011). Important studies indicate that burst generation of endogenous NO is one of the immediate responses to pathogen attack on the plant (Delledonne et al. 1998). NO plays a role in plant disease resistance and in the development of hypersensitive cell death (Falak et al. 2021). Not only in plants, but also in fungi and fungi-like oomycetes, NO regulates several processes during their growth, reproduction, development, and immune responses to biotic interactions (Jedelská et al. 2021). NO plays a key role in the molecular mechanisms of pathogen recognition and plant-mediated local and systemic defense. NO participates in the hypersensitive response (HR), thus inducing plant defense against pathogen infection (Delledonne et al. 1998). NO is also a key component of pathogen recognition leading to the activation of local and systemic immune responses. These orchestrated mechanisms involve ROS/RNS generation, activation of antioxidant enzyme activities, changes in cellular redox state, post-translation modifications of thiol-containing proteins (S-nitrosation), protein nitration, and interactions with phytohormones (Sedlarova et al. 2015). Indeed, to cope with biotic stress conditions, plants have developed smart approaches for precise balance of important stress regulators and plant growth, such as NO, ROS, RNS, and phytohormones (Sánchez-Vicente et al. 2019). Importantly, NO-mediated plant defense is linked to cellular redox changes caused by nitrosation of cysteine thiol moieties of targeted proteins (S-nitrosation reaction) under physiological and pathophysiological conditions leading to the formation of S-nitrosoproteins and disulfide bridges. S-nitrosation is one of the features of NO controlling several cellular events by modulating the activity of enzymes, protein-protein interactions, and protein localization (Tada et al. 2008). In plants, two enzymatic systems regulate the protein S-nitrosation: (i) thioredoxin (Trx)/thioredoxin reductase system, and (ii) indirect denitrosation performed by S-nitrosoglutathione reductase (GSNOR). GSNOR modulates S-nitrosoprotein in plant response in plant defense (Jedelská et al. 2021). These protein modifications might inhibit the activity of pathogen effectors in challenged plant cells, disarming the virulence machinery of the attacking pathogen (Sedlářová et al. 2015). Immediately after pathogen perception by plant, plant defense system generates NO and ROS/RNS, inducing HR response via S-nitrosation of transcripton factor proteins. This signaling response significantly changes the protein structures, affecting their ability to bind at their specific sites in the promotion of target genes (Falak et al. 2021). Moreover, it has been demonstrated that different NO-responsive genes have been identified in plants upon application of NO donors, indicating a plant adaptative response to biotic stress conditions (Mata-Pérez et al. 2016). The formation of NO/ROS/RNS burst as a defense plant mechanism enhances the expression of genes associated with plant immunity and defense (Levine et al. 1994). In this context, modulation of gene expression by transcription factors, hypersensitive response (HR), and production of pathogenesis-related proteins occur (Sedlářová et al. 2015).

Selected examples of NO/NO donors in biotic stress in plants

This section presents and discusses the recent progress in the investigation of NO signaling effects on plant immunity to cope with pathogens.

The NO-dependent post-infection signaling related to defense against virulent and avirulent races of *Phytophthora infestans* (a water mold or an oomycete that causes potato blight, a serious disease) was evaluated in potato leaves (Abramowski et al. 2015). The blight disease control was related to the NO signaling pathways through protein S-nitrosation and redox homeostasis. In fact, the nitrosation of free thiol groups (cysteine residues) in proteins (S-nitrosation), leading to the formation of S-nitrosoproteins by NO is linked to the generation of reactive oxygen (ROS) and nitrogen (RNS) species, boosting the immune responses in potato leaves. Many plant proteins are reported to be S-nitrosated; however, the mechanisms are still not completely understood and further studies are required. In fact, the authors reported the generation of superoxide ($O_2^{\bullet-}$) that reacts with NO in a diffusion-limited reaction leading to the harmful peroxynitrite ($ONOO^-$), in addition to the production of hydrogen peroxide (H_2O_2), allied to the activity of superoxide dismutase (SOD) (Abramowski et al. 2015).

In another strategy, the enhancing effect of NO in the efficacy of biotic elicitor (BE) in the improvement of the innate immune system of tomato infected with *Fusarium oxysporum* Schltdl was evaluated (Chakraborty et al. 2021). Tomato is very important from the perspective of nutritional values, and *Fusarium* wilt is related to yield loss under field conditions. Tomato plants were treated for 48 h with BE (2%), sodium nitroprusside (SNP, a NO donor), and BE along with NO-modulators. As expected, the disease incidence was significantly impaired with BE treatment, improving the NO production. Moreover, the activity of antioxidant enzymes and total phenol and flavonoid levels were increased, compared to the control. A two-fold increase in endogenous plant NO production was observed in plants treated with BE; moreover, the expression of defense genes was upregulated upon SNP and BE treatments, after 48 h (Chakraborty et al. 2012).

Fusarium crown rot is an important disease worldwide that infects wheat, and the lack of effective fungicides and resistant cultivars make it difficult to control the disease. In this sense, biocontrol is an environmentally friendly approach to control the spread of diseases caused by *Fusarium*. The root endophytic fungus *Piriformospora indica* hosts plants mitigating biotic and abiotic stresses. The ability of *Piriformospora indica* to protect wheat seedlings against *Fusarium pseudograminearum*, which leads to crown rot and the functions of the signaling molecules NO and polyamines (PA) in the induction of biotic stress responses, was investigated (Dehghanpour-Farashah et al. 2019). Overall, in greenhouse experiments, the results demonstrated that *Piriformospora indica* significantly impaired the disease progress on wheat seedlings and leaves, while increasing several plant growth parameters, in comparison with the control. Administration of the NO donor SNP and spermidine (a PA donor) enhanced the immunity of the plant. *Piriformospora indica* induced plant resistance by increasing H_2O_2 levels, increasing the activity of catalase (CAT) and guaiacol peroxidase (GPX), enhancing the membrane stability index (MSI), relative water content (RWC), and callose deposition, compared to control groups (i.e., uninoculated controls and plants inoculated with *Fusarium pseudograminearum*. Thus, PA and NO enhanced the basal plant resistance to *Fusarium* crown rot (Dehghanpour-Farashah et al. 2019).

Recently, hydrogen sulfide (H_2S) has been applied not only in biomedical applications but also in agriculture. Like NO, H_2S is a signaling molecule found in plants with potent protective effects on stress responses (Yamasaki and Cohen 2016). It has been demonstrated that application of H_2S donor (sodium hydrosulfide, NaHS at a concentration of 5 mM for 12 h) induces resistance in apples against *Penicillium expansum* by regulating NO and H_2O_2 generation of phenylpropanoid metabolism (Deng et al. 2021). The authors observed that NaHS application significantly reduced the diameter of lesions in apples colonized with *Penicillium expansum* (Figure 1).

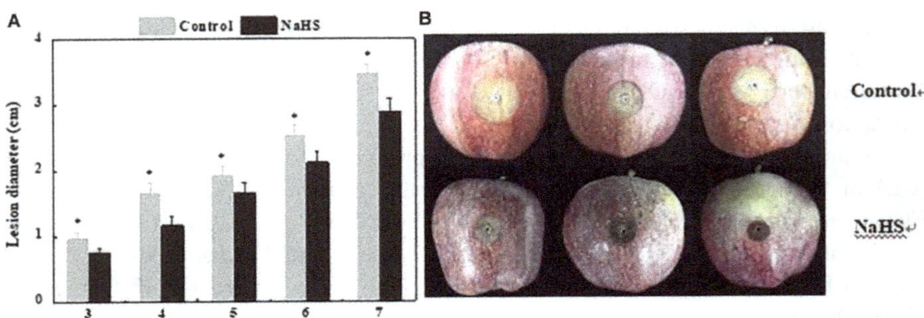

Figure 1: Inhibition of *Penicillium expansum*-induced decay in apples by NaHS. The bars represent the standard error of the mean (±SE). The lesion diameter (A) and representative photographs were taken after 5 days (B). *Significant differences at $P < 0.05$. Reproduced from reference Deng et al. 2021 under the terms of the Creative Commons Attribution License (CC BY).

Interestingly, NaHS treatment induces the endogenous production of H_2S, NO, and H_2O_2 in the fruits, modulating the host defense mechanism. In addition, NaHS treatment increases the activity of antioxidant enzymes and the accumulation of caffeic, coumaric, sinapic acids, ferulic, and total phenolic compounds (Deng et al. 2021). Recently, the crosstalk between NO and H_2S has been extensively investigated and further studies are required.

The involvement of ROS/RNS and NO in the mechanism of tomato resistance to the pathogen *Salmonella enterica* was demonstrated (Ferelli et al. 2020). The NO donor SNP (168 μM) was used to treat *Larix olgensis* seedlings to elucidate the regulatory role of NO at the transcriptional level (Hu et al. 2015) (Figure 2). *Larix olgensis* is a common coniferous species found in China, in plantation forests, and vulnerable to pathogens. SNP upregulated important genes related to plant-pathogen infection (MAPKKK, WRKY33, PR1, and FLS2). These results suggest the involvement of NO in the plant's resistance to pathogens and offer a contribution to the genomic and transcriptomic resources of *Larix olgensis* (Hu et al. 2015).

NO donors can also be used in postharvest (Machado et al. 2022). The effects of NO on the control of postharvest anthracnose, one of the most common postharvest diseases in citrus fruit, caused by *Colletotrichum gloeosporioides* in citrus fruits were reported (Zhou et al. 2015). Nowadays, the postharvest control of anthracnose in citrus fruit is based on synthetic fungicides, which impose toxicity to humans and

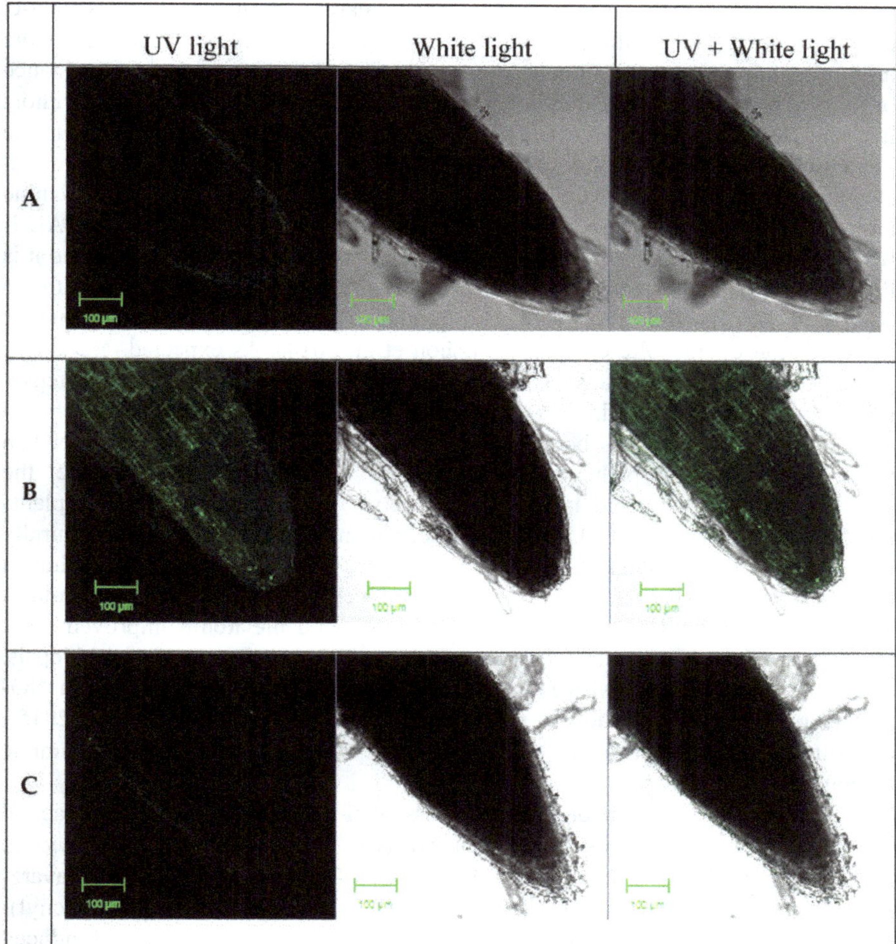

Figure 2: Effects of SNP (NO donor) on NO generation in *L. olgensis* seedlings. NO fluorescence was detected using the NO-specific 4-amino-5-methylamino-2`,7`-dichlorodihydrofluorescein diacetate (DAF-FM-DA) green fluorescent dye, and viewed by confocal laser scanning microscopy. (A) Root of control seedlings; (B) Root of seedlings treated with 167.8 μM sodium nitroprusside (SNP, NO donor) for 5 h; (C) Root of seedlings treated with 167.8 μM SNP and 100 μM 2-4-(carboxyphenyl)-4,4,5,5-tetramethyl-imidazoline-1-oxyl-3-oxide (cPTIO) for 5 h. Reproduced from Hu et al. 2015 under the terms of the Creative Commons Attribution License (CC BY).

to the environment, in addition to the development of resistance by pathogens. Thus, great effort has been made for the development of new strategies for control of the disease. The NO donor SNP (50 μM) significantly reduces the disease incidence and lesion size of citrus fruit contaminated with *Colletotrichum gloeosporioides* at 20°C under storage. SNP improved the titratable acidity content and the ascorbic acid and decreased the total soluble solid levels in the fruit during storage. Moreover, SNP modulated hydrogen peroxide content enhanced the synthesis of phenolic compounds, and induced peroxidase, phenylalanine ammonia-lyase, catalase, polyphenol oxidase activities, and the ascorbate-glutathione cycle. Thus, NO donors might find important applications in the food industry by inducing disease resistance of fruits to pathogenic fungi, thus enhancing the postharvest life of fruits. The authors hypothesized that NO has a critical role in modulating defense responses and in the induction of resistance of plants to fungal pathogens (Zhou et al. 2015). Important enzymes upregulated by NO in the combat of biotic stress are phenylalanine ammonia-lyase (PAL), peroxidase (POD), and polyphenol oxidase (PPO). PAL is responsible for the plant synthesis of lignums, phytoalexins, and phenolics, and it is the first enzyme in the phenylpropanoid pathway. POD is involved in lignin synthesis (Liu et al. 2010), while PPO generates quinines (that are toxic to pathogens) by the oxidation of phenolic compounds (Zhou et al. 2014). As expected, NO donor generated H_2O_2, which is a second messenger and further activates plant defense-related genes (Knörzer et al. 1999).

Although progress has been made, the exact mechanism of action of NO in plant defense against pathogens is still under investigation. For instance, the crosstalk between NO and melatonin (*N*-acetyl-5-methoxytryptamine) in plants under biotic stress (Shi et al. 2015) has been demonstrated. Melatonin, naturally found in plants, is a secondary messenger in plant response to biotic stress. Pathogen bacterial infection (Pst DC3000) increases the levels of both NO and melatonin in *Arabidopsis* leaves. The elevated contents of NO and melatonin improved plant resistance to *Pseudomonas syringe* pv. tomato (Pst) DC3000 infection in Arabidopsis. Interestingly, in *Pseudomonas syringe* pv. tomato (Pst) DC3000 infection in NO-deficient mutants of Arabidopsis, melatonin lost its effectiveness (Shi et al. 2015). These results suggest crosstalk between NO and melatonin in the innate immune response of *Arabidopsis*.

The NO donor that belongs to the class of S-nitrosothiol, S-nitroso-*N*-acetyl *D*-penicillamine (SNAP) was used to induce resistance of bean plant to *Pseudomonas fluorescent* (CHA0) and basal resistance against *Rhizoctonia solani* (Keshavarz-Tohid et al. 2016). In contrast, application of the NO scavenger 2-(4-carboxyphenyl)-4,4,5,5-tetramethyl-imidazoline-1-oxyl-3-oxide (cPTIO) reduced basal and induced plant resistance to the pathogen. Again, the activities of key enzymes (peroxidase) (POX), catalase (CAT), and ascorbate peroxidase (APX) increased along with the levels of H_2O_2. Overall, NO donor enhanced H_2O_2 formation and regulated the redox state of the host plant, inducing plant defense against the pathogen (Keshavarz-Tohid et al. 2016).

The role of NO and ROS in the plant defense against pathogens was studied in the *Lactuca* spp.- *Bremia lactucae* pathosystem by using a leaf disc model (Sedlářová et al. 2011). The NO donor SNP decreased the development of *Bremia lactucae* infection structures within 48 hours post-inoculation. This effect was followed by retardation of sporulation. Accumulation of NO in the penetrated cells, suggesting the hypersensitive reaction, was demonstrated by scanning microscopy (Sedlářová et al. 2011).

In another study, the involvement of NO/ROS and heat shock proteins (HSP) produced in response to biotic stress to prevent cell damage was reported (Piterková et al. 2013). Indeed, the accumulation of Hsp70 proteins in tomato leaves induced by biotic stress was studied by using a model system of leaf discs with two tomato genotypes *Solanum lycopersicum* cv. *Amateur* and *Solanum chmielewskii*, which differ in their resistance to fungal pathogen *Oidium neolycopersici*. Treatment with NO donor (S-nitrosoglutathione, GSNO at a concentration of 0.1 mM) modulated the content of Hsp 70 in a manner related to the genotype resistance. Accordingly, levels of endogenous NO production correlated with Hsp 70 accumulation. The authors conclude that NO and ROS are related to defense mechanisms in plants (Piterková et al. 2013).

The defense mechanism of NO is also related to plant hormones. In this direction, plant treatment with brassinosteroids (BRs), growth-promoting phytohormones, increased levels of accumulated NO resulting in a significant reduction of virus accumulation in *Arabidopsis thaliana* (Zou et al. 2018). In addition to NO levels, photosystem parameters and antioxidant system in the plant were up-regulated in brassinolide (BL)-treated plant, whereas NO scavenging downregulated it, suggesting that NO might be involved in BRs-induced virus resistance in *Arabidopsis* (Zou et al. 2018). It has been reported that NO action in plant defense against pathogens is associated with plant hormones such as BRs, cytokinin, abscisic acid, and ethylene (Deng et al. 2016). BRs activate virus resistance by producing NO and NIA1-dependent NO production is responsible for BR-induced defense response in *Arabidopsis* (Zou et al. 2018).

It should be noted that most of the research to evaluate the effect of NO/NO donors in plants under biotic stress conditions has been performed under controlled experimental conditions, which might be completely different from the "real world". The observed responses can be different by changing the variable conditions in the environment. Importantly, plant fitness and plant ability to activate immune response are dependent on nitrogen nutrition (Jedelská et al. 2021). In this direction, further studies to investigate how NO signaling crosstalk with nitrogen metabolism and thus plant defense is a prospective research topic that should be further explored in the plant application of NO.

Conclusions

In summary, recent evidences have demonstrated that the endogenously found signaling molecule NO, which is a free radical and gaseous molecule, plays a pivotal role in plant growth, development, and defense. Indeed, NO is a vital player not only in animals but also in plant cells. In this sense, over the last decades, various important publications have been describing and exploring the functional aspects of NO in plants, since NO is known to be involved in several biological processes in plants, such as plant defense against pathogens (biotic stress), iron homeostasis, mitigation to abiotic stress, and seed signaling pathways of plant hormones. Although several signs of progress have been achieved in the understanding of NO actions in plant tissues, the exact mechanism of action needs to be further investigated. Under biotic stress, NO bursts plant immunity by the generation of ROS and RNS, in addition to posttranslational modifications of cysteine residues of plant proteins (S-nitrosation) and tyrosine nitration. S-nitrosation of vital proteins (S-NO attachment to cysteine residues of proteins) has been reported to be a key event in plant immunity, including modifications in protein-protein interactions, enzyme pathways, and protein localization, among others. As NO is a free radical, NO donors, such as SNP and S-nitrosothiols, have been used in plants to mitigate biotic stresses. However, the combination of NO donors and nanomaterials has been poorly explored in agriculture applications, especially in biotic stress, as it has been extensively used in biomedical applications and, to a lesser extent, to cope with plants to mitigate abiotic stress. Thus, further studies are required to better understand the roles of NO/NO in plant defense and in the application of NO-releasing nanomaterials in plants under biotic stress conditions. This chapter will open new avenues in chemistry/biochemistry of NO, agriculture/food production, and nanotechnology.

References

Abramowski, D., Arasimowicz-Jelonek, M., Izbiańska, K., Billert, H. and Floryszak-Wieczorek, J. 2015. Nitric oxide modulates redox-mediated defense in potato challenged with *Phytophthora infestans*. Eur. J. Plant Pathol. 143: 237–260.

Corpas, F.J., Gonzalez-Gordo, S. and Palma, J.M. 2020. Nitric oxide: A radical molecule with potential biotechnological applications in fruit ripening. J. Biotechnol. 324: 211–219.

Dehghanpour-Farashah, S., Taheri, P. and Falahati-Rastegar, M. 2019. Effect of polyamines and nitric oxide in *Piriformospora* indica-induced resistance and basal immunity of wheat against *Fusarium pseudograminearum*. Biol. Control 136: 104006.

Delledonne, M., Xia, Y., Dixon, R.A. and Lamb, C. 1998. Nitric oxide functions as a signal in plant disease resistance. Nature 394: 585–588.

Deng, H., Wang, B., Liu, Y., Ma, L., Zong, Y., Prusky, D. and Bi, Y. 2021. Sodium hydrosulfide induces resistance against *Penicillium expansum* in apples by regulating hydrogen peroxide and nitric oxide activation of phenylpropanoid metabolism. Front. Microbiol. 12: 720372.

Deng, X.G., Zhu, T., Zou, L.J., Han, X.Y., Zhou, X., Xi, D.H., Zhang, D.W. and Lin, H.H. 2016. Orchestration of hydrogen peroxide and nitric oxide in brassinosteroid-mediated systemic virus resistance in Nicotiana benthamiana. Plant J. 85: 478–493.

Falak, N., Imran, Q.M., Hussain, A. and Yun, B.-W. 2021. Transcription factors as the "Blitzkrieg" of plant defense: A pragmatic view of nitric oxide's role in gene regulation. Int. J. Mol. Sci. 22: 522.

Ferelli, A.M.C., Bolten, S., Szczesny, B. and Micallef, S.A. 2020. *Salmonella enterica* elicits and is restricted by nitric oxide and reactive oxygen species on tomato. Front. Microbiol. 11: 391. Doi: 10.3389/fmicb.2020.00391

Gimenez, E., Salinas, M. and Manzano-Agugliaro F. 2018. Sustainability 10: 391.

González de Molina, M., Soto Fernández, D., Infante-Amate, J., Aguilera, E., Vila Traver, J. and Guzmán, G.I. 2017. Decoupling food from land: The evolution of Spanish agriculture from 1960 to 2010. Sustainability 9: 2348.

Hu, X., Yang, J. and Li, C. 2015. Transcriptomic response to nitric oxide treatment in *Larix olgensis* Henry. Int. J. Mol. Sci. 16: 28582–28597.

Jedelská, T., Luhová, L. and Petřivalský, M. 2021. Nitric oxide signalling in plant interactions with pathogenic fungi and oomycetes. J. Exp. Botany 72: 848–863.

Keshavarz-Tohida, V., Taheria, P., Taghavi, S.M. and Tarighi, S. 2016. The role of nitric oxide in basal and induced resistance in relation with hydrogen peroxide and antioxidant enzymes. J. Plant Physiol. 199: 29–38.

Knörzer, O.C., Lederer, B., Durner, J. and Böger, P. 1999. Antioxidative defense activation in soybean cells. Physiol. Plant 107: 294–302.

Kolbert, Z., Barroso, J.B., Brouquisse, R., Corpas, F.J., Gupta, K.J., Lindermayr, C., Loake, G.J., Palma, J.M., Petrivalský, M., Wendehenne, D. and Hancock, J.T., 2019. A forty year journey: The generation and roles of NO in plants. Nitric Oxide 93: 53–70.

Levine, A., Tenhaken, R., Dixon, R. and Lamb, C. 1994. H_2O_2 from the oxidative burst orchestrates the plant hypersensitive disease resistance response. Cell 79: 583–593.

Liu, F.J., Tu, K., Shao, X.F., Zhao, Y., Tu, S.C. and Su, J. 2010. Effect of hot air treatment in combination with *Pichia guilliermondii* on postharvest anthracnose rot of loquat fruit. Postharvest Biol. Technol. 58: 65–71.

Lopes-Oliveira, P.J., Gomes, D.G., Pelegrino, M.T., Bianchini, E., Pimenta, J.A., Stolf-Moreira, R., Seabra, A.B. and Oliveira, H.C. 2019. Effects of nitric oxide-releasing nanoparticles on neotropical tree seedlings submitted to acclimation under full sun in the nursery. Sci. Rep. 9: 17371.

Machado, M.R., Veiga, J.C., Silveira, N.M., Seabra, A.B., Pelegrino, M.T., Boza, Y.E.A.G., Cia, P., Valentini, S.R.T. and Bron, I.U. 2022. Nitric oxide supply reduces ethylene production, softening and weight loss in papaya fruit. Bragantia 81: e1222.

Mata-Pérez, P., Begara-Morales, J.C., Luque, F., Padilla, M.N., Jiménez-Ruiz, J., Sánchez-Calvo, B., Fierro-Risco, J. and Barroso, J.B. 2016.Transcriptomic analyses on the role of nitric oxide in plant disease resistance. Curr. Issues Mol. Biol. 19: 121–128.

Muthamilarasan, M. and Prasad, M. 2013. Plant innate immunity: An updated insight into defense mechanism. J. Biosci. 38: 433–449.

Nabi, R.B.S., Tayade, R., Imran, Q.M., Hussain, A., Shahid, M. and Yun, B.-W. 2020. Functional insight of nitric-oxide induced DUF genes in *Arabidopsis thaliana*. Front. Plant Sci. 11: 1041.

Oerke, E.C. 2006. Crop losses to pests. J. Agric. Sci. 144: 31–43.

Oliveira, H.C., Gomes, B.C., Pelegrino, M.T. and Seabra, A.B. 2016. Nitric oxide-releasing chitosan nanoparticles alleviate the effects of salt stress in maize plants. Nitric Oxide 61: 10–19.

Palmgren, M.G., Edenbrandt, A.K., Vedel, S.E., Andersen, M.M., Landes, X., Østerberg, J.T., Falhof, J., Olsen, L.I., Christensen, S.B. and Sandøe, P. 2015. Are we ready for back-to-nature crop breeding? Trends Plant Sci. 20: 155–164.

Piterková, J., Luhová, L., Mieslerová, B., Lebeda, A. and Petrivalsky, M. 2013. Nitric oxide and reactive oxygen species regulate the accumulation of heat shock proteins in tomato leaves in response to heat shock and pathogen infection. Plant Sci. 207: 57–65.

Rasul, S., Dubreuil-Maurizi, C., Lamotte, O., Koen, E., Poinssot, B., Alcaraz, G., Wendehenne, D. and Jeandroz, S. 2012. Nitric oxide production mediates oligogalacturonide-triggered immunity and resistance to *Botrytis cinerea* in *Arabidopsis thaliana*. Plant Cell Environ. 35: 1483–1499.

Sánchez-Vicente, I. Fernández-Espinosa, M.G. and Lorenzo, O. 2019. Nitric oxide molecular targets: Reprogramming plant development upon stress. J. Exp. Bot. 70: 4441–4460.

Sedlářová, M., Kubienová, L., Trojanová, Z.D., Luhová, L., Lebeda, A. and Petřivalský, M. 2015. The role of nitric oxide in development and pathogenesis of biotrophic phytopathogens e downy and powdery mildews. Adv. Bot. Res. 77: 264–283.

Sedlářová, M., Petřivalský, M., Piterková, J., Luhová, L., Kočířová, J. and Lebeda, A. 2011. Influence of nitric oxide and reactive oxygen species on development of lettuce downy mildew in *Lactuca* spp. Eur. J. Plant Pathol. 129: 267–280.

Shi, H., Chen, Y., Tan, D.X., Reiter, R.J., Chan, Z. and He, C. 2015. Melatonin induces nitric oxide and the potential mechanisms relate to innate immunity against bacterial pathogen infection in *Arabidopsis*. J. Pineal. Res. 59: 102–108.

Silveira, N.M., Prataviera, P.J.C., Pieretti, J.C., Seabra, A.B., Almeida, R.L., Machado, E.C. and Ribeiro, R.V. 2021. Chitosan-encapsulated nitric oxide donors enhance physiological recovery of sugarcane plants after water deficit. Environ. Exp. Bot. 190: 104593.

Tada, Y., Spoel, S.H., Pajerowska-Mukhtar, K., Mou, Z., Song, J., Wang, C., Zuo, J. and Dong, X. 2008. Plant immunity requires conformational charges of NPR1 via S-nitrosylation and thioredoxins. Science 2008: 321.

Taiz, L. and Zeiger, E. 2006. Secondary metabolites and plant defense. pp. 315–344. *In*: Taiz, L. and Zeiger, E. (eds.). Plant Physiology; Sinauer Associates: Sunderland, UK. Volume 4. ISBN 9780878935659.

Yamasaki, H. and Cohen, M.F. 2016. Biological consilience of hydrogen sulfide and nitric oxide in plants: Gases of primordial earth linking plant, microbial and animal physiologies. Nitric Oxide 55: 91–100.

Zhou, Y., Lia, S. and Zeng, K. 2015. Exogenous nitric oxide-induced postharvest disease resistance in citrus fruit to *Colletotrichum gloeosporioides*. J. Sci. Food Agric. 96: 505–512.

Zhou, Y.H., Ming, J., Deng, L.L. and Zeng, K.F. 2014. Effect of *Pichia membranaefaciens* in combination with salicylic acid on postharvest blue and green mold decay in citrus fruits. Biol. Control 74: 21–29.

Zou, L.J., Denga, X.G., Zhang, L., Zhu, T., Tan, W.R., Muhammada, A., Zhub, L.J., Zhanga, C., Zhanga, D.W. and Lin, H.H. 2018. Nitric oxide as a signaling molecule in brassinosteroid-mediated virus resistance to *Cucumber mosaic virus* in *Arabidopsis thaliana*. Physiol. Plant. 163: 196–210.

Index

Editors' Biography

Dr. Graciela Avila-Quezada is a Professor at the Faculty of Agrotechnology Science, Universidad Autónoma de Chihuahua, in México. She has been conducting research related to plant pathogens for 30 years, beginning with her undergraduate thesis work focused on the control of *Puccinia graminis* f. sp. *avenae*. She was a visiting scientist at the Volcani Center in Israel; Colegio de Postgraduados, Texcoco, Mexico; SENASICA Laboratories, Ministry of Agriculture in Tecamac, State of Mexico; the California Polytechnic State Univ-San Luis Obispo California (Calpoly); Institut de Recherche pour le Développement IRD Montpellier, France; Universidad Autonoma Chapingo, Mexico State; Universidad de Córdoba, Spain. Dr. Avila was President of the Mexican Society of Phytopathology; Vice-president of the National Phytosanitary Advisory Council (CONACOFI); Executive Secretary of the National System for Research and Technology Transfer (SNITT) in the Ministry of Agriculture (Sagarpa); and Director of Academic Cooperation, in the Coordination of Science and Technology, Presidency of Mexico. Besides, she has conducted more than 12 research projects and published more than 70 research papers in national and international journals focused on plant pathogens.

Dr. Mahendra Rai is presently a visiting Professor at the Department of Microbiology, Nicolaus Copernicus University, Torun, Poland. Formerly, he was Professor and Head of the Department of Biotechnology, SGB Amravati University, Maharashtra, India. He was a visiting scientist at the University of Geneva, Debrecen University, Hungary; University of Campinas, Brazil; VSB Technical University of Ostrava, Czech Republic, National University of Rosario, Argentina, and the University of Sao Paulo. He has published more than 425 research papers in national and international journals. In addition, he has edited/authored more than 69 books and 6 patents. Recently, he has been featured in Stanford's list of top 2% scientists under Nanoscience and Nanotechnology.

For Product Safety Concerns and Information please contact our EU representative GPSR@taylorandfrancis.com
Taylor & Francis Verlag GmbH, Kaufingerstraße 24, 80331 München, Germany

www.ingramcontent.com/pod-product-compliance
Lightning Source LLC
Chambersburg PA
CBHW060357220326
41598CB00023B/2945

* 9 7 8 1 0 3 2 3 8 3 2 8 6 *